2nd International Conference on Novel Functional Materials (ICNFM2019)

IOP Conference Series: Materials Science and Engineering
Volume 733

Shanghai, China
8 - 9 November 2019

ISBN: 978-1-7138-2592-0
ISSN: 1757-8981

Printed from e-media with permission by:

Curran Associates, Inc.
57 Morehouse Lane
Red Hook, NY 12571

Some format issues inherent in the e-media version may also appear in this print version.

Institute of Physics
Dirac House, Temple Back
Bristol BS1 6BE UK

Phone: 44 1 17 929 7481
Fax: 44 1 17 920 0979

techtracking@iop.org

Additional copies of this publication are available from:

Curran Associates, Inc.
57 Morehouse Lane
Red Hook, NY 12571 USA
Phone: 845-758-0400
Fax: 845-758-2633
Email: curran@proceedings.com
Web: www.proceedings.com

TABLE OF CONTENTS

IOP Conf. Series: Materials Science and Engineering **733** (2020) 011001 doi:10.1088/1757-899X/733/1/011001

Preface

Dear Distinguished Authors and Guests,

The Organizing Committee warmly welcomes you to 2019 2nd International Conference on novel functional materials (ICNFM2019), held on 8th-9th, November, 2019, Shanghai, China.

ICNFM2019 aims to bring together leading academic scientists, researchers and research scholars to exchange and share their experiences and research results on all aspects of advanced functional materials. It also provides a premier interdisciplinary platform for researchers, practitioners and educators to present and discuss the most recent innovations, trends, and concerns as well as practical challenges encountered and solutions adopted in the fields of advanced functional materials.

The proceeding records the fully refereed papers presented at the conference. The main conference themes and tracks are Advanced Materials Design and Mechanics and other related topics. Hopefully, all participants and other interested readers benefit scientifically from the proceedings and also find it stimulating in the process.

We would like to thank the program chairs, organization staff and the members of the program committees for their work. On behalf of the organizing committee, I would like to thank all the editors for their great support to ICNFM2019. Without their excellent editorial work, ICNFM2019 will not be published so timely and successfully. We are grateful to all those who have contributed to the success of ICNFM2019.

Finally, we wish all the authors and attendees of ICNFM2019 a unique, rewarding and enjoyable memory at ICNFM2019 in Shanghai, China. We look forward to your participation in the 3rd ICNFM in 2020.

With our warmest regards,
Bin Liu
Conference Organizing Chair
November 27th 2019

IOP Conf. Series: Materials Science and Engineering **733** (2020) 011002 doi:10.1088/1757-899X/733/1/011002

ICNFM2019 Committee

General Chair

Prof. Bin Liu, North University of China

Technical Program Committee Chairs

Prof. An Baigang, University of Science and Technology LiaoNing, China
Prof. Luo Xudong, University of Science and Technology LiaoNing, China
Prof.Zhao Hongyang, University of Science and Technology LiaoNing, China
Prof. Li Xianchun, University of Science and Technology LiaoNing, China
Prof. He Zhongliang, University of Science and Technology LiaoNing, China
Prof. Zhou Yanwen, University of Science and Technology LiaoNing, China
Prof.Hu Zhizhi, University of Science and Technology LiaoNing, China
Prof. Hasebe, Yasushi, Saitama University of Technology, Japan
Prof. Zhang Run, Australian Institute of Bioengineering and Nanotechnology(AIBN), Australian
Prof. Hideki Okamoto, Okayama University, Japan
Prof. Stephen C. Bergmeier, Ohio State University, USA
Prof. Ramesh Singh, University of Malaya, Malaysia
Prof. Mohd Hamdi Bin Abd Shukor, University of Malaya, Malaysia
Prof. Ahmed El-Shafie, University of Malaya, Malaysia
Prof. Stefan Talu， Technical University of Cluj-Napoca， Romania
Dr. Beddiaf ZAIDI， University of Batna， Algeria
Dr. ANAND VYAS， Hong Kong Polytechnic University， Hong Kong
Prof. RAJIV KUMAR， Goswami Ganesh Dutt Sanatan Dharam College， Indian
Prof. Mamoun FELLAH， Khenchela University， Algeria
Prof. Achanai Buasri， Chulalongkorn University, Thailand
Prof. Feng-Huei Lin， National Health Research Institutes (NHRI),Taiwan
Dr. Yili Li， University of Washington, Seattle

First name	Surname
Jeng Jong	Lin
Yuan	Lu
Yuan	Lu
Yuanyuan	Ding
Qiangcai	Liu
Zhenye	Zhu
Zhenye	Zhu
Fei	Zhang
Manoon	Masniyom
Ruixia	Liu
Chongyi	Wei
Xiaohua	Liu
Xiyuan	Yao
Shiling	Zhang
Dachuan	Qin
Tian	Lan
Yi	Yu
Rumyana	Lyubenova Lazarova
Yuanzhe	Yao
Yongqiang	Qin
Jiajia	Wang
Jiajia	Wang
Xiaorui	Zheng
Xiaorui	Zheng
Tianling	Du
Zhihua	Zhu
Jinghui	Zhang
Guanglei	Lv
Fan	Zhang
Vygandas	Vaitkus
Heping	Liu
Zongbo	Li
Jinxiang	Chen
Yanru	Li
Kadyirov	Aidar iLdusovich
Huiping	Liu
Guangliang	Song
Zhengcun	Zhou
Qinyi	Li
Zhiyong	Wang
Tianyang	Du
Xinhui	Li
Yifan	Chen

IOP Conf. Series: Materials Science and Engineering **733** (2020) 011003 doi:10.1088/1757-899X/733/1/011003

Skripkina	Tatiana
Skripkina	Tatiana
Panida	Charnkeitkong
Zhu	Chen
Geping	Bi
Tianchi	Wang

IOP Conf. Series: Materials Science and Engineering **733** (2020) 011003 doi:10.1088/1757-899X/733/1/011003

Institution
Vanung University
Xi'an Special Equipment Inspection Institute
Xi'an Special Equipment Inspection Institute
College of Mechanical and Electronic Engineering, Weifang Univers
Qingdao University of Science and Technology
Harbin Institute of Technology
Harbin Institute of Technology
Chengdu Fine Optical Engineering Research Center
Prince of Songkla University
Institute of Mechanics, Chinese Academy of Sciences
Yanshan University School of Mechanical Engineering
Sinopec Beijing Research Institute of Chemical Industry
Northwestern Polytechnical University
State Grid Chongqing Electric Power Company Chongqing Electric
Huadian electric power research institute Co. , LTD.
Aerospace Institute of Advanced Materials & Processing Technolog
Aerospace Institute of Advanced Materials & Processing Technolog
Institute of Metal Science Equipment and Technologies with Hydro
University of Electronic Science and Technology of China
Hefei University of Technology
Beijing University Of Chemical Technology
Beijing University Of Chemical Technology
Xi' an Modern Chemistry Research Institute
Xi' an Modern Chemistry Research Institute
Research Institute of Highway Ministry of Transport
Guangdong Mechanical & Electrical Polytechnic
Beijing University Of Chemical Technology
CNOOC Safety Technology Services Co., Ltd
China University of Petroleum (East China)
Kaunas University of Technology
North University of China
yunnan university
Peking university Shenzhen graduate school
Naval Logistical Academy
IPEAT FRC Kazan Scientific Center of RAS
Institute of Modern Physics, Chinese Academy of Sciences
Donghua University
Suzhou Vocational University
Huzhou University
Yanshan university
Shenyang JianZhu University
Shenyang Jianzhu University
Shenyang Jianzhu University

IOP Conf. Series: Materials Science and Engineering **733** (2020) 011003 doi:10.1088/1757-899X/733/1/011003

Institute of Solid State Chemistry and Mechanochemistry SB RAS, N
Institute of Solid State Chemistry and Mechanochemistry SB RAS, N
Department of Chemical Engineering, College of Engineering, Rang
Northeastern University, China
China National Quality Supervision and Inspection Center for Knife,
Nanjing University of Science and Technology, China

IOP Conf. Series: Materials Science and Engineering **733** (2020) 011003 doi:10.1088/1757-899X/733/1/011003

Email address
jjlin@vnu.edu.tw
luyuan19801104@163.com
luyuan19801104@163.com
deeyuanyuan@163.com
1913077247@qq.com
zhuzy@hit.edu.cn
zhuzy@hit.edu.cn
zfchunri@163.com
manoon.ma@psu.ac.th
liuruixia@imech.ac.cn
weichongyi@outlook.com
654175383@qq.com
yaoxiyuan@nwpu.edu.cn
526793305@qq.com
kindlitbear0@163.com
lantiansky2016@163.com
ccinyuyi@163.com
R.Lazarova@ims.bas.bg
yzyao@tsinghua.edu.cn
albon@hfut.edu.cn
2805156949@qq.com
2805156949@qq.com
zhxrui@qq.com
1042736537@qq.com
184823174@qq.com
184823174@qq.com
591690385@qq.com
lvgl2@cnooc.com.cn
zhangfan51@qq.com
vvaitkus@ktu.lt
peace666@126.com
xiongkai@ynu.edu.cn
chenjinxiang1@tcl.com
jianghaibo022@126.com
aidarik@rambler.ru
liuhp2008@impcas.ac.cn
sgl_tong@dhu.edu.cn
1355326160@qq.com
tangps@zjhu.edu.cn
zhiyongwhrb@qq.com
136964904@qq.com
2712126752@qq.com
356374012@qq.com

IOP Conf. Series: Materials Science and Engineering **733** (2020) 011003 doi:10.1088/1757-899X/733/1/011003

urazovatanya@mail.ru
urazovatanya@mail.ru
panida.s@rsu.ac.th
tianmei613@sina.com
bigeping@sina.com
tianchiwang@aliyun.com

Peer review statement

All papers published in this volume of *IOP Conference Series: Materials Science and Engineering* have been peer reviewed through processes administered by the proceedings Editors. Reviews were conducted by expert referees to the professional and scientific standards expected of a proceedings journal published by IOP Publishing.

IOP Conf. Series: Materials Science and Engineering **733** (2020) 012001 doi:10.1088/1757-899X/733/1/012001

Effects of Friction Time on microstructures and mechanical properties of friction welded T92/Super304H dissimilar steel weld joints

Yuan Lu*, **Debiao Li, Cheng Bi, Fuyang Sun, Jianlong Zhang, Bo Cao**

Xi'an Special Equipment Inspection Institute, Xi'an, 710065, China

*Corresponding author's e-mail: luyuan19801104@163.com

Abstract. In this paper, the T92 martensitic steel and the Super304H austenitic steel were welded by continuous drive friction welding technique. The friction welding joints exhibited good mechanical and metallurgical properties. The influence of friction time on the microstructure and mechanical properties of T92/Super304H dissimilar steel joints was investigated. With an increase in the friction time, the grains in the welding zone and the heat affected zone didn't grow up and the number of second-phase particles precipitates increased. The tensile fracture location was in the Super304H base metal and didn't change. With an increase in the friction time, the tensile strength didn't change, and the impact toughness decreased.

1. Introduction

Energy shortage and environmental pollution around the world urges people to develop new techniques of clear and high efficiency thermal power generation. Nowadays, the ultra-super critical power generation technique with much higher parameters including vapor temperature and pressure, has been generally accepted as the most feasible and reliable thermal power generation technology. In order to fulfill the tough parameters of the ultra-super critical power generation, traditional heat-resistant steels used in the subcritical and supercritical units have been replaced by those new heat-resistant steels. The recently developed Super304H (0.1C-18Cr-9Ni-3Cu-Nb-N) austenitic steel has shown considerable promise due to its high oxidation resistance, corrosion resistance and creep resistance. The Super304H has higher strength at elevated temperatures are required for super heater tubes in the boilers. This excellent creep rupture strength is based on finely precipitated particles such as Nb(C, N) and NbC. Thus, it is widely used for super heaters and reheaters, which have the abominable service environment in Ultra-super critical boilers. The T92 (9Cr-0.5Mo-2-W-V-Nb) martensitic steel approximately similar to T91 but with a little modification in chemical compositions for preferable higher temperature properties. The T92 martensitic steel will certainly have a broad application prospect in forth coming ultra-super boilers [1,2]. In this case, the welding between T92 and Super304H steels will be necessary.

At present, the T92 martensitic steel and Super304H austenitic steel are welded through a gas tungsten arc welding (GTAW) technique usually in engineering applications. The GTAW technique is mature and suitable for field welding. Nevertheless, the T92/Super304H joints welding through the conventional fusion welding still exhibit inferior mechanical properties, due to: (i) The base metal near the fusion line during the fusion welding process is prone to form the wide heat affected zone with coarse grains; (ii) A transition layer (fusion zone) will appear in the welding zone near the weld line;

The friction welding is a solid state welding process, and the welding temperature was under the melting temperature. This welding process nullifies adverse effects of the uncertainties in the filler metal selection by GTAW technique. The friction welding has a number of advantages over the conventional fusion welding process. The major advantage relies on the direct conversion of mechanical energy into thermal energy at the joint interface in contrast to fusion welding. The very high temperature gradient available in the joint area accounts for a very small heat affected zone. Owing to the narrow the heat affected zone, the welding distortion is kept minimum. Additionally, because it is a solid state process, the defects associated with melting solidification phenomenon, such as porosity and slag are not present [3-5]. The aim of this paper is to fabricate a combination with better mechanical properties between the Super304H austenitic steel and the T92 martensitic steel using continuous drive friction welding. The effect of friction time on the microstructure and mechanical properties of the welded joints were thus investigated.

2. Experimental procedures

The pipes of Super304H austenitic steel and T92 martensitic steel that in the sizes of Φ44.5mm x 9mm were used as base metals. The pipes of Super304H and T92 were fwelded by a continuous drive friction welding machine. In the process of friction welding, the friction speed was 1500 rpm, the friction pressure was 150 MPa and the upset pressure was 200 MPa. In the process of friction welding, the friction time was 6 S of Process 1, 10 S of Process 2 and 14 S of Process 3 respectively. The metallographs and fractographs of the welding joints were performed by an optical microscope and a scanning electron microscope respectively. The tensile samples were tested using a testing instrument. The impact samples were tested using an impacting instrument.

3. Results and Discussion

Figure 1, 2 and 3 shows the metallography of the welding joints with different friction time respectively. During the friction welding process, the welded zone produced agglutinate and shear tear behaviour, which led to the deformation of the grains in the welding zone, the dynamic recrystallization driving force and lattice distortion energy increased. The thermoplastic deformation temperature and unit volume free energy of recrystallized grains decreased, which produce a large number of recrystallized nucleations. The welding zone included a welding interface that was straight and clear, the areas (about 50μm from the welding interface) on both sides of the welding interface. The Super304H side of the welding zone was mainly composed of fine equiaxed austenite grains because of grain refinement. The martensite grains which in the T92 side of the welding zone were austenitized due to the higher temperature generated in the friction process. Than it produced a large number of fine equiaxed austenitic grains by dynamic recrystallization. In the cooling process, the recrystallization of the equiaxed austenitic grains takes place, resulting in the formation of the fine plate martensite grains. The welding zone of T92/Super304H dissimilar steel welded by fusion welding process had the typical coarse δ-phase grains. The coarse δ-phase grains were not formed in the welding zone by the friction welding. Because that the welding interface is well combined, and the austenite grains and plate martensite grains on both sides of the welding interface was fine, so the welding zone of the welding joints exhibit excellent mechanical properties. The heat input during the friction welding process was much less than the heat input during the fusion welding process. The width of the heat affected zone caused by friction welding process were much smaller than the width of the heat affected zone caused by the fusion welding process. The tendency of grains growth during the friction welding process was less obvious than the tendency of grains growth during the fusion welding process. So the heat affected zone of the welding joints with the fine grains showed excellent microstructure. [6, 7]. So the friction welded Super304H/T92 joints with the fine grains in the welding zone and heat affected zone showed excellent microstructure and mechanical properties.

IOP Conf. Series: Materials Science and Engineering **733** (2020) 012001 doi:10.1088/1757-899X/733/1/012001

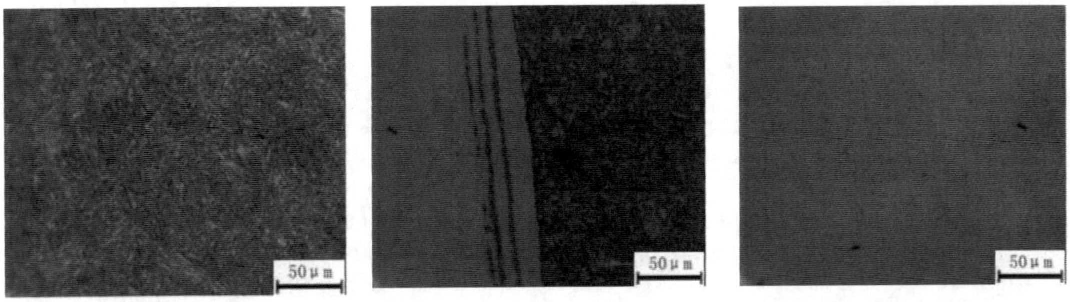

(a) (b) (c)

Figure 1 The metallography of welded joints for Process 1 (a) the T92 heat affected zone, (b) the welding zone, (c) the Super304H heat affected zone

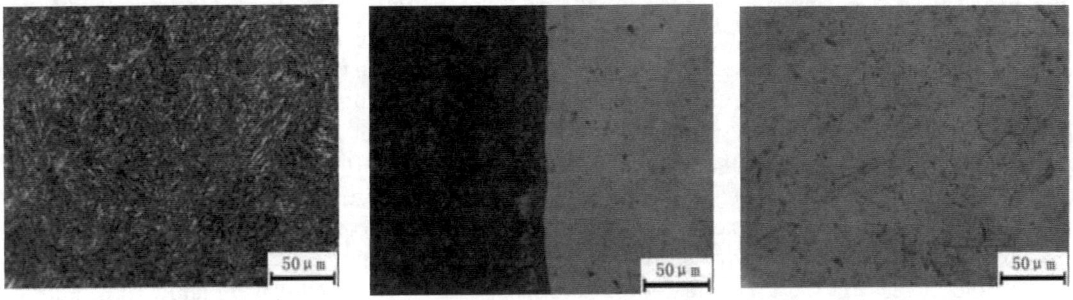

(a) (b) (c)

Figure 2 The metallography of welded joints for Process 2 (a) the T92 heat affected zone, (b) the welding zone, (c) the Super304H heat affected zone

(a) (b) (c)

Figure 3 The metallograph of welded joint for Process 3 (a) the T92 heat affected zone, (b) the welding zone, (c) the Super304H heat affected zone

The surface temperature of the welding joints was equal to the temperature of the welding heat source in friction welding process. So the surface temperature of the welding joints had a great influence on the microstructure and mechanical properties of the welding joints. The surface temperature of the welding joints was calculated according to the Eq (1)

$$T(t) = qt/(\pi\lambda C) \quad (1)$$

T was referred to the surface temperature of the welding joints， *t* was referred to friction time and *q* was referred to friction heating power. When friction heating power was constant, as the friction time increased, the surface temperature of the welding joints increased correspondingly. Because that melting process did not occur in the welding zone, although the surface temperature of the welding joints increased, the size change of the grains in the welding zone was not detectable. As the friction time increased, the heat input during the friction welding process increased in the heat affected zone, and the temperature of the heat affected zone increased accordingly. The size change of the grains which in the Super304H heat affected zone and in the T92 heat affected zone was not detectable, but the amount of the precipitated phases in the both heat affected zone increased obviously [8,9].

Table 2 depicted the variation of the yield strength and tensile strength as a function of the friction time. It was detected that the fracture position of the welding joints was in the Super304H base metal for the different friction time. The fractographs of the tensile strength samples were shown in figure 4. The fracture morphology was characterized by typical ductile fracture, and a lot of fine and shallow dimples distributed on the facture surface. As the friction time increased, the tensile strength change of the welding joints was not obviously. The tensile strength change is certainly corresponding to the microstructure evolution. Table 2 depicted the variation of the impact toughness as a function of the friction time. The grains size had a great influence on the impact toughness of each zone of the welded joints. When the grains size was smaller, the impact toughness was higher. The grain size of the heat affected zone was the larger, so the impact toughness of the heat affected zone was the lower. As the friction time increased, the dislocation density and lattice distortion energy in the welding zone increased, and the impact toughness of the welding zone decreased. The precipitation of the $Cr_{23}C_6$ carbides were main reason that the impact toughness of the heat affected zone decreased with the friction time increased. The fractographs of the impact toughness samples were shown in figure 5, 6 and 7. The fracture mode of the impact toughness samples was ductile fracture with the aggregation dimples on the facture surface. In the welding zone, the dimples were distributed densely and deeply, and the shapes of dimples were round. In the heat affected zone, the dimples were distributed sparsely and shallow, and the shapes of dimples were flat.

Table 1 The mechanical properties of Super304H/T92 welded joints

	Tensile strength /MPa			Impact toughness /J	
	$R_{p0.2}$	R_m	Fracture position	Welding zone	Heat affected zone (Super304H/T92)
Process 1	454.77	670.79	Super304H base metal	83.8	78.7/65.2
Process 2	432.62	689.05	Super304H base metal	74.3	61.6/49.9
Process 3	475.82	712.43	Super304H base metal	71.8	52.8/43.4

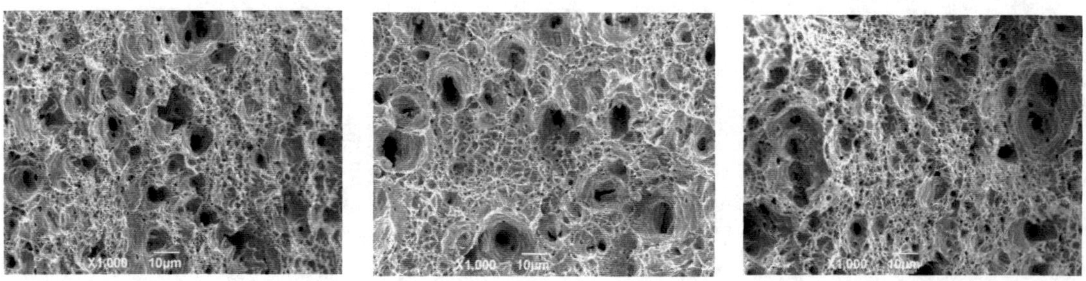

(a) (b) (c)

Figure 4 The fracture appearance of the tensile strength samples (a) Process 1, (b) Process 2, (c) Process 3

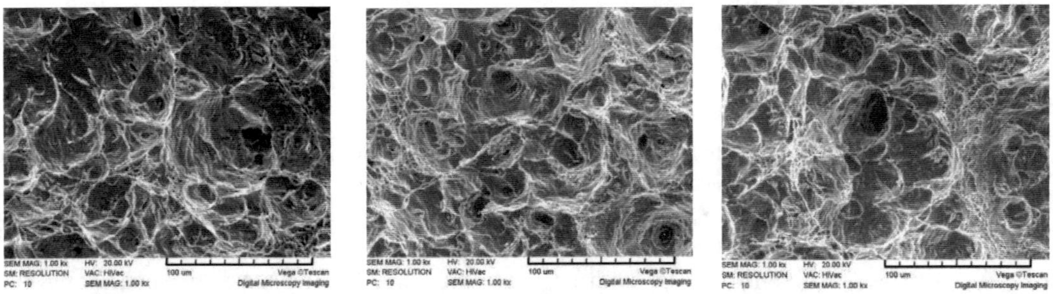

(a) (b) (c)

Figure 5 The fracture appearance of the impact toughness samples for Process 1 (a) the T92 heat affected zone, (b)the welding zone, (c) the Super304H heat affected zone

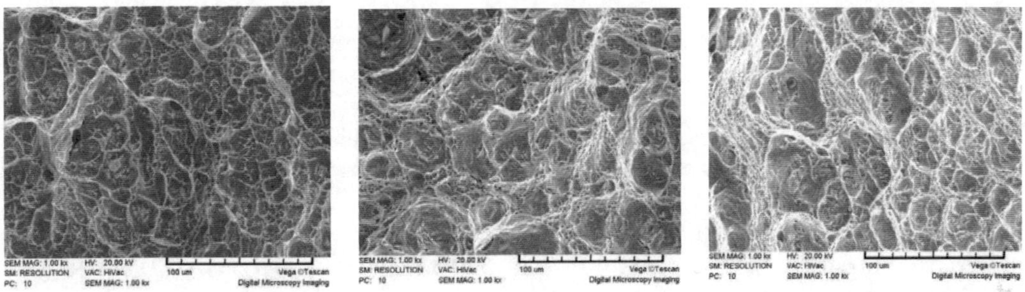

Figure 6 The fracture appearance of the impact toughness samples for Process 2 (a) the T92 heat affected zone, (b)the welding zone, (c) the Super304H heat affected zone

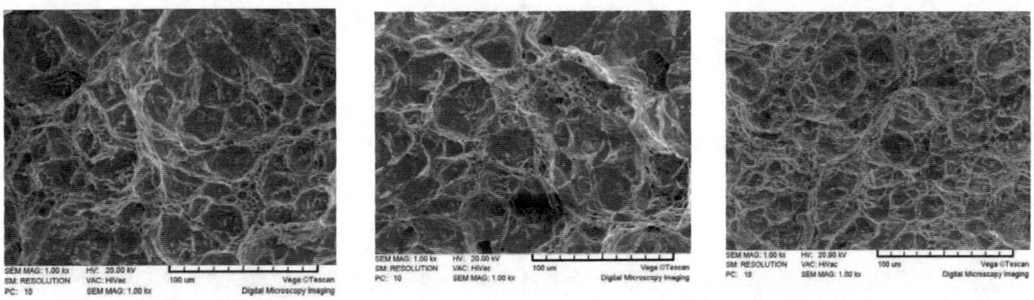

(a) (b) (c)

Figure 7 The fracture appearance of the impact toughness samples for Process 3 (a) the T92 heat affected zone, (b)the welding zone, (c) the Super304H heat affected zone

IOP Conf. Series: Materials Science and Engineering **733** (2020) 012001 doi:10.1088/1757-899X/733/1/012001

Conclusions

The T92/Super304H dissimilar steel welded joints with excellent properties were welded by continuous drive friction welding technique. The welding interface was well combined, and the austenite grains and plate martensite grains in the welding zone was fine. The fracture position of the welding joints was in the Super304H base metal. The influence of friction time on the microstructure and properties of T92/Super304H dissimilar steel joints was investigated. With the friction time increased, the grains of weld zone didn't grow up and the number of second-phase particles precipitates increased. The impact toughness decreased, the tensile fracture didn't change.

References

[1] X M Li, Y Zou and Z W Zhang. (2010) Microstructure Evolution of a Novel Super304H Steel Aged at High Temperatures. Mater. Trans, 608: 164-173

[2] J Cao, Y Gong and Z G Yang. (2011) Creep fracture behavior of dissimilar weld joints between T92 martensitic and HR3C austenitic steels. Int. J. Press. Ves. Pip, 88: 94–98.

[3] P Li, J L Li and M Salman. (2014) Effect of friction time on mechanical and metallurgical properties of continuous drive friction welded Ti6Al4V/SUS321 joints. Materials. &. Design, 56: 649-656.

[4] P Li, J Li, X Li. (2015) A study of the mechanisms involved in initial friction process of continuous drive friction welding. J. Adhes. Sci. Technol, 29: 1246-1257.

[5] X Li, J Li, Z Liao . (2018) Effect of rotation speed on friction behavior and radially non-uniform local mechanical properties of AA6061-T6 rotary friction welded joint. J. Adhes. Sci. Technol, 32: 1987-2006.

[6] G H Chen, Y M Song, J Q Wang. (2012) High-temperature short-term tensile test and creep rupture strength prediction of the T92/TP347H dissimilar steel weld joints. Eng. Fail. Anal, 26:220-229.

[7] J Cao, Y Gong and Z G Yang. (2011) Microstructural analysis on creep properties of dissimilar materials joints between T92 martensitic and HR3C austenitic steels. Mater. Sci. Eng. A, 528:6103-6111.

[8] J Cao, Y Gong and K Zhu. (2011) Microstructure and mechanical properties of dissimilar materials joints between T92 martensitic and S304H austenitic steels. Materials. &. Design, 32:2763-2770

[9] G H Chen, Q Zhang and J J Liu. (2013) Microstructures and mechanical properties of T92/Super304H dissimilar steel weld joints after high-temperature ageing. Materials. &. Design, 44:469-475

IOP Conf. Series: Materials Science and Engineering **733** (2020) 012002 doi:10.1088/1757-899X/733/1/012002

Effects of Friction Pressure on Microstructures and Mechanical properties of friction welded Super304H austenitic welding joints

Yuan Lu*, Debiao Li, Jianlong Zhang, Cheng Bi, Fuyang Sun, Yang Yang

Xi'an Special Equipment Inspection Institute, Xi'an, 710065, China

*Corresponding author's e-mail: luyuan19801104@163.com

Abstract. In this paper, the Super304H austenitic steel were welded by continuous drive friction welding technique. The friction welding joints exhibited good mechanical and metallurgical properties. The influence of friction pressure on the microstructure and mechanical properties of Super304H welding joints was investigated. With an increase in the friction pressure, the grains in the welding zone and the heat affected zone didn't grow up and the number of the precipitated phases increased. The fracture location of tensile strength samples was in the heat affected zone. With an increase in the friction pressure, the Vickers hardness and the tensile strength increased.

1. Introduction

Increased heat efficiency is always the innovative driving forces in the development of ultra-supercritical boilers for fossil power plants, whose steam temperature is up to 600°C and pressure exceeds 25MPa. Under this ultra-supercritical condition, the heat efficiency can rise to around 45%. Many ultra-supercritical boilers have been built and successfully operated all over the world. At present, the worldwide aspirations for ensuring energy safety with simultaneous tightening of the requirements of the act on permissible emission of pollutants into the atmosphere have contributed to the development of materials that allow the improvement of steam working parameters resulting in units designed for ultra-supercritical steam parameters. The construction of boilers with ultra-supercritical steam parameters between 550 and 760°C requires the use of new materials with higher mechanical parameters and more complex manufacturing process for making the components of boiler pressure section out of them [1].

In order to fulfill the tough parameters of the ultra-super critical power generation, traditional heat-resistant steels used in the ultra-supercritical units have been replaced by those new heat-resistant steels. To meet the critical demands for the ultra-supercritical critical power generation, a new austenitic steel Super304H (0.1C-18Cr-9Ni-3Cu-Nb-N) has been developed. The Super304H has higher strength at elevated temperatures are required for super-heater tubes in fossil fired boilers. The conventional fusion welding joints of austenitic steel exhibit inferior mechanical properties due to the formation of intermetallic compounds at the joint interface, and the excessive residual stresses. The friction welding can be more suitable than fusion ones since many problems associated with melting are eliminated or reduced [2, 3]. The aim of this paper is to fabricate a combination the Super304H austenitic steel by using continuous drive friction welding with better mechanical properties. The effect of friction pressure on the microstructure and mechanical properties of the welded joints were investigated.

2. Experimental procedures

The pipes of Super304H austenitic steel that in the sizes of Φ44.5mm x 9mm were used as base metals. The pipes of Super304H were friction welded by a continuous drive friction welding machine. In the process of friction welding, the friction speed was 1500 rpm, the upset pressure was 150 MPa and the friction time was 5s. The friction pressure was 100 MPa of Process 1, 150 MPa of Process 2 and 200 MPa of Process 3 respectively. The metallography and fractographs of the welding joints were performed by an optical microscope and a scanning electron microscope respectively. The X-ray XRD patterns were examined by X-ray Diffraction. The tensile samples were tested using a testing instrument, and Vickers hardness values were tested using a Vickers hardness testing instrument.

3. Results and Discussion

Figure 1 showed the X-ray diffraction profiles for the Super304H welding joints welded by Process 2. It could be sure that the peaks with the highest intensity correspond to the γ-Fe phase and the precipitated phases of $Cr_{23}C_6$ and NbC. Therefore, the main phases were the γ-Fe phase and the precipitated phases of $Cr_{23}C_6$ and NbC. The Cr element had a tendency to form carbides, some C element and Cr element were precipitated from matrix and formed the precipitated phases $Cr_{23}C_6$ at the grain boundary preferentially. Some N element replaces the element C, and a part of the NbC were converted into Nb(C, N). It was difficult to index the Nb(C, N) due to lower intensities peaks. Therefore, $Cr_{23}C_6$ and NbC were denoted precipitated phases in the X-ray diffraction profiles [4, 5].

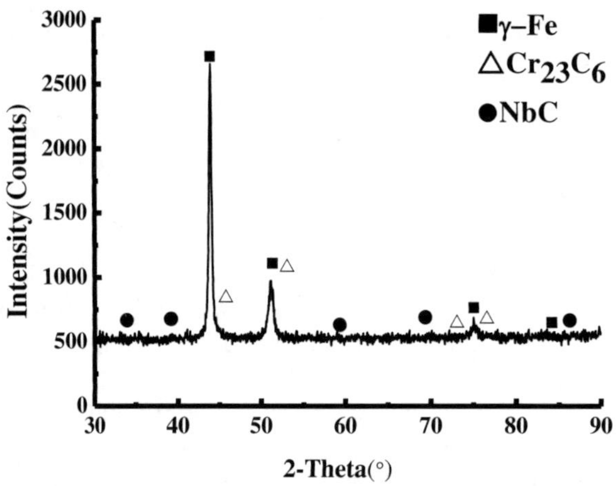

Figure 1 the XRD patterns of for the Super304H welding joints

Figure 2, 3 and 4 shows the metallography of the welding joints with different friction pressure respectively. During the friction welding process, the welded zone produced agglutinate and shear tear behaviour, which led to the deformation of the austenitic grains in the welding zone, and the dynamic recrystallization driving force and lattice distortion energy increased. The thermoplastic deformation temperature and unit volume free energy of recrystallized grains decreased, which produce a large number of recrystallized nucleations that leading to the fine grained dynamic recrystallization structure of the weld zone. The welding interface that was straight and clear, and the welding interface is well combined. The welding zone of austenitic steel welded by fusion welding was typical dendritic morphology with large grains. Compared with the conventional fusion welding joints, the welding zone were mainly composed of fine equiaxed austenite grains because of grain refinement. So the welding zone exhibit excellent microstructure and mechanical properties [6]. The friction welding was solid state welding techniques, the melting process did not occur in the welding zone and the heat affected zone. So the heat input during the friction welding process was much less than the heat input during the fusion welding process. The width of the heat affected zone caused by friction welding process were much smaller than the width of the heat affected zone caused by the fusion welding process[6]. The tendency of grains growth during the friction welding process was less obvious than the tendency of grains growth during the fusion welding process. So the friction welded Super304H joints with the fine grains in the welding zone and heat affected zone showed excellent microstructure and mechanical properties.

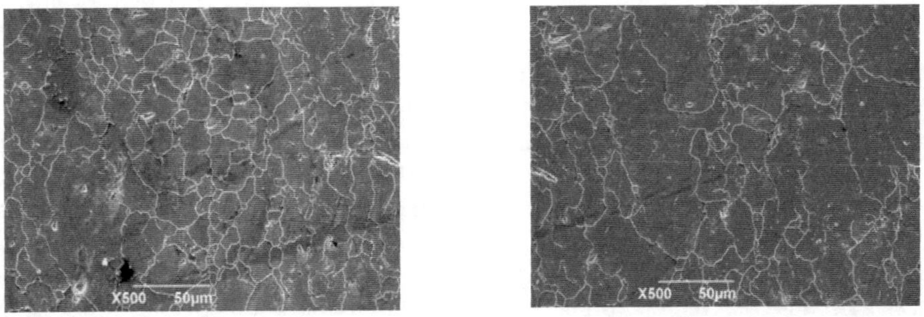

(a) the welding zone (b) the heat affected zone

Figure 2 The metallography of Super304H welding joints for Process 1

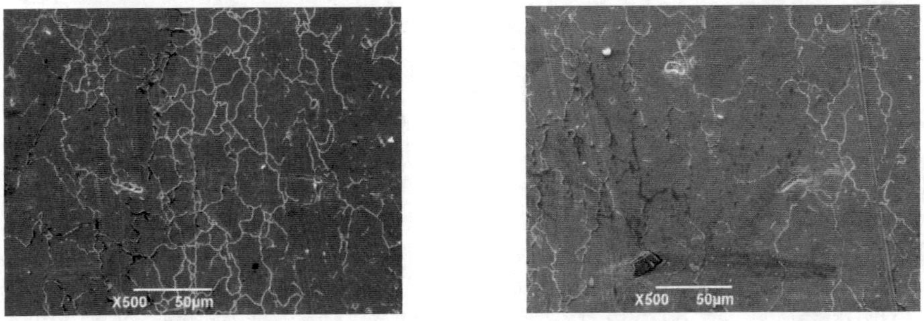

(a) the welding zone (b) the heat affected zone

Figure 3 The metallography of Super304H welding joints for Process 2

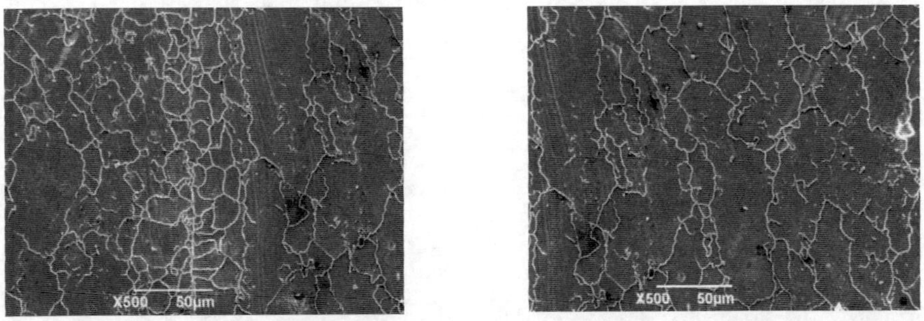

(a) the welding zone (b) the heat affected zone

Figure 4 The metallography of Super304H welding joints for Process 3

The surface temperature of the welding joints was equal to the temperature of the welding heat source in friction welding process, so the surface temperature of the welding joints had a great influence on the microstructure and mechanical properties of the welding joints. The surface temperature of the welding joints was calculated according to the Eq (1)

$$T(t) = qt/(\pi\lambda C) \quad (1)$$

T was referred to the surface temperature of the welding joints, t was referred to friction time and q was referred to friction heating power. It was calculated according to the Eq (2)

$$q = (\pi nT)/30 \quad (2)$$

T was referred to the friction torque, n was referred to the friction speed, and q was referred to the friction heating power. The friction heating power was mainly depended on the friction torque and the friction speed. During the friction welding process, when the friction speed was constant, as the welding torque increased, the friction heating power and the surface temperature of the welding joints increased correspondingly. The friction torque was mainly depended on the friction pressure and the coefficient of friction, so the friction torque was proportional to the friction pressure. Therefore, as the friction pressure increased, the friction heating power and the surface temperature of the welding joints increased. Because that melting process did not occur in the welding zone, although the surface temperature of the welding joints increased, the size change of the grains in the welding zone was not detectable. As the friction pressure increased, the heat input during the friction welding process increased and the temperature of the heat affected zone increased accordingly. The size change of the austenite grains which in the heat affected zone was not detectable, but the amount of the precipitated phases $Cr_{23}C_6$ and NbC in the both heat affected zone increased obviously [7, 8].

Table 1 depicted the variation of the yield strength and tensile strength as a function of the friction pressure. It was detected that the fracture position of the welding joints was in the heat affected zone for the different friction pressure. The fractographs of the tensile strength samples were shown in figure 5. As the friction pressure increased, the tensile strength of the welding joints increased gradually. The friction heating power and the temperature of the heat affected zone increased during the friction welding process with the friction pressure increased. The amount of the precipitated phases $Cr_{23}C_6$ and NbC increased obviously in the heat affected zone, the dispersion strengthening effect of the $Cr_{23}C_6$ and NbC carbides was main reason of the tensile strength increased. The fracture morphology was characterized by typical ductile fracture, and a lot of fine and shallow dimples distributed [9]. Table 1 depicted the variation of the Vickers hardness as a function of the friction pressure. When the grains size was smaller, the Vickers hardness was higher. The grains size of the welding zone with fine grains was the smaller, so the Vickers hardness of the welding zone was the higher. As the friction pressure increased, the dislocation density and lattice distortion energy in the welding zone increased, and the Vickers hardness of the welding zone increased. The precipitation of the $Cr_{23}C_6$ carbides was the reason that the Vickers hardness of the heat affected zone increased.

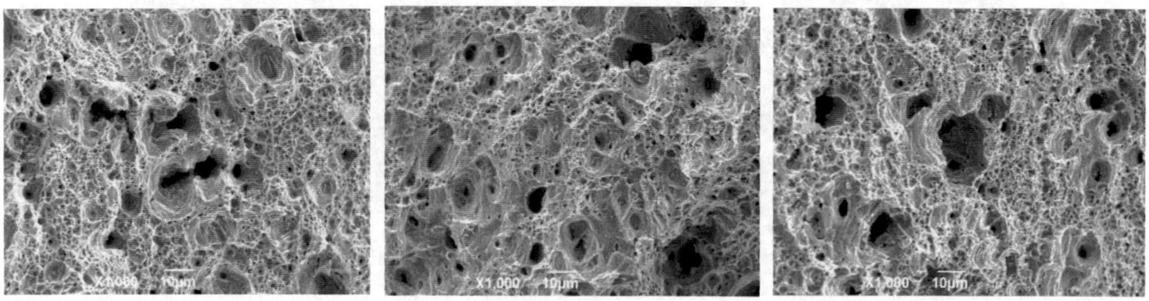

(a) Process 1 (b) Process 2 (c) Process 3

Figure 5 The fracture appearance of the tensile strength samples

Table 1 The mechanical properties of Super304H welded joints

	Tensile strength /MPa			Vickers hardness / Hv	
	$R_{p0.2}$	R_m	Fracture position	Welding zone	Heat affected zone
Process 1	456.29	736.66	heat affected zone	333.71	265.92
Process 2	493.14	764.51	heat affected zone	367.80	282.22
Process 3	518.99	788.29	heat affected zone	402.21	305.71

4. Conclusions

The Super304H austenitic steel welded joints with excellent properties were welded by continuous drive friction welding technique. The welding interface was well combined, and the austenite grains in the welding zone was fine. The fracture position was in the heat affected zone. The influence of friction pressure on the microstructure and properties of Super304H joints was investigated. With an increase in the friction pressure, the grains of weld zone didn't grow up and the number of second-phase particles precipitates increased. With an increase in the friction pressure, the tensile strength and the Vickers hardness increased.

References

[1] X M Li, Y Zou and Z W Zhang. (2010) Microstructure Evolution of a Novel Super304H Steel Aged at High Temperatures. Mater. Trans, 608: 164-173

[2] P Li, J Li, X Li. (2015) A study of the mechanisms involved in initial friction process of continuous drive friction welding. J. Adhes. Sci. Technol, 29: 1246-1257.

[3] X Li, J Li, Z Liao . (2018) Effect of rotation speed on friction behavior and radially non-uniform local mechanical properties of AA6061-T6 rotary friction welded joint. J. Adhes. Sci. Technol, 32: 1987-2006.

[4] K MyungYeon, K Sukchul and C InSuk. (2014) High-temperature tensile and creep deformation of cross-weld specimens of weld joint between T92 martensitic and Super304H austenitic steels. Mater. Charact. 97: 161-168.

[5] J Cao, Y Gongan and K Zhu. (2011) Microstructure and mechanical properties of dissimilar materials joints between T92 martensitic and S304H austenitic steels. Materials. &. Design, 32:2763-2770

[6] G H Chen, Q Zhang and J J Liu. (2013) Microstructures and mechanical properties of T92/Super304H dissimilar steel weld joints after high-temperature ageing. Materials. &. Design, 44:469-475

[7] A Zielinski. (2012) Structure and properties of Super 304H steel for pressure elements of boilers with ultra-supercritical parameters. Journal of Achievements in Materials and Manufacturing Engineering, 55:403-409

[8] S Indrani, E Amankwaha, S Kumara. (2011) Microstructure and mechanical properties of annealed SUS 304H austenitic stainless steel with copper. Mater. Sci. Eng. A, 528: 4491-4499.

IOP Conf. Series: Materials Science and Engineering **733** (2020) 012003 doi:10.1088/1757-899X/733/1/012003

Application Research of Quantitative Deductive Kansei Engineering in APP Interface Optimization Design

Yuanyuan Ding

Lecturer, College of Mechanical and Electronic Engineering, Weifang University, Weifang, China
E-mail: deeyuanyuan@163.com

Abstract. In order to better explore people's emotional appeals in the design of Application interface, this paper applies the quantitative deductive method of Kansei Engineering to the research of its optimization design. Firstly, questionnaire survey, cluster analysis and other methods are used to determine the design elements of APP interface and representative perceptual words, and then the semantic difference method is used to establish the relationship between the two, so as to carry out perceptual evaluation on representative APP interface samples. On this basis, qualitative analysis is used to explore the corresponding relationship between perceptual words and design elements. Finally, the validity of the conclusions obtained from the qualitative analysis is verified by the orthogonal analysis method, and the corresponding relationship between the consumer's emotional demands and APP interface design elements is thus established.

1. Introduction

As the mainstream media of modern social information dissemination, mobile phones have been integrated into all aspects of people's lives. The mobile APP interface provides us with a virtual visual environment that transcends physical material. How to utilize mobile phone software interface design with limited design space and visual elements has become an important task faced by the designers.

In order to consider more about the user's intention in the design, this paper applies the quantitative deductive method of Kansei Engineering to the research of APP interface optimization design. By the means of computer and statistical analysis, this paper aims to effectively obtain the corresponding relation between consumer's psychological and emotional needs and APP interface design elements, thus improving the emotional and spiritual care of the APP interface for consumers.

2. Application and Implementation of Quantitative Deductive Kansei Engineering in Research

2.1. Analysis of APP Interface Samples

In the research, APP interface samples were collected, cluster analysis was used to cluster the samples, and the samples were divided into 8 categories. Then k-means clustering was applied and the clustering number was set to 8. The coordinate values were analyzed, the distance between each sample and the clustering center was calculated, and the closest interface sample to the cluster center is a representative sample of the class. Taking category 1 as an example, the classification results are shown in Table 1. The sample (21) is the closest to the clustering center and is the representative sample of category 1. Similarly, the eight most representative APP interface samples are finally determined.

Table 1. Distance between Samples in Category 1 and the Clustering Center.

Samples	Distance to the cluster center
Sample（**21**）	0.63812
Sample（**19**）	0.77527
Sample（**3**）	0.82738
Sample（**7**）	0.97437
Sample（**33**）	1.19268

2.2. *Emotional evaluation of semantic differential method（SD）*

A 7-level semantic difference scale (SD scale) was established, as shown in Table 2. The subjects were asked to make perceptual evaluation on 8 representative APP interface separately.

Table 2. Example of SD scale for samples.

Complicated	-3	-2	-1	0	1	2	3	Simple
Messy	-3	-2	-1	0	1	2	3	Unitary
Monotonous	-3	-2	-1	0	1	2	3	Colorful
Ornate	-3	-2	-1	0	1	2	3	Plain
Conventional	-3	-2	-1	0	1	2	3	Modern
Cautious	-3	-2	-1	0	1	2	3	Lively
Hard-to-use	-3	-2	-1	0	1	2	3	Easy-to-use
Inconspicuous	-3	-2	-1	0	1	2	3	Eye-catching
With low commercial value	-3	-2	-1	0	1	2	3	With high commercial value

2.3. *Establishment of corresponding relationship between design elements and kansei vocabulary*

2.3.1. *Qualitative analysis*

The perceptual data of APP interface samples obtained from the above SD scale evaluation test were collated and analyzed, and the average scores of kansei vocabulary corresponding to each sample were calculated, the questionnaire results analysis table as shown in Figure 1 was generated through EXCEL, and then the data was qualitatively analyzed according to the chart.

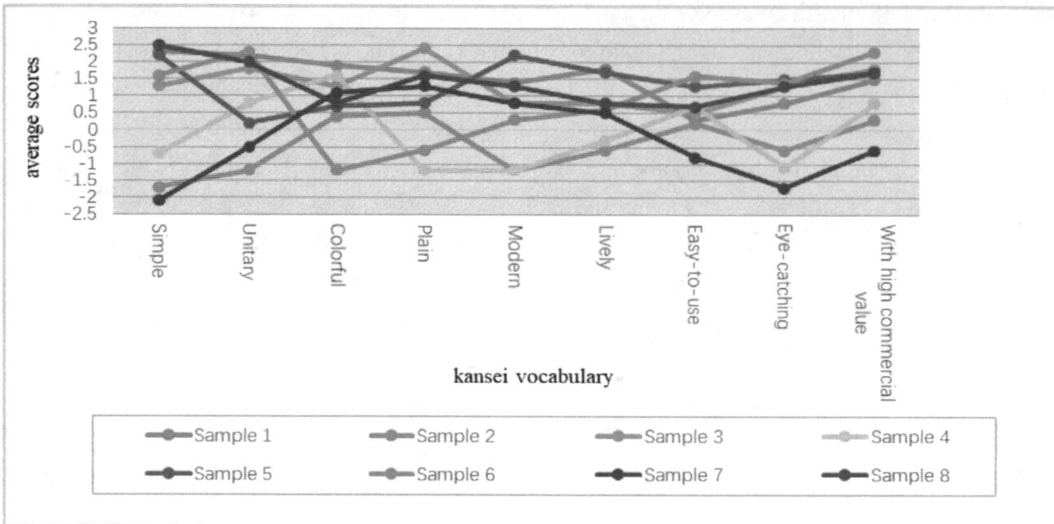

Figure1. Survey results analysis of SD scale.

According to the content shown in Figure 1, the scores of each APP interface sample and the corresponding kansei vocabulary are analyzed, and the highest and lowest scores corresponding to each kansei vocabulary can be found. In addition, with further comprehensive analysis of each kansei vocabulary and sample design element, a certain correspondence law between the two can be found, and the corresponding relation between the kansei vocabulary and the design elements is obtained as shown in Table 3. In general, the four factors of the layout, color, picture and navigation bar of the APP interface have a greater influence on the emotional appeals of the subjects, while other factors have a less obvious influence on the emotional appeals.

Table 3. The corresponding relation between kansei vocabulary and design elements.

Design elements \ Kansei appeals		Simple	Unitary	Colorful	Plain	Modern	Lively	Easy-to-use	Eye-catchin	With high commercial value
Layout	Regional style		√					√	√	
	Freestyle				√					√
Color	Similar color		√						√	
	Contrast color				√			√		√
Text	Same font color	√					√	√	√	
	Different font colors		√	√	√					√
Picture	Picture scrolling		√					√		
	Picture fixed	√			√		√		√	√
Navigation bar	In the middle of the interface	√					√	√	√	
	Below the interface									
Link	Color differentiation	√	√	√	√	√				√

	No obvious link instructions							√		
Animation	With animation				√			√		
	Without animation		√			√				√

2.3.2. Verifying the Conclusion by Orthogonal Analysis

In order to verify the conclusions of the above qualitative analysis and conduct in-depth analysis, this part will adopt orthogonal analysis method to conduct research.

In the orthogonal test, various experimental conditions are called "factors", and the state in which each factor is desirable is called "level", and the quantity to evaluate and measure the experimental effect is called "index". When different levels of each factor are taken, the index will change, but the change range of the index caused by different factors is different. When the index variation range (range R) caused by different levels of factors is large, the factor has a great influence on the index, which is the main factor. Therefore, by analyzing the range of the orthogonal test, the primary and secondary degree of the influence of different factors (elements) and levels (design elements) on the index (emotional appeal) can be obtained, and the conclusion of "which design elements combination should be selected to achieve a certain sensibility of the APP" can be drawn to guide the emotional design.

In this study, the commonly used $L_8(2^7)$ orthogonal table was used for orthogonal test, indicating that 8 experiments were needed according to 8 samples, and 7 factors could be observed at most, each factor at 2 level.

As some of the 8 representative APP interface samples selected by k-means clustering method before did not meet the requirements of $L_8(2^7)$ orthogonal table and the analysis of APP interface design features, 8 representative samples were re-selected through evaluation to make the factors and levels of each sample meet the requirements of $L_8(2^7)$ orthogonal table.

By using the 9 groups of typical kansei vocabulary selected in 1.2.2, the SD perceptual evaluation of the 8 newly selected representative APP interface samples was re-examined according to the experimental steps of the semantic differential method. Then the evaluation data were sorted out and the orthogonal test table of L8(27) was established. The sum of each index at different levels of each factor was calculated, i.e. T1, T2. The processing results of orthogonal test data obtained were shown in Table 4.

Table 4. Orthogonal Test Data Processing Results.

Factor / Index		Layout	Color	Text	Picture	Navigation bar	Link	Animation
Simple	T1	3.35	3.86	1.35	0.93	4.31	3.38	4.15
	T2	1.09	0.51	-0.08	0.44	0.96	-0.55	3.39
Unitary	T1	3.87	2.61	1.43	2.05	2.21	4.15	1.53
	T2	-0.04	-1.71	-1.95	1.40	-0.42	1.92	1.27
Colorful	T1	2.91	2.09	1.52	1.53	2.08	0.92	3.17
	T2	2.08	-1.84	-1.33	-0.68	0.25	-0.65	-0.02
Plain	T1	2.21	2.04	1.23	3.03	1.52	3.62	1.21
	T2	1.24	0.37	-1.03	2.06	0.37	3.27	0.44
Modern	T1	1.91	2.75	2.63	2.07	1.09	3.87	1.93

	T2	1.36	2.22	1.91	0.10	0.91	3.61	1.10
Lively	T1	0.92	0.53	2.25	2.33	0.24	2.47	2.65
	T2	-1.03	-1.74	-1.78	1.08	-1.69	0.81	-1.67
Easy-to-use	T1	3.19	4.43	3.38	2.64	3.17	3.63	3.32
	T2	-0.05	2.88	2.69	2.21	-0.20	-0.40	2.77
Eye-catching	T1	3.73	1.23	1.93	2.05	3.19	3.42	3.87
	T2	0.72	-1.03	0.97	1.48	-0.07	-0.41	0.30
With high commercial value	T1	4.59	2.92	2.61	1.13	0.95	4.02	0.65
	T2	4.3	2.54	2.34	-1.40	-1.32	0.78	-2.33

According to the orthogonal test data processing results in Table 8, the range R (| T1 -T2 |) was calculated, and the obtained range calculation results are shown in Table 5.

Table 5. Orthogonal Test Data Range Calculation Result.

Factor / Index	Layout	Color	Text	Picture	Navigation bar	Link	Animation
Simple	2.26	3.35	1.43	0.49	3.35	3.93	0.76
Unitary	3.91	4.32	3.38	0.65	2.63	2.23	0.26
Colorful	0.83	3.96	2.85	2.21	1.53	1.57	3.19
Plain	0.97	1.67	2.26	0.97	1.15	0.35	0.77
Modern	0.55	0.53	0.72	1.97	0.18	0.26	0.83
Lively	1.95	2.27	4.03	1.25	1.93	1.66	4.32
Easy-to-use	3.24	1.55	0.69	0.43	3.37	4.03	0.55
Eye-catching	2.65	2.62	0.96	0.57	3.26	3.83	3.57
With high commercial value	0.29	0.38	0.27	2.53	2.27	3.24	2.98

According to the standard of "the greater the variation range (range R) of the index caused by taking different levels of the factor, the greater the influence of the factor on the index and this factor is the main influence factor", the following conclusions can be drawn through the range analysis and factor analysis in Table 5:

(1) The influence of different factors (elements) and levels (design elements) on the index (emotional appeal) is exemplified by the influence of factors and levels on the four indicators of simplicity, integrity, applicability and conspicuousness, to illustrate the influence of design elements on emotional appeal, as shown in Table 6.

Table 6. The Influence of Design Elements on Emotional Appeal.

Index	The influence of factors on indicators	The influence of level on indicators
Simplicity	Color, navigation bars, and links have the greatest influence on simplicity; layout and text have a greater influence on it; other factors have little influence on it.	The APP interface design consisting of similar colors, navigation bars located at the bottom of the interface, color-differentiated links and other elements is considered to be simpler.

Integrity	Layout, color and text have the greatest influence on integrity; navigation bars and link have a greater influence on it; other factors have little influence on it.	The APP interface design consisting of regional layout, similar color, text with the same font color and other elements is considered to be more integrated.

Similarly, the influence of factors and levels on the seven indicators of color, applicability, conspicuousness, simplicity, modernity, liveliness and high commerciality can be obtained, indicating the degree of influence of design elements on these emotional appeals.

(2) Based on the above analysis of the influence of the design elements on emotional appeal, the four factors of APP interface, i.e. layout, color, picture and navigation bar have the greatest influence on consumers' emotional appeals, while other factors have a slightly weaker or less obvious influence on emotional appeal.

(3) The conclusions obtained by the orthogonal analysis test are basically consistent with the conclusions obtained from the previous qualitative analysis.

2.4. Interpret the Analysis Results and Provide Design Guidance

Through the analysis of Table 3 and Table 6, it can be seen that in the optimization design process of APP interface, the ideal kansei vocabulary can be selected, so as to obtain the APP interface design elements that can best reflect the vocabulary. For example, the combination of design elements of "integrated APP interface" is roughly as follows: regional type (layout) + similar color (color) + same font color (text) + being located at the bottom of the interface (navigation bar) + color differentiation type (link). In the same way, other combinations of APP interface design elements that reflect the user's emotional appeal can be obtained.

3. Conclusions

In this paper, the implementation process of the quantitative deductive method of Kansei Engineering in the optimization design of APP interface is discussed in detail. In addition, qualitative analysis and orthogonal analysis are used to verify each other to establish the relationship between relevant kansei vocabulary and design elements, and transform people's vague psychological and emotional needs for APP interface into specific design elements. The future research will also discuss the rules of the influence of different configuration of design elements on users. As the research progresses further, it will provide more scientific and feasible references for the optimization design of APP interface.

References:

[1] Ming Changchun and Kong Zhaolin. *2018J.Advanced Engineering Informatics*, Vol.**38** pp.826-839

[2] Wu Qiong. *2010.Art & Design*. Vol.**201** pp.34-37.

[3] Cheng-Ta Yeh, Mu-Chen Chen. 2018J.Computers & Industrial Engineering.Vol.120, pp.401-417

[4] Xia Bozhong. 2008.*Orthogonal Test.* (Jilin :Jilin People's Publishing House). p 133

[5] Wei Xu and Dov Furie. *2019J. Ergonomics*, Vol.**62 (7)**, pp.954-980

IOP Conf. Series: Materials Science and Engineering **733** (2020) 012004 doi:10.1088/1757-899X/733/1/012004

Analysis of waveguide with special resonant window which can reduce the energy consumption of waste rubber pyrolysis

Qiangcai Liu[1], Xin Su[1], Lihao Hu[1], Xianbing Zhao[1] and Zhihua Li[1,2]

1 Qingdao University of Science and Technology, 99 Songling Road, Laoshan District, Qingdao, 266601,China
2 E-mail address: 1913077247@qq.com

Abstract. Microwave pyrolysis is a new technology for waste rubber treatment. In this paper, the microwave energy transfer characteristic working at a frequency of 2.45 GHz in different types of rectangular waveguides are investigated with HFSS simulation software. The results show that although the quartz sealing plate of the rectangular waveguide can prevent the combustible gas generated by waste rubber pyrolysis from entering the microwave generator, it also causes some microwave reflection at the same time, which will reduce the microwave energy transmission efficiency, shorten the service life of the microwave generator. While the resonant window of the rectangular waveguide, that is, a rectangular window with a certain size in the centre of the copper sealing plate, and with a quartz plate embedded in the window, can avoid such microwave reflection, not only enhance the sealing performance, but also improve the waste rubber pyrolysis efficiency and reduce the pyrolysis energy consumption.

1. Introduction

With the extensive use of rubber products, the amount of waste rubber is also increasing at an alarming rate and the "Black pollution" caused by waste rubber is becoming more and more serious. How to treat waste rubber effectively, especially to realize the recycling of waste rubber, has become an important research subject for environmental protection and resource conservation[1,2]. In recent years, microwave pyrolysis as a new approach for waste rubber treatment has attracted widespread attention in the industry, and is considered to be the most effective and sustainable process for waste rubber recycling due to its eco-friendly process and its advantages over the conventional methods such as its strong penetrability, features of rapidly and uniformly cracking, high energy utilization efficiency and so forth. The commonly used microwave frequencies for microwave pyrolysis equipment is 2.45GHz. In the microwave pyrolysis process, the waste rubber is cracked as high-quality oil, combustible gas and carbon black, thereby realizing high-value recycling of waste rubber[3].

In order to prevent the combustible gas from entering the microwave generator through the waveguide, a quartz sealing plate is set in the waveguide when designing the microwave pyrolysis equipment. Quartz is a material with excellent wave transmission performance. However, it has been confirmed that when the microwave is transmitted in the waveguide, there is some of the microwave energy fail to pass through the quartz sealing plate, but be reflected back to the microwave generator, causing the microwave generator to heat up, it not only shortens the service life of microwave generator, but also increases the energy consumption of pyrolysis[4].

HFSS is a full-wave three-dimensional electromagnetic simulation software, by adopting the finite element method, it can calculate the electromagnetic characteristics of various RF/microwave passive components rapidly and accurately, thereby through gaining the S-parameter, propagation constant, and electromagnetic property to optimize the components performance[5]. In this paper, through establishing reasonable three-dimensional waveguide models and setting reasonable boundary conditions and excitations, then the HFSS software can accurately accomplish the simulation of the electromagnetic field distribution of the waveguide, and the corresponding S parameter value can be obtained, furthermore, based on this, the influence of quartz sealing plate and resonant window on pyrolysis performance can be analysed.

2. The rectangular waveguide with quartz sealing plate

2.1. Theoretical basis

Scattering parameter, or S parameter for short, is a network parameter based on the relationship between incident wave and reflected wave, and it describes the circuit network in terms of the reflected signal from another device port and the transmitted signal from this port to another port. The scattering matrix equation can be expressed as equation (1).

$$\begin{pmatrix} b_1 \\ b_2 \\ \cdots \\ b_n \end{pmatrix} = \begin{pmatrix} S_{11} & S_{12} & \cdots & S_{1n} \\ S_{21} & S_{22} & \cdots & S_{2n} \\ \cdots & \cdots & \cdots & \cdots \\ S_{n1} & S_{n2} & \cdots & S_{nn} \end{pmatrix} \begin{pmatrix} a_1 \\ a_2 \\ \cdots \\ a_n \end{pmatrix} \quad (1)$$

Here, a and b are the matrix representation of the normalized incident waves and normalized reflected waves, respectively, and both contains n normalized incident waves and n normalized reflected waves; Sij is the transfer coefficient from port j to port i under the circumstances that all other ports are connected to the load.

In figure 1, a two-port transmission network is taken as an example. Assuming that Port 1 and Port 2 are the signal input port and output port, respectively, according to the beforementioned basis, S_{11} and S_{21} served as two more important S parameters is obtained compared to other S parameters, where S_{11} represents the return loss, that is, how much energy is reflected back to the source port(Port 1); S_{21} represents the insertion loss, that is, how much energy is transmitted to the destination port (Port 2). Therefore, the S parameter can be used to indicate the reflected and transmitted power, and calculate the efficiency of microwave energy transmission, and thus the microwave pyrolysis effect can be analysed according to the efficiency of microwave transmission. Using the terms P, P_1 and P_2 to represent the microwave incident power, the microwave transmission power from port 1 to port 2 and the microwave reflected power, respectively, and expressing S parameter in logarithmic form, then the internal correspondences of which can be illustrated by equation (2) and equation (3).

$$S_{11} = 10\lg(\frac{P_2}{P}) \quad (2)$$

$$S_{21} = 10\lg(\frac{P_1}{P}) \quad (3)$$

According to equation (2) and equation (3), the calculation equations of energy reflection and transmission efficiency were deduced as follows:

$$\eta_{11} = (\frac{P_2}{P}) \times 100\% = 10^{\frac{S_{11}}{10}} \times 100\% \quad (4)$$

$$\eta_{21} = (\frac{P_1}{P}) \times 100\% = 10^{\frac{S_{21}}{10}} \times 100\% \quad (5)$$

Where η_{11} is energy reflection efficiency and η_{21} is energy transmission efficiency.

2.2. Simulation analysis

The waveguide model BJ22 is determined by the standard rectangular waveguide selection specification[6]. The three-dimensional dimensions of the hollow and rectangular metal waveguide are 109.22mm,54.61mm,200mm in length, width and height, and the microwave frequency is 2.45 GHz. Set the side boundary condition of the rectangular waveguide to Perfect E, and set the excitations of the two end faces of the rectangular waveguide to Wave Ports. The waveguide model with a quartz sealing plate is shown in figure 2. The sealing plate is made of quartz material with a relative dielectric constant of 3.78 and a thickness of 4mm. Its length and width are parallel with the size of the rectangular waveguide section size. In addition, as a control, an empty waveguide model that has the same parameters but without the sealing plate is also analyzed.

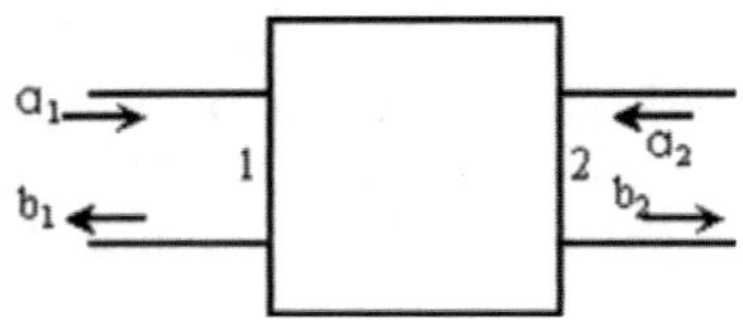

Figure 1. The schematic diagram of a two-port transmission network.

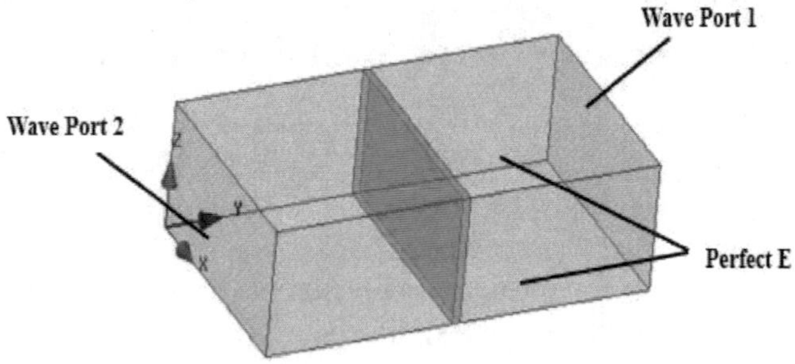

Figure 2. The waveguide model with a quartz sealing plate.

The S parameter values obtained by the simulation of the two waveguide models were listed in table 1. Based on the data in the table, the energy reflection efficiency (η_{11}) and energy transmission efficiency (η_{21}) can be calculated by equation (4) and equation (5) mentioned before. Finally, the conclusions can be displayed as follows: For the empty waveguide, the energy reflection efficiency is zero, while the transmission efficiency is 100%. In other words, under ideal conductor boundary conditions, microwave energy can be completely transmitted from port 1 to port 2 through the waveguide if there is no quartz sealing plate. For the waveguide with quartz sealing plate, the energy reflection efficiency is about 10.18%, and the energy transmission efficiency is about 89.82%. According to the conclusions and the comparison of the calculations, it can be seen that the quartz sealing plate will cause some of energy reflection in the process of energy transmission, and as the thickness of the quartz seal increases, the reflection will become more serious. As a consequence, the reflection of microwave energy will reduce the energy transmission efficiency and increase the energy consumption of pyrolysis. In addition, if the reflected microwave energy is absorbed by the microwave generator, it's service life will be shortened in some way.

Table 1. The S parameters of empty waveguide and waveguide with a quartz sealing plate.

Model	Frequency (GHZ)	S11(dB)	S21(dB)
Empty waveguide	2.45	-70.569268	0
Waveguide with a quartz sealing plate	2.45	-9.922160	-0.466770

3. The rectangular waveguide with a resonant window

3.1. Working principle of the resonant window

Combining the capacitive diaphragm and the inductive diaphragm together will form a rectangular window-shaped diaphragm, and its equivalent circuit is a parallel-resonance circuit called resonant window[7].

The schematic diagram of the resonant window and its equivalent circuit are shown in figure 3.

As is demonstrated in figure 3(a), the resonant window can be regarded as the small waveguide with length a′, width b′ and height t′, while the waveguide with length a, width b, and height t is considered as the main waveguide. If the characteristic impedance of the small waveguide is equal to the characteristic impedance of the main waveguide, it is equivalent to the state where the inductance and the capacitance of the equivalent circuit are in a state of parallel resonance, and the relation can be illustrated by equation (6).

$$\frac{b}{a}\cdot\eta_1\cdot\left(1-\left(\frac{\lambda_1}{2a}\right)^2\right)^{-\frac{1}{2}}=\frac{b'}{a'}\cdot\eta_2\cdot\left(1-\left(\frac{\lambda_2}{2a'}\right)^2\right)^{-\frac{1}{2}} \tag{6}$$

In the expression above,

$$\eta_1=\left(\frac{\mu_1}{\varepsilon_1}\right)^{\frac{1}{2}},\lambda_1=\lambda_0\cdot(\varepsilon_{r1})^{-\frac{1}{2}};\quad \eta_2=\left(\frac{\mu_2}{\varepsilon_2}\right)^{\frac{1}{2}},\lambda_2=\lambda_0\cdot(\varepsilon_{r2})^{-\frac{1}{2}}.$$

Here, μ_1, ε_1 and ε_{r1} are the permeability, dielectric coefficient and relative dielectric constant of the packing medium in the main waveguide respectively. μ_2, ε_2 and ε_{r2} are the permeability, dielectric coefficient and relative dielectric constant of the packing medium in the small waveguide respectively.

The characteristics of the rectangular waveguide with a resonant window are as follows:(1) when the operating frequency is exactly equal to its resonant frequency, the microwave can pass through the resonant window without reflection;(2) when the operating frequency is not equal to its resonant frequency, microwave radiation reflection occurs due to the inductive or capacitive nature of the resonant window. Therefore, if the rectangular waveguide has a resonant window with appropriate size and its resonant frequency is 2.45GHz, theoretically, the energy transmission efficiency of the waveguide can reach 100% and with no reflection occurs.

3.2. Simulation analysis

The form of the sealing plate is changed from the quartz plate shown in figure 2 to the resonant window shown in figure 4. And the sealing plate is composed of a copper sealing plate and a quartz plate, both have a thickness of 4 mm. Specifically, the length and width of the copper sealing plate are the same as those of the rectangular waveguide, and in the center of the copper sealing plate opened a rectangular window, the quartz plate is embedded exactly in the rectangular window, in fact, it is equivalent to setting a quartz window in the center of the copper sealing plate figure 4 shows the rectangular waveguide model with a resonant window (quartz window).

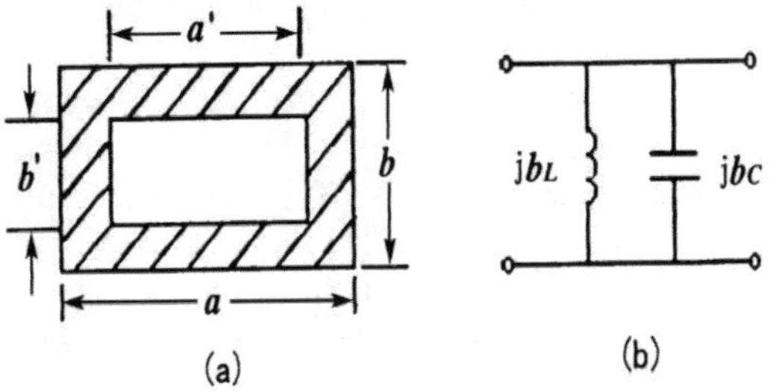

Figure 3. The schematic diagram of the resonant window and its equivalent circuit.

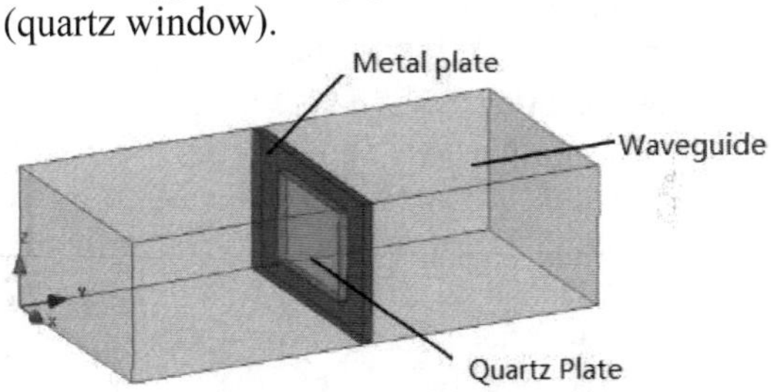

Figure 4. The rectangular waveguide model with a resonant window.

To obtain the appropriate size of the resonant window, the simulations targeting at different rectangular quartz windows with different lengths and widths were performed. The simulation results indicate that when the length and width of the rectangular quartz window are 55mm and 35mm respectively, the transmission efficiency of the waveguide is the highest. And in this case, the S-parameter curve and the electric field distribution of the waveguide were obtained and they have been provided in figure 5 and figure 6.

It can be seen from figure 5 that the rectangular waveguide with resonant window has higher energy transmission efficiency while the microwave frequency is between 2.4GHz and 2.5GHz. It should be noted that when the microwave working frequency is 2.45GHz, the return loss reached its peak close to zero, that is, the energy transmission efficiency is about 100%, almost the same as the transmission efficiency of the empty waveguide. So, it is reasonable to believe that the arrangement of the resonant window does significantly reduce the generation of reflection and improve the transmission efficiency of the microwave energy provided by the microwave generator working at 2.45GHz frequency. From figure 6, it can be seen that the waveguide electric field distribution is uniform, indicating that the microwave energy distribution is uniform. Hereby it is easy to draw a conclusion that the resonant window set in the rectangular waveguide solves the microwave energy reflection problem well, improves the transmission efficiency, and reduces the energy consumption. The results of the subsequent experiment are in agree with the results of the analysis conclusions.

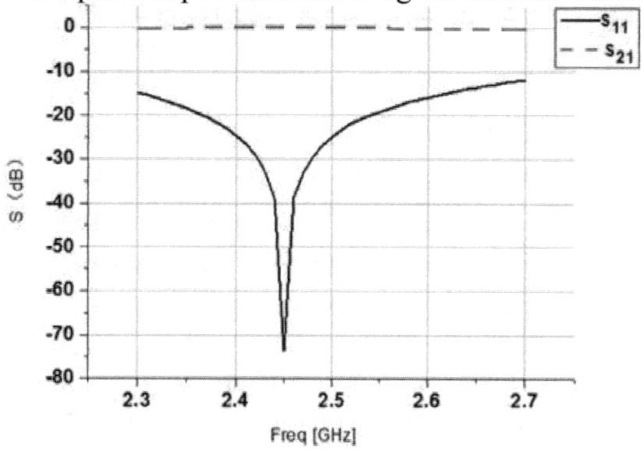

Figure 5. The S parameter curve of the resonant window.

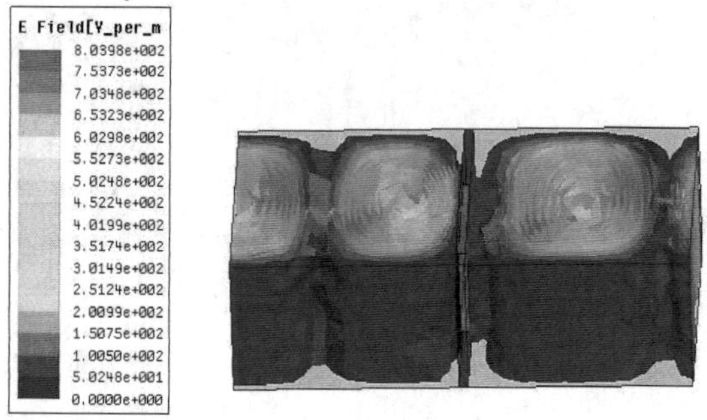

Figure 6. The electric field distribution of rectangular waveguide with a resonant window.

4. Conclusions

Aiming at the microwave energy reflection problem caused by the quartz sealing plate set in the waveguide of the microwave pyrolysis equipment, simulation about the microwave transmission behavior in different types of rectangular waveguides is performed through HFSS software in the paper. According to the performance of the simulation, the following conclusions were drawn, and it's been verified to be correct by experiments:

(1) Although the quartz sealing plate of the rectangular waveguide can prevent the combustible gas from entering the microwave generator, it will also cause microwave energy reflection, accounting for 10.18% of the total microwave energy, which will result in a decrease in microwave energy transmission efficiency, shorten the service life of the microwave generator.

(2) The microwave energy transmission efficiency of the waveguide can be improved while a resonant window is set inside the waveguide. And when the length and width of the resonant window are 55mm and 35mm, respectively, actually, that means the length and width of the embedded quartz window are 55mm and 35mm, respectively, the waveguide can achieve a microwave energy transmission efficiency of about 100%. In consequence, there is reason to believe that the resonant window can not only enhances the sealing performance, but also improves the waste rubber pyrolysis efficiency and reduces the pyrolysis energy consumption.

5. References

[1] Chen S N 2018 Investigation and analysis of the development status of rubber products industry *HEBEI QIYE* **(8)** 7-8

[2] Li Z 2018 Current situation, problems and countermeasures of recycling of waste tires in China *China Tire Resources Recycling* **(12)** 41-3

[3] Li Z H,Li Y,Song Y S,Liu Q C and Chen S Z 2018 Simulation and experimental study on continuous microwave pyrolysis for waste rubber *China Rubber Industry* **65(5)** 495-8

[4] Li Z H and Guo N 2018 Influence of tuning screw position in waveguide on pyrolysis energy consumption of waste rubber *China Rubber Industry* **65(5)** 573-5

[5] Li M Y and Liu M 2013 *HFSS Electromagnetic Simulation Design from Entry to Proficiency* （Beijing:Post & Telecom Press） chapter 2 pp 101-13

[6] Huachang Machinery Plant.GB/T11450.2-1989 *Relevant Specification for Ordinary Rectangular Waveguides* (Beijing:Standards Press of China) p 5

[7] Liao C E 1994 *Fundamentals of Microwave Technology* (Xian:XIDIAN University Press) pp 274-5

IOP Conf. Series: Materials Science and Engineering **733** (2020) 012005 doi:10.1088/1757-899X/733/1/012005

Nano layered-spinel $0.8Li_2MnO_3 \cdot 0.2LiMn_2O_4$ as high-performance cathode for Li-ion batteries

Zhenye Zhu[1] and Sang Zhuo[2]

1 School of Materials Science and Engineering, Harbin Institute of Technology, Shenzhen Graduate School, Shenzhen, 518055, China
2 The Eighth Affiliated Hospital, Sun Yat-Sen University 518033, PR China
Telephone:+86-0755-26032709, E-mail: zhuzy@hitsz.edu.cn

Abstract. The nano layered-spinel $0.8Li_2MnO_3 \cdot 0.2LiMn_2O_4$ cathode material has been synthesized by a simple solvothermal method. X-ray diffraction studies show that the as-prepared material has two compositions, including Li_2MnO_3 and $LiMn_2O_4$. Moreover, the surface morphology of the precursor and cathode material were characterized by scanning electron microscopy. Meanwhile, the EDS graph reflects homogeneous mixing of the two compositions due to the novel precursor synthesizing method. Consequently, as a cathode material for LIBs, the resulting nano layered-spinel $0.8Li_2MnO_3 \cdot 0.2LiMn_2O_4$ achieves a high discharge capacity of 248 mA h g^{-1} at 25 mA g^{-1} and 162 mA h g^{-1} even at 1250 mA g^{-1}. In addition, the cathode material shows a high reversible capability with 180 mA h g^{-1} after 50 cycles, which delivers medium cycling performance.

1. Introduction

Nowadays, lithium ion batteries(LIB) have widely used in many areas closely related to our lives since it is commercialized by Sony corporation. However, there is ever-increasing demand for electronic device with high energy density and excellent cycle performance. Among all issues, cathode materials may be the key point, such as low capacity, inferior rate capability and so on. As common cathode material of LIB, $LiCoO_2$ has a high theoretical capacity of 274 mA h g^{-1}, but only half of it is available in the practical application.[1] Therefore, cathode materials for lithium ion batteries with high performance are urgently needed to be developed. Fortunately, lithium-rich manganese oxide materials are discovered to be the most promising candidates for cathode materials of lithium ion battery which can provide very large specific capacity more than 250 mA h g^{-1} when charged over 4.6 V or higher, though it has several limits, such as relatively poor rate capability, low first cycle efficiency and significant voltage decay.[2-9]

The research on lithium-rich manganese oxide cathode materials indicates that high capacity benefits from the component of Li_2MnO_3 in the material and the ideal capacity of which can amount to 460 mA h g^{-1}.[10,11] However, pure Li_2MnO_3 is nearly inactive and has no reversible capacity. Therefore, additional components have been mixed in Li_2MnO_3, such as $LiMnO_2$, $LiCoO_2$, $LiNiO_2$, to improve it electrochemical activity. But this method decreases the ratio of Li_2MnO_3 which reduces the capacity correspondingly. We design another method by decreasing the size of Li_2MnO_3 particle to form nanoparticle of Li_2MnO_3 to improve the electrochemical activity.[12-15] At present, the preparation of the anode material by the solvothermal method can significantly reduce the path of the lithium ion insertion/deintercalation process. In addition, another problem to be resolved is poor rate capability. In previous studies, the spinel structure of $LiMn_2O_4$ has been used to improve the rate performance. In the crystal structure of $LiMn_2O_4$, Li ions locates on the tetrahedral sites, and the transition metal ions locates on the octahedral sites in an oxygen cubic close packed network. This structure provides fast 3D lithium diffusion frame which is useful to improve the rate capability. Up to now, many researches are focussed on mixing spinel $LiMn_2O_4$ into original cathode.[16,17] The original cathode may be $LiMnO_2$. But most don't achieve expected goals of increasing rate performance. The reason could be spinel $LiMn_2O_4$ doesn't mix into the original cathode material homogeneously. So we are expected to synthesize a homogeneous mixture of spinel $LiMn_2O_4$ and original material to improve the rate capability. However, conventionally, the original cathode is first synthesized and then enters the spinel $LiMn_2O_4$ in a specific manner, which usually causes uneven distribution of $LiMn_2O_4$. Herein, we propose a novel approach which put the step of mixing the two components forward to precursor preparation. By this method, a homogeneous precursor containing both components is first obtained, and then the final cathode material is obtained by annealing.[18-21]

To achieve the goal of high capacity and rate performance, we choice nanoscale Li_2MnO_3 as original cathode. In this work, we have successfully synthesized nano layered-spinel $0.8Li_2MnO_3 \cdot 0.2LiMn_2O_4$ by solvothermal method, which make spinel $LiMn_2O_4$ and layered Li_2MnO_3 mix homogeneously. In addition, characterization methods are utilized to characterize the morphology and crystal structure of the composite. Meanwhile, the electrochemical properties of the composites were investigated, including initial capacity, cycle performance and rate performance of the cathode material.

2. Experimental

2.1. Preparation of Sample

The precursor of $0.8Li_2MnO_3 \cdot 0.2LiMn_2O_4$ cathode material was synthesized by solvothermal method. Raw materials contain $CH_3COOLi \cdot 2H_2O$, $Mn(CH_3COO)_2 \cdot 4H_2O$ and urea. Firstly, 1542.6 mg $CH_3COOLi \cdot 2H_2O$(5 mol% excesses), 2352.9 mg $Mn(CH_3COO)_2 \cdot 4H_2O$ and 2473.1 mg urea were dissolved in 70 mL ethanol at room temperature. After stirring for 2 hours, ethanol was added into the solution until the volume reached 80 mL. Then the solution was transferred into 100 mL autoclave and heated at 200 °C for 20 hours in muffle. After cooling to room temperature, the precursor was collected by using a centrifuge for 3 times. After sufficient drying, the precursor was calcined at 600 °C for 8 hours in the air. Finally, powder was washed several times with DI water and then dried at 80 °C for 10 hours to obtain final cathode materials.

2.2. Material Characterizations

The microstructures morphology and size of powders were characterized by scanning electron microscopy(SEM, Hitachi).The structure and phase composition of synthesized samples were characterized by X-ray diffraction (XRD, Rigaku RINT2000 with Cu-Kα radiation) in the range of 10°-90°(2θ) with a scan rate of 8 $°min^{-1}$.

2.3. Electrochemical Measurements

The electrochemical properties of the synthesized samples were tested using coin cells (CR2032). The positive electrode was prepared by mixing prepared active material (80 wt%), carbon conducting additive (Super 10 wt%) and poly(vinylidene fluoride) (PVDF 10 wt%) and dissolved them in N-methyl-2-pyrrolidone (NMP) then magnetic stirring over night. After that spread the uniformed mixture onto a smooth aluminum foil. The positive electrodes were dried at 75 °C for 3 h in the air and then at 120 °C for 10 h in a vacuum oven. Last diced the electrode to the needed size. Battery cells were then fabricated in an Ar-filled glove box with a lithium foil and porous polypropylene film(Celgard2400) as anode material and separator, and 1M $LiPF_6$ (EC+DMC, 1: 1 in volume) was used as the electrolyte respectively. The charge-discharge tests were performed in the voltage range of 2.0-4.8 V (1 C=250 mA g^{-1}).

3. Results and Discussion

3.1. Structure and morphology

The XRD patterns of the synthesized precursor and cathode material are shown in Figure 1. Peaks appearing in the precursor can be indexed to $MnCO_3$ and Li_2CO_3, implying that lithium element and manganese element have been mixed on precursor preparation. The pattern for cathode material has a number of peaks that can be well-indexed to Li_2MnO_3 (JCPDS no. 27-1252) and $LiMn_2O_4$ (JCPDS no. 35-0782), indicating that the materials synthesized have achieved our desired results.

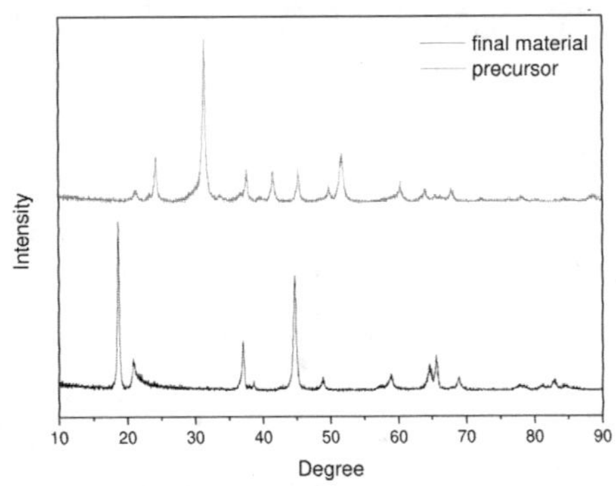

Figure 1. XRD patterns of precursor and cathode material.

Morphology of precursor and final cathode were characterized by scanning electron microscopy (SEM). Figure 2 shows the SEM images of the precursor and final cathode samples. The SEM image of precursor shows spherical particle with appropriate interval each other on the magnification time of 2500. Meanwhile, the morphology of the particle shows perfectly spherical with little fragmentary on the magnification time of 10000. In addition, SEM imaging also showed essentially that the spherical particle is agglomerated by smaller secondary particles with a particle size of 30-100 nm about 80000 times magnification, indicating the formation of nanoscale structure. Above all, homogeneous mixing is the key factor whether this cathode can demonstrate high rate capability. Moreover, the elemental contents of cathode material were analyzed through the energy-dispersive spectrometer in scanning electron microscopy (SEM-EDS). Figure 3 shows the EDS graph of the cathode. These results reveal that the ratio of Li and Mn in Li_2MnO_3 and $LiMn_2O_4$ has a distinct difference. Meanwhile, for the reason light atomic mass of Li that shows inferior effect on EDS, Mn is used to reflect the composition of Li_2MnO_3 and $LiMn_2O_4$. The content of Mn in Li_2MnO_3 and $LiMn_2O_4$ can be estimated to be 47 and 30.4 wt%, respectively. Additionally, the SEM-EDS result indicates the same brightness in the material which reflects homogeneous mixing of Li_2MnO_3 and $LiMn_2O_4$.

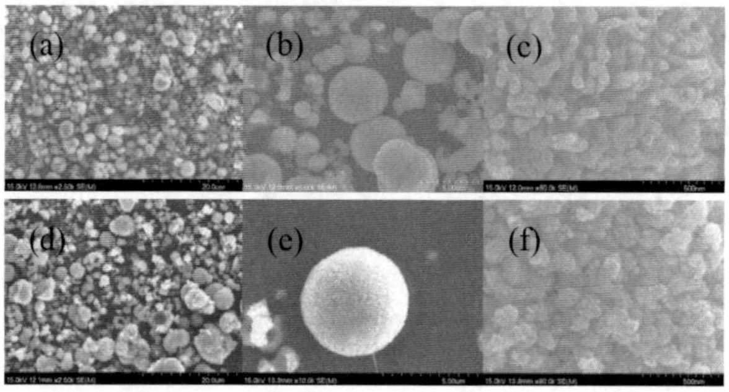

Figure 2. SEM of precursor and cathode material in different magnification. (a) precursor, 2.5k (b) precursor, 10k (c) precursor,80k (d) cathode,2.5k (e) cathode,10k (f) cathode,80k.

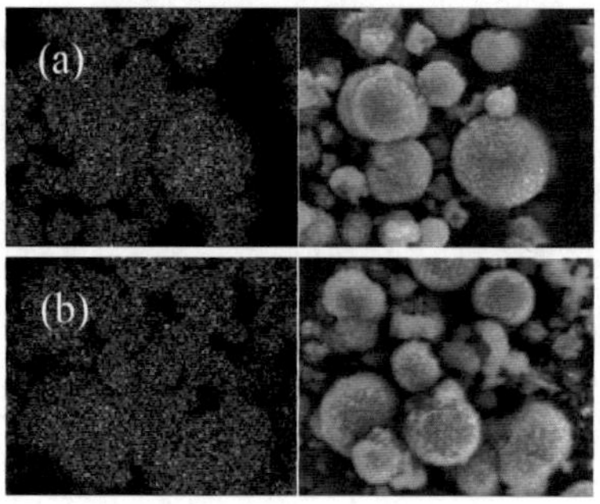

Figure 3. EDS graph of precursor. (a) and cathode material (b) on Mn element.

3.2. Electrochemical Performances

Figure 4 shows galvanostatic profiles of charge and discharge processes of $0.8Li_2MnO_3 \cdot 0.2LiMn_2O_4$ at 250 mA g^{-1} in the potential range of 2.0-4.8 V vs. Li/Li^+ at room temperature. The cathode electrode delivers an initial charge specific capacity of 299 mA h g^{-1} and a discharge specific capacity of 248 mA h g^{-1}, higher than that synthesized by conventional method. High capacity attributes to nano Li_2MnO_3 which activates by nanostructure. From the graph, we can see the first charge curve is different from subsequent cycles. There are two plateaus located at 3.7-4.3 V and above 4.5 V. The first plateau is attributed to lithium ion extraction from $LiMn_2O_4$ phase. The second plateau is considered to be the Li_2O taking off from the layered Li_2MnO_3 structure, which results in high irreversible charge capacity of the cathode during the initial charge process. The subsequent cycles only show one plateau correlated to $LiMn_2O_4$ phase.

The most desirable result is the elevation on rate performance. Figure 5 shows the rate performance of synthesized material at varied current densities ranging from 0.1 to 5.0 C. As can be seen, the discharge specific capacities of $0.8Li_2MnO_3 \cdot 0.2LiMn_2O_4$ are 248, 236, 220,

202, 177 and 162 mA h g^{-1} at current densities of 0.1 C, 0.2 C, 0.5 C, 1 C, 2 C and 5 C, respectively. These result indicates that active material structure of uniformly mixed Li_2MnO_3 and $LiMn_2O_4$ as cathode has significantly improved the rate performance. Meanwhile, the structure is decided by the preparation method. Acetate in the raw material is divided into two parts, one is acetic part which has strong alcohol-affinity, another is metal ions which are alcohol-resistant. Urea would decompose to CO_2 and NH_3 when the temperature reaches 130 °C. Therefore, there are many tiny bubbles existing in alcohol. Alcohol-resistant part metal ions reject alcohol and aggregate in the bubbles, however, the alcohol-affinity part dissolves in alcohol. Therefore, Li ions and Mn ions are aggregated together homogeneously in the bubbles. The metal ions react with CO_2 to generate $MnCO_3$ and Li_2CO_3 under 200 °C for 20 hours. From the SEM graph of precursor, we can see the spherical particles. Last, the precursor is annealing at 800 °C for 10 hours to generate the mixture of Li_2MnO_3 and $LiMn_2O_4$.

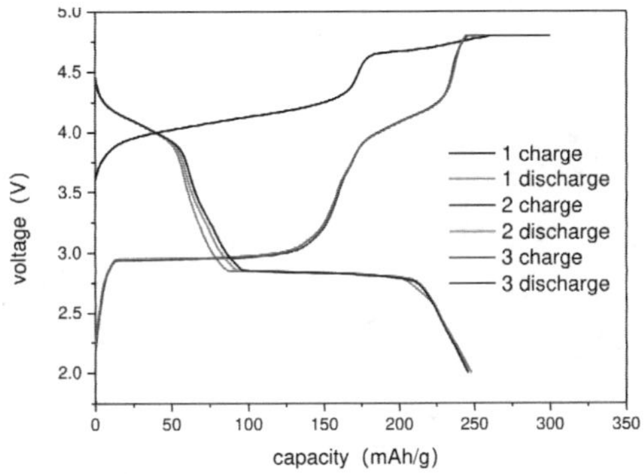

Figure 4. Charge/discharge curves for the initial three cycles of synthesized cathode under a current density of 25 mA g^{-1}.

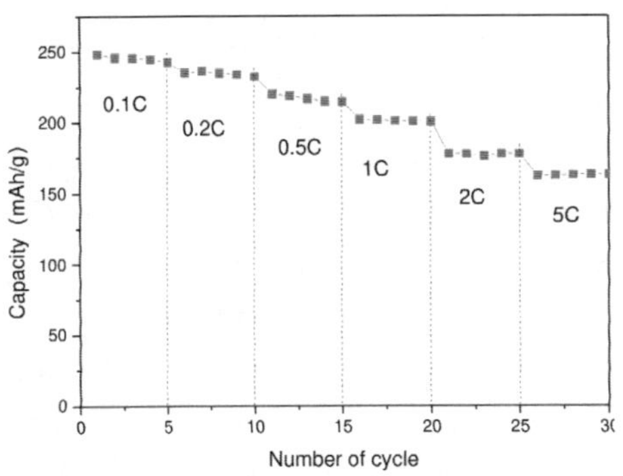

Figure 5. Rate capability of synthesized cathode.

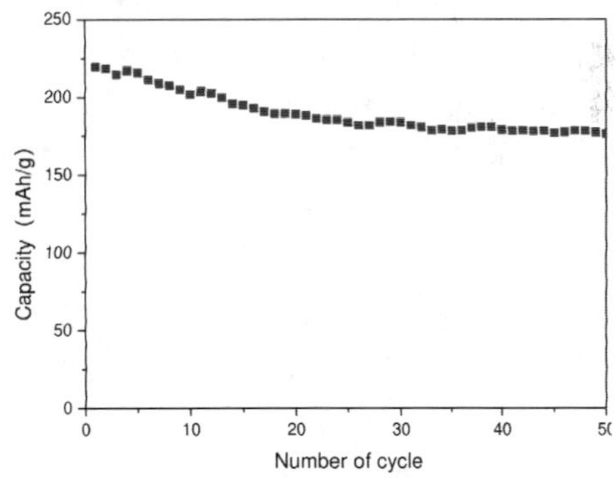

Figure 6. Cycling performance of synthesized cathode.

The cycling performance of $0.8Li_2MnO_3 \cdot 0.2LiMn_2O_4$ at a current density of 0.5C (125 mA g^{-1}) is shown in Figure 6. As a cathode material for LIBs, the reversible capacity attenuates 29 mA h g^{-1} from the initial cycle to 30th cycle and maintains 180 mA h g^{-1} after 50 cycles. The cycling performance doesn't get the ideal effect. The reason may be the jahn-teller effect which leads to the structural distortion of $LiMn_2O_4$. We plan to solve this problem by doping and coating in the following research.

4. Conclusions

In summary, we have provided a novel method to fabricate layered-spinel cathode material for lithium-ion battery. The nanoscale structure improves the activity of Li_2MnO_3 which is related to the capacity of the cathode. When tested as a cathode for LIBs, the $0.8Li_2MnO_3 \cdot 0.2LiMn_2O_4$ exhibited excellent specific capacity which could achieve 248 mA h g^{-1}. Meanwhile, the rate performance has been improved remarkably. The discharge capacity can surpass 200 mA h g^{-1} even under a current density of 1 C. At a higher current density of 5C, the discharge capacity can also maintain 162 mA h g^{-1}. This work provides a successful material with combination property of high specific capacity and excellent rate capability for the development of cathode materials for LIBs, which shows broad prospects in future LIBs applications.

Acknowledgements
This work is supported by Shenzhen Basic Research Free Exploration Project (Nos.JCYJ20180306171650007).

References

[1] Luo S, Wang K, Wang J, Jiang K, Li Q and Fan S 2012 *Advanced materials.* **24** 2294-8.
[2] Jeon H J, Monim S A, Kang C S and Son J T 2013 *Journal of Physics and Chemistry of Solids.* **74** 1185-1195.
[3] Jin X, Xu Q, Yuan X, Zhou L and Xia Y 2013 *Electrochimica Acta.* **114** 605-610.
[4] Lanz P, Villevieille C and Novák P 2014 *Electrochimica Acta.* **130** 206-212.
[5] Wei X, Zhang S, Du Z, Yang P, Wang J and Ren Y 2013 *Electrochimica Acta.* **107** 549-554.
[6] Yu X, Lyu Y, Gu L, Wu H, Bak S M, Zhou Y, Amine K, Ehrlich S N, Li H, Nam K W and Yang X Q 2013 *Advanced Energy Materials.* 4. doi: 10.1002/aenm. 201300950.
[7] Zhang X, Lengyel M and Axelbaum R L 2014 *AIChE Journal.* **60** 443-450.
[8] Zhu Z and Zhu L 2014 *Journal of Power Sources.* **256** 178-182.
[9] Chen Z, Ren Y, Lee E, Johnson C, Qin Y and Amine K 2013 *Advanced Energy* Materials. **3** 729-736.
[10] Rana J, Stan M and Kloepsch R 2014 *Advanced Energy Materials.* **4(5)** 1300998.
[11] Koga H, Croguennec D and Michel M 2013 *Journal of Power Sources.* **236** 250-258.
[12] Choi A, Palanisamy K and Kim Y 2014 *Journal of Alloys and Compounds.* **591** 356-361.
[13] Xiao W, Li H and Fei H 2013 *New Journal of Chemistry.* **12** 207890.
[14] Xue L, Zhang S and Li S 2013 *Journal of Alloys and Compounds.* **577** 560-563.
[15] Zhao W, Harada S and Furuya Y 2014 *Journal of Power Sources.* **261** 324-331.
[16] Kim J, Kim K and Cho W 2012 *Nano Letters.* **12 (12)** 6358-6365.
[17] Kim D, Sandi G and Croy J 2012 *Journal of the Electrochemical Society.* **160(1)** A31-A38.
[18] Croy J, Kang S and Balasubramanian M 2011 *Electrochemistry Communications.* **13(10)** 1063-1066.
[19] Lee E, Huq A and Manthiram A 2013 *Journal of Power Sources.* **240** 193-203.
[20] Rajarathinam S, Mitra S and Petla R 2013 *Electrochimica Acta.* **108** 135-144.
[21] Wu F, Li N and Su Y 2013 *Advanced Materials.* **25(27)** 3722-3726.

Enhanced electrochemical performance of F-doped $Li_{1.2}Mn_{4/6}Ni_{1/6}Co_{1/6}O_{2.35}$ cathode materials for Li-ion batteries

Zhenye Zhu [1*] **Zhuo Sang** [2] **Hui Li** [1]
1 School of Materials Science and Engineering, Harbin Institute of Technology (Shen zhen), Shen zhen 518055, PR China
2 The Eighth Affiliated Hospital, Sun Yat-Sen University 518033, PR China
Corresponding author： Zhenye Zhu zhuzy@hit.edu.cn

Abstract. F-doped cathode material $Li_{1.2}Mn_{4/6}Ni_{1/6}Co_{1/6}O_{(2.35-x)}F_x$ (x=0,0.05,0.10,0.15) were synthesized successfully by two-step co-precipitation method. X-ray diffraction pattern shows that fluorine doping does not bring impurities, but the lattice parameters increase slightly with fluorine content increasing in $Li_{1.2}Mn_{4/6}Ni_{1/6}Co_{1/6}O_{(2.35-x)}F_x$. At the current density of 0.5C (125mAhg^{-1}), $Li_{1.2}Mn_{4/6}Ni_{1/6}Co_{1/6}O_{2.30}F_{0.05}$ has the highest capacity retention (94.56%) and discharge capacity(177.4mA•h/g) after 50th cycles, while capacity retention and discharge capacity after 50th cycles of $Li_{1.2}Mn_{4/6}Ni_{1/6}Co_{1/6}O_{2.35}$ material are only 86.19% and 164.7 mA•h/g. And $Li_{1.2}Mn_{4/6}Ni_{1/6}Co_{1/6}O_{2.30}F_{0.05}$ material exhibits a noticeable higher discharge capacity than other $Li_{1.2}Mn_{4/6}Ni_{1/6}Co_{1/6}O_{(2.35-x)}F_x$ (x=0, 0.10,0.15) materials at high current density. The initial discharge specific capacity of $Li_{1.2}Mn_{4/6}Ni_{1/6}Co_{1/6}O_{2.30}F_{0.05}$ material is increased to 121.1 mA•h/g and 82.3 mA•h/g compared to 91.1 mA•h/g and 66.8 mA•h/g of $Li_{1.2}Mn_{4/6}Ni_{1/6}Co_{1/6}O_{2.35}$ material at 2C and 5C. It can be concluded that electrochemical performances of $Li_{1.2}Mn_{4/6}Ni_{1/6}Co_{1/6}O_{2.35}$ material are improved due to fluorine doping.

1. Introduction

Significant attention has been paid to lithium-ion batteries for their unmatchable combination of high energy and power density[1]. In recent years, lithium-ion batteries have been successfully used in all kinds of portable devices, and even in electric vehicles (EVs)[2-4]. But lithium-ion batteries still have some shortcoming, such as low capacity under high current density, instability, insecurity and so on. Based on these problems, researchers need to find new and effective ways to improve the comprehensive performance of lithium-ion batteries [5].

Li-rich layered oxides, $xLi_2MnO_3\cdot(1-x)LiMO_2$ (M=Ni, Mn, Co), are the promising alternate cathode material for its much higher capacity (>200mAhg^{-1} at lower rate), lower cost, excellent cycling performance[6]. For example, $LiNi_{1/3}Co_{1/3}Mn_{1/3}O_2$ has been extensively studied as cathode material for Li-ion battery, since Ohzuku and Makimura reported it[7]. Layer-structured oxides

$(1-z)Li[Li_{1/3}Mn_{2/3}]O_2 \cdot zLi[Mn_{0.5-y}Ni_{0.5-y}Co_{2y}]O_2$ (y=1/6, and 0.25≤z≤0.75) deliver capacities as high as 250 $mAhg^{-1}$ when charged to 4.8V [8]. Doping modification has been proved an effective method to improve electrochemical performance of electrode materials, such as fluorine substitution [9-16].

It is reported that fluorine substitution catalyzes the growth of the primary particles of $Li[Li_{1/6}Ni_{1/4}Mn_{7/12}]O_{2-x}F_x$ (x=0,0.025, 0.05, 0.075, 0.1) cathode materials[17]. Although the initial discharge capacity decreases as the fluorine content increasing, the fluorine substituted materials present significant improvement in the cycling performance. Yue and his colleagues found that Fluorine substituted $LiNi_{0.8}Co_{0.1}Mn_{0.1}O_2$ material maintained stable layered structure after 100 cycles and still presented smooth surface without any structure transformation [18-20]. And They attributed the phenomenon to fluorine substitution in protecting the electrode from HF attack and maintaining the structural stability. Codoped (Li^+ and F^-) $Li_{1+x}Ni_{0.5}Mn_{1.5}O_{4-x}F_x$ spinel powders were synthesized by Li, et al[21], the codoped particles formed a smooth and uniform surface film, which mitigated the interaction of active materials and electrolyte solution as well as improved stability of cycling. Kim and his colleagues reported that Fluorine was substituted to facilitate the movement of lithium ions in $0.3Li_2MnO_3 \cdot 0.7Li[Mn_{0.60}Ni_{0.25}Co_{0.15}]O_{1.975}F_{0.025}$ cathode materials[22]. The introduction of fluorine forms a more compact SEI layer and reduces the dissolution of transition metals. The fluorine gradient-doped $LiNi_{0.5}Mn_{1.5}O_4$ spinels were synthesized by Luo, et al[23], the fluorine gradient-doped $LiNi_{0.5}Mn_{1.5}O_4$ sample exhibited an improved long-term cycling stability and high rate performance, due to the suppression of the reaction between electrolyte and cathode, resulting in a decrease in the total resistance and the formation of a thin, uniform and smooth film on the surface. $Li[Li_{0.2}Mn_{0.54}Ni_{0.13}Co_{0.13}]O_{2-x}F_x$ (x = 0, 0.05 and 0.10) were synthesized by Zheng and his colleagues[24], and they found that fluorine incorporation stabilized the electrode/electrolyte interface by suppressing the formation of poorly conducting LiF in the SEI layer and thus maintained stable interfacial resistances. As a result, cycling stability of $Li[Li_{0.2}Mn_{0.54}Ni_{0.13}Co_{0.13}]O_{2-x}F_x$ was significantly improved. It was reported by Li that the discharge voltage plateau of $Li_{1.2}Mn_{0.54}Ni_{0.13}Co_{0.13}O_{2-x}F_x$ (x = 0, 0.02, 0.05, 0.08) materials were stabilized by F doping due to the retarded layered-to-spinel phase transition [25], thus a high energy density output of the cell could be maintained. $Li[Li_{1/6}Ni_{1/6}Co_{1/6}Mn_{1/2}]O_{2-x}F_x$(x = 0.00 to 0.07) materials were synthesized by Song, et al[26]. Due to the addition of fluorine, the reversible capacity significantly increased as the irreversibility was suppressed during the first cycle. It is also found that the presence of fluorine facilitated the reduction of cobalt and manganese ions in Li-rich layered oxide, and that the reduced transition metal ions suppressed structural changes.

In this study, F-doped material $Li_{1.2}Mn_{4/6}Ni_{1/6}Co_{1/6}O_{2.35-x}F_x$ (x=0, 0.05, 0.10, 0.15) was synthesized by two-step co-precipitation method. Phase analysis was implemented on high-resolution X-ray diffractometer (Rigaku,Japan) with K_α radiation. Scanning electron microscopic(SEM) images were operated with a S4700 Field emission scanning electron microscope(Hitachi,Japan).The electrochemical measurements were carried out by using LAND batteries testing system.

2. Experimental

2.1. Synthesis: two-step temperature co-precipitation method

A $[Mn_{4/6}Ni_{1/6}Co_{1/6}]CO_3$ precursor reagent was synthesized by two-step co-precipitation method. $NiSO_4 \cdot 7H_2O$ (Shanghai Aladdin, 99.9%), $CoSO_4 \cdot 7H_2O$ (Shanghai Aladdin,99.999%) and $MnSO_4 \cdot H_2O$ (Shanghai Aladdin, 99%) were used as raw materials. Figure 1. shows the preparation procedure of the precursor $[Mn_{4/6}Ni_{1/6}Co_{1/6}]CO_3$ and the cathode material $Li_{1.2}Mn_{4/6}Ni_{1/6}Co_{1/6}O_{2.35-x}F_x$. Firstly, the appropriate amount of $NiSO_4 \cdot 7H_2O$ and $CoSO_4 \cdot 7H_2O$ is dissolved in distilled water at a molar ratio of 3:1 and put into the reaction tank together. Then, add the aqueous solution of sodium carbonate slowly at 40 °C. Raise the reaction temperature to 80°C after the dropwise-addition is completed, and then the mixed solution ($NiSO_4 \cdot 7H_2O$, $CoSO_4 \cdot 7H_2O$ and $MnSO_4 \cdot H_2O$ molar ratio is 0.5:1.5:8) is added into the continuous stirred (900r/min) solution in the reaction tank, while the required Na_2CO_3 is also dropped into the reactor. After the completion of the precipitation reaction, the resultant slurry solution is stirred for 3h then aged at 80°C for 12h. After that, the $[Mn_{4/6}Ni_{1/6}Co_{1/6}]CO_3$ precipitate is separated by the vacuum filtration procedure and then washed several times with the distilled water and dried in vacuum at 110°C for 12h. Then, in order to prepare cathode material $Li_{1.2}Mn_{4/6}Ni_{1/6}Co_{1/6}O_{(2.35-x)}F_x$, 5% excess $LiCO_3$ and 3% excess LiF are mixed thoroughly with the obtained $[Mn_{4/6}Ni_{1/6}Co_{1/6}]CO_3$. The mixture is first calcined at 500°C for 5h and then 850°C for 15h in air with a heating rate of 5°Cmin^{-1}.

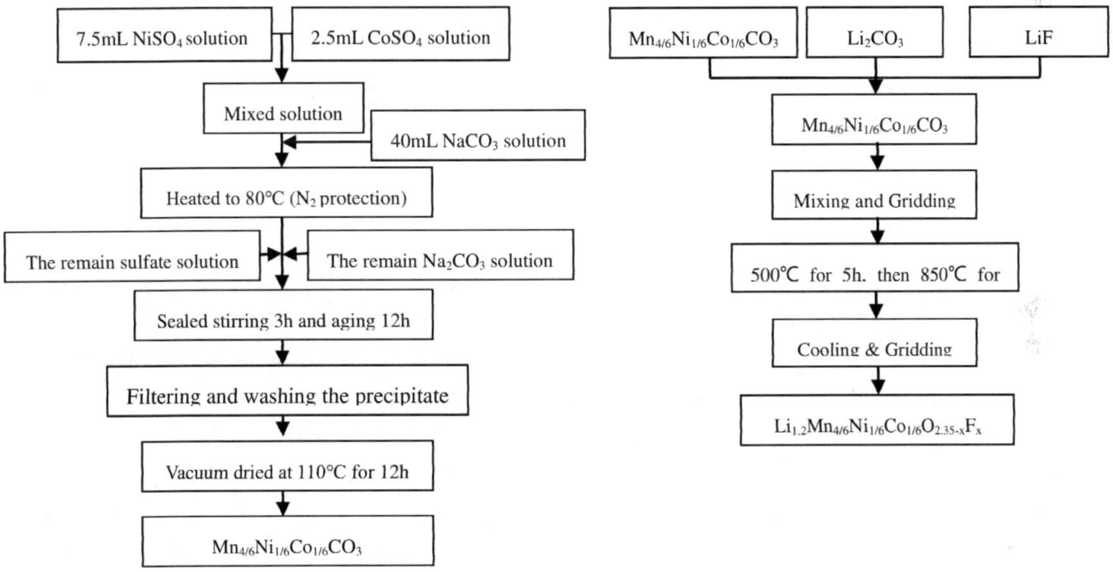

Figure 1. Schematic diagram of the synthesis of precursor $[Mn_{4/6}Ni_{1/6}Co_{1/6}]CO_3$ and cathode material $Li_{1.2}Mn_{4/6}Ni_{1/6}Co_{1/6}O_{2.35-x}F_x$.

2.2. *Structure and morphology analysis*

X-ray diffraction (XRD, Rigaku RINT2000 with Cu-Kα radiation) in a 2θ range of 10°-80° with a scan rate of 6°min^{-1} is employed to analyze the structure and phase composition of synthesized samples. The scanning electron microscopy (SEM, Hitachi) images are taken to get the morphology and size of the samples.

2.3. Electrochemical measurements

The electrochemical properties of $Li_{1.2}Mn_{4/6}Ni_{1/6}Co_{1/6}O_{2.35-x}F_x$ composites are examined by using a coin cells (CR2032) assembled in an Ar-filled glove-box. The positive electrode is prepared by spreading the mixture of prepared active material (80wt%), carbon conducting additive (Super-P, 10wt%) and polyvinylidene fluoride (PVDF,10wt%) dissolved in N-methyl-2-pyrrolidone (NMP) onto a smooth aluminum foil. The positive electrode is allowed to dry at 75°C for 3h in the air and then at 120°C for 10h in a vacuum oven. The cathode film is punched into appropriate size to assemble coin cells. Lithium foil is used as negative electrode, Celgard 2400 membrane is used as the separator, and the electrolyte is 1 M $LiPF_6$ dissolved in a mixture of ethylene carbonate (EC) and dimethyl carbonate (DMC) with volume ratio 1:1.

3. Results and discussion

3.1. Structure and morphology of F-doped $Li_{1.2}Mn_{4/6}Ni_{1/6}Co_{1/6}O_{2.35-x}F_x$ composites

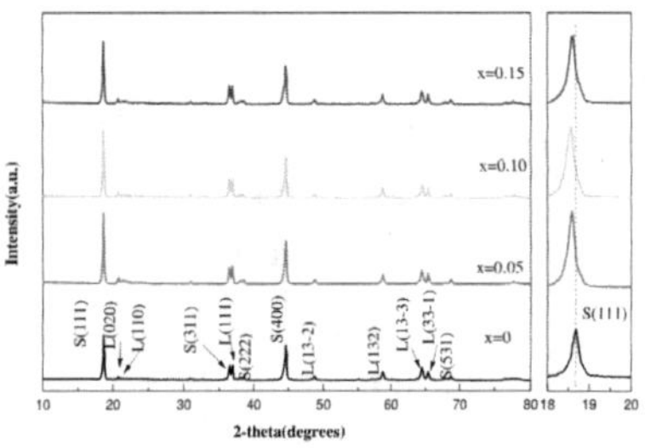

Figure 2. XRD patterns of $Li_{1.2}Mn_{4/6}Ni_{1/6}Co_{1/6}O_{2.35-x}F_x$ (x=0, 0.05, 0.10, 0.15) materials.

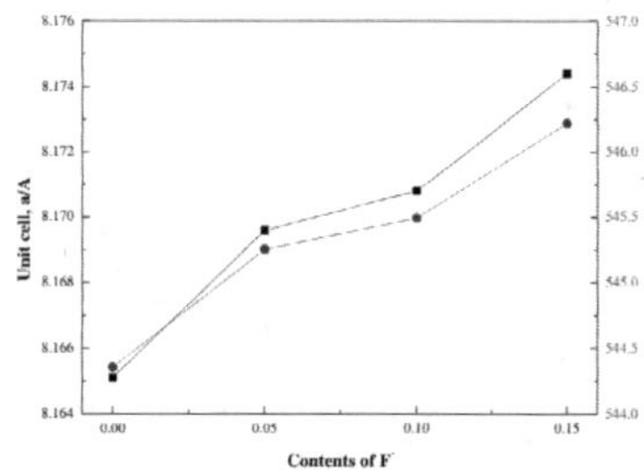

Figure 3. Variation of lattice parameters as a function of F amount in $Li_{1.2}Mn_{4/6}Ni_{1/6}Co_{1/6}O_{2.35-x}F_x$ (x=0, 0.05, 0.10, 0.15) materials.

XRD patterns of synthesized $Li_{1.2}Mn_{4/6}Ni_{1/6}Co_{1/6}O_{2.35-x}F_x$ (x=0, 0.05, 0.10, 0.15) composites are presented in Figure 2. All diffraction peaks of materials can be indexed to cubic structure spinel $LiNi_{0.5}Mn_{1.5}O_4$ (Space group: Fd-3m), layer structure Li_2MnO_3 (Space group: C2/m) and layer structure $LiMn_{1/3}Ni_{1/3}Co_{1/3}O_2$ materials. Compared with pristine $Li_{1.2}Mn_{4/6}Ni_{1/6}Co_{1/6}O_{2.35}$, there is no new peaks or impurities in the fluorine-doped materials, which demonstrates that fluorine substitution does not alter the crystal structure of $Li_{1.2}Mn_{4/6}Ni_{1/6}Co_{1/6}O_{2.35}$. By observing the patterns carefully, the S (111) peaks of the F-doped materials shift to small angle obviously, which is caused by the increase of the lattice parameter of the modified material [27].

The variation of lattice parameters as a function of F amounts are shown in Figure 3. The lattice parameter was calculated by JAD 6.0. As indicated in Figure 3, with increasing fluorine content, the lattice parameters gradually increase from a= 8.165(1)Å and v=544.355(9)$Å^3$ to a=8.174(4)Å and v=546.222(1)$Å^3$. The increase of lattice parameters may be induced by two reasons. The first reason is that partial substitution of oxygen by fluorine leads to an unbalance of valence, which requires partial reduction of transition metal ions Mn^{4+}(r=0.53Å) to Mn^{3+}(r=0.645Å), Ni^{4+}(r=4.8Å) to Ni^{3+}(r=5.6Å) and Co^{3+} (r=0.545Å) to Co^{2+} (r=0.65Å) for charge compensation, and the second one is that the chemical bond between lithium and fluorine is stronger than that between lithium and oxygen resulting from the charge compensation by the fluorine substitution [28,29].

The particle morphologies of $Li_{1.2}Mn_{4/6}Ni_{1/6}Co_{1/6}O_{2.35-x}F_x$ composites observed by SEM are shown in Figure 4 $Li_{1.2}Mn_{4/6}Ni_{1/6}Co_{1/6}O_{2.35}$, $Li_{1.2}Mn_{4/6}Ni_{1/6}Co_{1/6}O_{2.30}F_{0.05}$ and $Li_{1.2}Mn_{4/6}Ni_{1/6}Co_{1/6}O_{2.25}F_{0.10}$ samples have good spherical secondary particles with an average diameter about 5μm, and the spherical secondary particles is comprised of a great number of small primary particles with a diameter in range of 200-800nm. The morphology of $Li_{1.2}Mn_{4/6}Ni_{1/6}Co_{1/6}O_{2.2}F_{0.15}$ in Figure 4 (d) shows that its spherical boundary is not smooth, some particles are broken and particle size is not uniform, which may be attribute to the excess of fluorine damaging the structure of the material. In Figure 4 (b), all the primary particles for $Li_{1.2}Mn_{4/6}Ni_{1/6}Co_{1/6}O_{2.3}F_{0.05}$ have smooth surfaces.

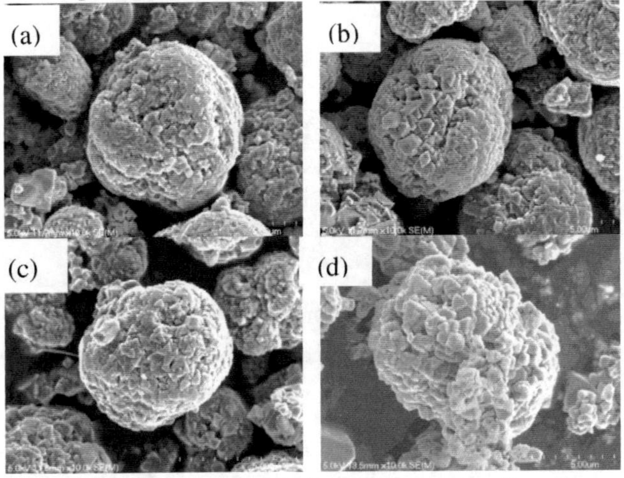

Figure 4. SEM images of $Li_{1.2}Mn_{4/6}Ni_{1/6}Co_{1/6}O_{2.35-x}F_x$: (a)x=0, (b)x=0.05, (c)x=0.10, (d)x=0.15.

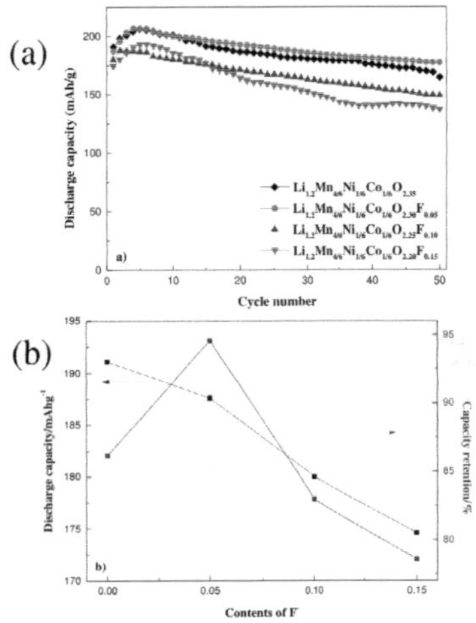

Figure 5. (a) Cycle performance of $Li_{1.2}Mn_{4/6}Ni_{1/6}Co_{1/6}O_{2.35-x}F_x$ (x=0, 0.05, 0.10, 0.15) at a discharge rate of 0.5C (1C=250mAhg^{-1}), and (b) the corresponding initial discharge capacity and capacity retention values.

3.2. Electrochemical performance

The electrochemical measurements were carried out by using LAND batteries testing system. The measurement temperature was 25 °C, the current density for 1 C rate was 250mAh·g^{-1}.

Figure 5 indicates that the discharge capacities of $Li_{1.2}Mn_{4/6}Ni_{1/6}Co_{1/6}O_{2.35-x}F_x$ composites for 50 cycles at the current density of 0.5C (125mAhg^{-1}) at a voltage range of 2.0-4.8V. As shown in Figure 5 (a), though the F-free $Li_{1.2}Mn_{4/6}Ni_{1/6}Co_{1/6}O_{2.35}$ has the highest initial discharge of 191.1mAhg^{-1}, its capacity decreases gradually in 10-50 cycles, leading to a capacity retention of only 86.19% after 50 cycles. Meanwhile, the capacity retention of $Li_{1.2}Mn_{4/6}Ni_{1/6}Co_{1/6}O_{2.35-x}F_x$ was 94.56% for x=0.05, 82.94% for x=0.10 and 78.58% for x=0.15, respectively. It is shown, after 11 cycles the discharge capacity of $Li_{1.2}Mn_{4/6}Ni_{1/6}Co_{1/6}O_{2.3}F_{0.05}$ is at its largest value. But $Li_{1.2}Mn_{4/6}Ni_{1/6}Co_{1/6}O_{2.25}F_{0.10}$ and $Li_{1.2}Mn_{4/6}Ni_{1/6}Co_{1/6}O_{2.20}F_{0.15}$ materials did not show better cycle performance than $Li_{1.2}Mn_{4/6}Ni_{1/6}Co_{1/6}O_{2.35}$ material, which may be the excess F damaging the structure of the phases and lead to the decrease of the discharge capacity. Figure5 (b) shows the corresponding initial discharge capacity and capacity retention values. With the content of F increasing, initial discharge capacity of

$Li_{1.2}Mn_{4/6}Ni_{1/6}Co_{1/6}O_{2.35-x}F_x$ cathode material decreases gradually, and capacity retention after 50 cycles first rises and then declines. When x=0.05, $Li_{1.2}Mn_{4/6}Ni_{1/6}Co_{1/6}O_{2.35-x}F_x$ cathode material delivers the largest capacity retention (94.56%) and discharge capacity after 50th cycles (177.4mA·h/g).

It can be concluded from the above analysis that fluorine doping positively impact on the structure stability and cycling performance of the cathode materials. The improved cycling performance of $Li_{1.2}Mn_{4/6}Ni_{1/6}Co_{1/6}O_{2.3}F_{0.05}$ composite should be due to the existence of the bonding of Li-F (stronger than the bonding of Li-O) which stabilizes the surface structure and enhances the stability of the electrode materials, and thus avoid the abrupt change of the lattice distortion [30,31].

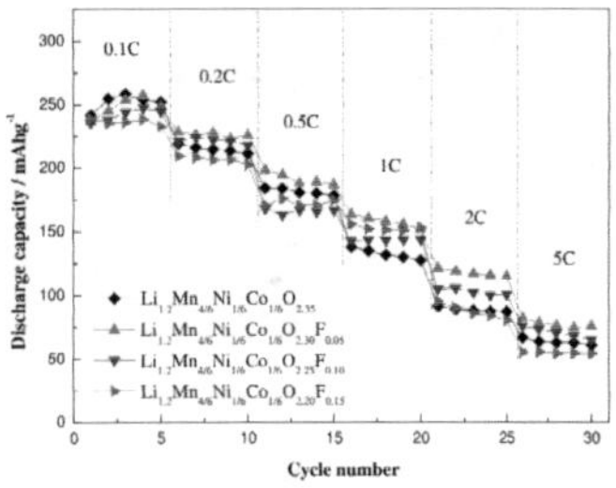

Figure 6. Rate capability of $Li_{1.2}Mn_{4/6}Ni_{1/6}Co_{1/6}O_{2.35-x}F_x$ (x=0, 0.05, 0.10, 0.15) in the voltage range of 2.0-4.8V.

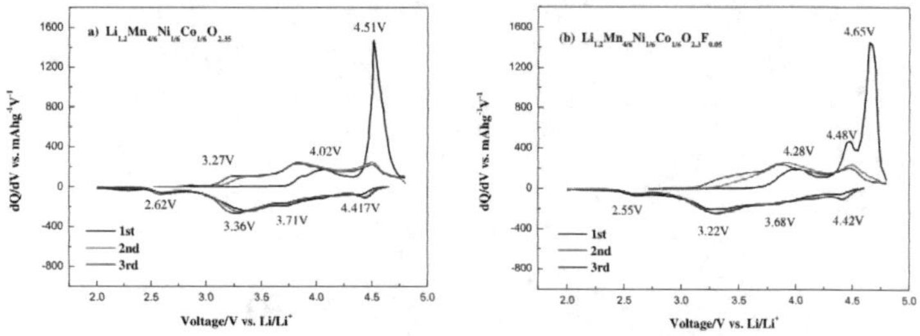

Figure 7. Initial charge/discharge curves and differential capacity vs. Voltage (dQ/dV) curves of (a) $Li_{1.2}Mn_{4/6}Ni_{1/6}Co_{1/6}O_{2.35}$ and (b) $Li_{1.2}Mn_{4/6}Ni_{1/6}Co_{1/6}O_{2.3}F_{0.05}$ electrodes.

Table 1. The first specific discharge capacity of $Li_{1.2}Mn_{4/6}Ni_{1/6}Co_{1/6}O_{2.35-x}F_x$ at different rates

F^- doping amount	First discharge specific capacity at different rates (mA·h/g)					
(x)	0.1C	0.2C	0.5C	1C	2C	5C
0	242	218.8	184.1	137.9	91.1	66.8
0.05	237.3	228.2	198.2	163.3	121.1	82.3
0.10	235.7	220.7	167.7	142.5	104.8	75.7
0.15	238.9	209.1	170.5	155.6	95.1	55.4

Figure 6. shows the rate capability of $Li_{1.2}Mn_{4/6}Ni_{1/6}Co_{1/6}O_{2.35-x}F_x$ composites (x=0,0.05, 0.10, 0.15) between 2.0V and 4.8V at charge-discharge rates from 0.1C to 5C. All composites deliver decreases capacities as the C rate is increased, but it is clearly observed that $Li_{1.2}Mn_{4/6}Ni_{1/6}Co_{1/6}O_{2.30}F_{0.05}$ exhibits a noticeable higher discharge capacity than $Li_{1.2}Mn_{4/6}Ni_{1/6}Co_{1/6}O_{2.35}$, $Li_{1.2}Mn_{4/6}Ni_{1/6}Co_{1/6}O_{2.25}F_{0.10}$ and $Li_{1.2}Mn_{4/6}Ni_{1/6}Co_{1/6}O_{2.20}F_{0.15}$ at high rates, and both cycle performance and rate capability of $Li_{1.2}Mn_{4/6}Ni_{1/6}Co_{1/6}O_{2.30}F_{0.05}$ are obviously better than those of other three materials. The corresponding data is given in Table 1, when the current density was 2C, the discharge specific capacity of F-doped $Li_{1.2}Mn_{4/6}Ni_{1/6}Co_{1/6}O_{2.30}F_{0.05}$ material is increased to 121.1mA·h/g, while that of $Li_{1.2}Mn_{4/6}Ni_{1/6}Co_{1/6}O_{2.35}$ material is only 91.1mA·h/g. It is found that $Li_{1.2}Mn_{4/6}Ni_{1/6}Co_{1/6}O_{2.30}F_{0.05}$ shows the best cycling performance according to our research. In other words, F^- substitution can

improve the cycling performance and rate capability of $Li_{1.2}Mn_{4/6}Ni_{1/6}Co_{1/6}O_{2.35}$ cathode materials.

The differential capacity curves (dQ/dV) of $Li_{1.2}Mn_{4/6}Ni_{1/6}Co_{1/6}O_{2.35}$ and $Li_{1.2}Mn_{4/6}Ni_{1/6}Co_{1/6}O_{2.3}F_{0.05}$ electrodes are depicted in Figure 7. The first potential slope in the initial charge curve between 3.8 and 4.4V of $Li_{1.2}Mn_{4/6}Ni_{1/6}Co_{1/6}O_{2.35}$ material is associated with lithium extraction from $Li_{1.2}Mn_{4/6}Ni_{1/6}Co_{1/6}O_{2.35}$-like component, accompanied by the oxidation process of Ni^{2+} to Ni^{4+} and Co^{3+} to Co^{4+} in the material [32], and these two processes correspond to peaks near 3.9V and 4.1V in the 1st differential capacity curve. In the two subsequent charging processes, the peaks in the differential capacity curve corresponding to the lithium removal from the $Li_{1.2}Mn_{4/6}Ni_{1/6}Co_{1/6}O_{2.35}$-like phase are separated from the peaks of the transition metal oxidation processes. The potential plateau of $Li_{1.2}Mn_{4/6}Ni_{1/6}Co_{1/6}O_{2.35}$ material near 4.51V corresponds to that of $Li_{1.2}Mn_{4/6}Ni_{1/6}Co_{1/6}O_{2.30}F_{0.05}$ material near 4.65V, corresponding to the loss of O^{2-} and de-intercalation of Li^{+} ($Li_2MnO_3 \rightarrow 2Li^{+}+2e^{-}+0.5O_2+MnO_2$), finally exists in the form of Li_2O[33]. And the oxidation peaks strength in differential capacity curve decreases and has a small voltage deviation, which is a characteristic of Li-rich oxide electrode. The weakening of oxidation peak is related to the gradual decrease of irreversible phase change of Li_2MnO_3 during the progress of charging and discharging cycle. In the first charging process of $Li_{1.2}Mn_{4/6}Ni_{1/6}Co_{1/6}O_{2.3}F_{0.05}$ material, the voltage of transition metal oxidation and the removal of Li_2O are both higher than those of $Li_{1.2}Mn_{4/6}Ni_{1/6}Co_{1/6}O_{2.35}$ material, and the corresponding peaks in the differential capacity curve are all shifted to higher voltage, which is due to F doping, while the reason need our further researches. In the discharging process, there are three obvious reduction peaks in the differential capacity curve of both $Li_{1.2}Mn_{4/6}Ni_{1/6}Co_{1/6}O_{2.35}$ material and $Li_{1.2}Mn_{4/6}Ni_{1/6}Co_{1/6}O_{2.3}F_{0.05}$ materials. The peak near 4.4V corresponds to the reduction process of Ni^{4+} to Ni^{3+}, and the peak near 3.7V corresponds to the reduction process of Ni^{3+} to Ni^{2+}, and the reduction process of Ni is also accompanied by the reduction of Co. The peak around 3.3V corresponds to the reduction process of Mn^{4+} to Mn^{3+}. In addition, reduction peaks appeared near 2.6V in the differential capacity curve of both $Li_{1.2}Mn_{4/6}Ni_{1/6}Co_{1/6}O_{2.35}$ material and $Li_{1.2}Mn_{4/6}Ni_{1/6}Co_{1/6}O_{2.3}F_{0.05}$ material, which corresponds to the lithium intercalation process of the spinel phase formed by the surface structure of the material [34].

4. Conclusion

F-doped $Li_{1.2}Mn_{4/6}Ni_{1/6}Co_{1/6}O_{2.35}$ material was successfully prepared by two-step temperature co-precipitation. With the doping of fluorine, the lattice parameter increases slightly but no impurity phase was formed. Electrochemical characterization demonstrated that F^{-} doping can improve high-rate capability and cycling performance of $Li_{1.2}Mn_{4/6}Ni_{1/6}Co_{1/6}O_{2.35}$ material. $Li_{1.2}Mn_{4/6}Ni_{1/6}Co_{1/6}O_{2.30}F_{0.05}$ composite has the highest capacity retention (94.56%) and discharge capacity after 50th cycles (177.4mA·h/g), compared with only 86.19% and 164.7 mA·h/g of $Li_{1.2}Mn_{4/6}Ni_{1/6}Co_{1/6}O_{2.35}$ material. Furthermore, the discharge capacity of $Li_{1.2}Mn_{4/6}Ni_{1/6}Co_{1/6}O_{2.30}F_{0.05}$ material was much higher than other $Li_{1.2}Mn_{4/6}Ni_{1/6}Co_{1/6}O_{(2.35-x)}F_x$ (x=0, 0.10, 0.15) materials at high current density. The initial discharge specific capacity of $Li_{1.2}Mn_{4/6}Ni_{1/6}Co_{1/6}O_{2.30}F_{0.05}$ material is increased from 91.1 mA·h/g to 121.1 mA·h/g compared to $Li_{1.2}Mn_{4/6}Ni_{1/6}Co_{1/6}O_{2.35}$ material at 2C. A conclusion can be drawn by comparing the properties of all kinds of $Li_{1.2}Mn_{4/6}Ni_{1/6}Co_{1/6}O_{2.35-x}F_x$ composites (x=0,0.05, 0.10, 0.15) that F^{-} doping can indeed improve the electrochemical properties of materials and optimal doping amount was 0.05 (mole ratio).

Acknowledgements

This work is supported by Shenzhen Basic Research Free Exploration Project (Nos.JCYJ20 180306171650007).

5. References

[1] Nitta N, Wu F, Lee J T, et al. *Materials Today*, 2015, 18(5):252-264.
[2] Wang H, Friedrich B. *Journal of Sustainable Metallurgy*, 2015, 1(2):168-178.
[3] Bruce Dunn, Haresh Kamath, Jean-Marie Tarascon. *Science*, 2011, 334: 928-935.
[4] Nam-Soon Choi, Zhonghai Chen, Stefan A. *Angewandte Reviews*, 2012, 5(1): 9994-10024.
[5] Kafle J, Harris J, Chang J, et al. *Journal of Power Sources*, 2018, 392:60-68.
[6] Nayak P K, Erickson E M, Schipper F, et al. *Advanced Energy Materials*, 2018, 8(8):1702397.
[7] WANG Hai-yan, TANG Ai-dong, et al. *Trans. Nonferrous Met. So*c, 2010, 20:803-808.
[8] Wu Y, Manthiram A. *Solid State Ionics*, 2009, 180(1):50-56.
[9] Kang S H, Amine K. *Journal of Power Sources*, 2005, 146(1):654-657.
[10] Kim G H, Kim M H, Myung S T, et al. *Journal of Power Sources*, 2005, 146(1-2):602-605.
[11] Kang S H, Belharouak I, Sun Y K, et al. *Journal of Power Sources*, 2005, 146(1-2):650-653.
[12] Oh S W, Park S H, Kim J H, et al. *Journal of Power Sources*, 2006, 157(1):464-470.
[13] Croguennec L, Bains J, Menetrier M, et al. *Journal of the Electrochemical Society*, 2009, 156(5):A349–A355.
[14] Luo Q, Muraliganth T, Manthiram A. *Solid State Ionics*, 2009, 180(9-10):703-707.
[15] Sha O, Tang Z, Wang S, et al. *Electrochimica Acta*, 2012, 77(none).
[16] Hagh N M, Amatucci G G. *Journal of Power Sources*, 2014, 256:457-469.
[17] Li H, Fan L Z. *Electrochimica Acta*, 2013, 113:407-411.
[18] Yue P, Wang Z, Guo H, et al. *Electrochimica Acta*, 2013, 92:1-8.
[19] Yue P, Wang Z, Wang J, et al. *Powder Technology*, 2013, 237:623-626.
[20] Yue P, Wang Z, Li X, et al. *Electrochimica Acta*, 2013, 95:112-118.
[21] Haiyan Li, Ying Luo, Jingying Xie. *Journal of Alloys and Compounds*, 2015, 639:346-351.
[22] Kim S M, Jin B S, Lee S M, et al. *Electrochimica Acta*, 2015, 171:35-41.
[23] Luo Y, Li H, Lu T, et al. *Electrochimica Acta*, 2017, 238:237-245.
[24] Zheng J, Wu X, Yang Y. *Electrochimica Acta*, 2013, 105(Complete):200-208.
[25] Li L, Song B H, Chang Y L, et al. *Journal of Power Sources*, 2015, 283:162-170.
[26] Song J H, Kapylou A, Choi H S, et al. *Journal of Power Sources*, 2016, 313:65-72.
[27] Croguennec L, Bains J, Menetrier M, et al. *Journal of the Electrochemical Society*, 2009, 156(5): A349–A355.
[28] Shin H S, Park S H, Yoon C S, et al. *Electrochem. Solid-State Lett*, 2005, 8(11): A559-A563.
[29] Woo S U, Park B C, Yoon C S. *Journal of The Electrochemical Society*, 2007, 154(7): A649.
[30] S.J. Shi, J.P. Tu, Y.Y. Tang, Y.Q. Zhang, X.Y. Liu, X.L. Wang, C.D. Gu, *Journal of Power Sources*. 2013, 225: 338.
[31] Y.S. He, L. Pei, X.Z. Liao, Z.F. Ma, *Journal of Fluorine Chemistry*, 2007, 128: 139.
[32] Deng, Ya-Ping, Fu F, Wu, Zhen-Guo, et al. *Chemelectrochem*, 2016, 3.
[33] Johnson C S, Kim J, Lefief C, et al. *Electrochemistry Communications*, 2004, 6(10):1085-1091.
[34] Hou P, Wang J, Song J, et al. *Chem. Commun*. 2015, 51(15): 3231-3234.

IOP Conf. Series: Materials Science and Engineering **733** (2020) 012007 doi:10.1088/1757-899X/733/1/012007

Comparative study on the effects of suspension way of sample holder on the large-aperture uniformity of refractive index of HfO_2 films

F Zhang[1,*], F Pan[2], Y W Wei[1], J Luo[2], Q Wu[3] and Z Wang[3]

1 Research Assistant, Chengdu Fine Optical Engineering Research Center, Chengdu 610041, China
2 Associate Researcher, Chengdu Fine Optical Engineering Research Center, Chengdu 610041, China
3 Engineer, Chengdu Fine Optical Engineering Research Center, Chengdu 610041, China
*E-mail: zfchunri@163.com

Abstract. For large-aperture dielectric film optics, the uniformity of film refractive index across the aperture is crucial to the uniformity of reflectance and transmittance. Herein, a comparative research has been performed on the refractive index uniformity of HfO_2 films across the Φ560 mm-aperture sample holder for inclined suspension and horizontal suspension. It's revealed that the refractive index of HfO_2 film at edge position of sample holder increased about 1.52% compared to that of HfO_2 film at center position for inclined suspension, while the refractive index varied only about 0.25% for horizontal suspension. The further analysis performed on film crystallization, morphology and microstructure show that the large variation of refractive index between the films at center and edge positions of inclined suspension sample holder was attributed to the different growth properties. The film at the edge crystallized more slightly and was denser, thus being of higher refractive index. And the different growth properties of the two films at center and edge positions can be explained in the light of the co-effects of atomic shadowing, reemission and diffusion of ad-atoms during film growth.

1. Introduction

The large-aperture dielectric film optics are indispensable to laser fusion facility [1], to which the spectrum performance of optics are especially important. The spectrum uniformity, as an assessment of spectrum performance of a large-aperture dielectric film optic, depends on the uniformity of thickness and refractive index (n) of dielectric films. And optimizing the large-aperture uniformity of film thickness is always a research focus for various film deposition techniques, because thickness inhomogeneity can result in the spectrum drifting significantly along the radial direction of a large-aperture optic. Although it's challenging to improve thickness uniformity of the large-aperture films, many strategies have been tried by researchers, for example, optimizing position of the evaporation source [2] and way of the substrate movement [3-6] in coating chamber, inserting a mask between substrate and evaporation source [6], and so on.

By contrast, the large-aperture uniformity of film refractive index attracts less attention from researchers, because it's generally better than that of film thickness and has slighter influence on the spectrum uniformity for common optics. However, it becomes vital to those special optics requiring constant transmittance or reflectance at specified light wavelength, thus should be carefully studied as well as film thickness uniformity. Herein, hafnia (HfO_2) films, which are widely used as high refractive index material in the field of high power laser due to their high damage threshold [7-9], had been deposited by reactive e-beam evaporation. And the effects of substrate suspension on the large-aperture uniformity of n of HfO_2 films had been investigated.

2. Experimental

HfO_2 films were deposited on fused silica substrates (Ф50 mm×5 mm) through the reaction of oxygen and hafnium gas evaporated by e-beam evaporation. The substrates were fixed into substrate holes of the sample holder in figure 1a. The substrate holes are equally spaced with interval of 70 mm along the radius of sample holder, covering a deposition area of Ф560 mm aperture. And half of the substrate holes are noted and named as *2-0#*, *2-1#*, *2-2#*, *2-3#* and *2-4#*, respectively, which are also the sample names of deposited HfO_2 films in the following description. As the sample holder rotates in a planetary way during film deposition, the properties of HfO_2 films deposited on these noted positions are adequate to represent the film properties across the whole deposition area of Ф560 mm aperture in consideration of the symmetry of planetary rotation. To maximize effects of the suspension way of sample holder on the n of HfO_2 films, 60°-inclined suspension (figure 1b) and horizontal suspension (figure 1c) had been adopted. It should be noted that HfO_2 films for 60°-inclined suspension and horizontal suspension had to be deposited in the coating machine of APS1504 and ZZS1350, respectively, due to the mechanical configuration of the coating machine. Nevertheless, it remains feasible to compare the effects of suspension way on HfO_2 films because the two coating machines (APS1504 and ZZS1350) are extremely similar with each other in the key aspects. For example, the location of the heater, e-gun, evaporating source, quartz crystal controller and evacuating outlet, as well as the configuration of evacuation system, are similar with each other for the two coating machines. As for 60°-inclined suspension, substrate positions of *2-1#* and *2-4#* had been selected to deposit HfO_2 films, while positions of *2-0#* and *2-3#* had been selected for horizontal suspension. The background pressure was controlled below 2.0×10^{-4} Pa prior to film deposition, and the deposition process was controlled automatically by quartz crystal controller. Detailed process conditions can be found in table 1.

The spectra of fabricated HfO_2 film samples had been measured by spectrophotometer

(PerkinElmer, Lambda 950). The *n* of samples had been derived from the spectra by the OptiLayer software [10]. The crystalline structure of samples had been analyzed by X-ray diffractometer (PANalytical, EMPYREAN). The surface morphology of samples had been measured by atomic force microscope (Park System, XE-150) with the non-contact mode. And the microstructure of samples had been characterized by field-emission scanning electron microscope (FEI, Helios NanoLab 650).

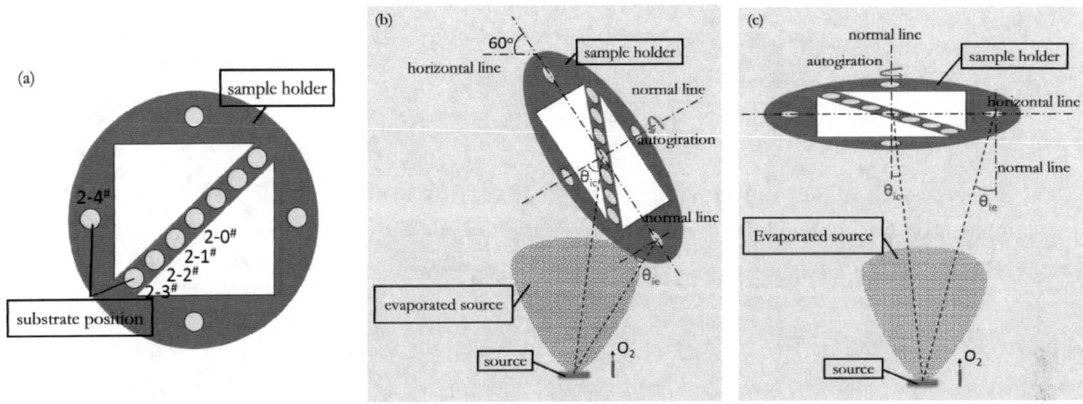

Figure 1. Schematic diagrams of sample holder (a) and its suspension ways (b, c).

Table 1. Detailed process conditions of HfO_2 film deposition for 60°-inclined suspension and horizontal suspension.

Suspension way of sample holder	Monitoring rate (nm/s)	Temperature (°C)	Oxygen pressure ($\times10^{-2}$ Pa)	Monitoring thickness (nm)
60°-inclined	0.1	210	1.8	50
horizontal	0.5	250	1.5	200

3. Results and discussion

Figure 2 shows the transmission spectra of deposited HfO_2 film samples for 60°-inclined suspension and horizontal suspension of sample holder. In the figure the spectra of sample *2-1#* and *2-4#* for 60°-inclined suspension differ obviously with each other. For example, the minimum transmittance of sample *2-1#* around λ=600nm varies about 1.18% with respect to sample *2-4#*, indicating that there is notable difference on *n* between the two samples. On the contrary, the spectra of sample *2-0#* and sample *2-3#* for horizontal suspension almost overlap with each other, suggesting that the two samples are of similar thickness and *n*. It's worth noting that some transmission maximums of samples have exceeded the transmittance of blank fused substrate, especially for sample *2-1#*, indicating that the *n* of the HfO_2 films are somewhat inhomogeneous and gradually decrease from the substrate to the film surface [11]. This n variation is opposite to that of HfO_2 films deposited by ion beam sputtering [12]. Moreover, the more the maximum exceeds, the more inhomogeneous the *n* is. Thus it can be concluded that *n* of sample *2-1#* is more inhomogeneous than that of sample *2-4#* for 60°-inclined suspension. Likewise, *n* of HfO_2 films for 60°-inclined suspension are more inhomogeneous than those for horizontal suspension.

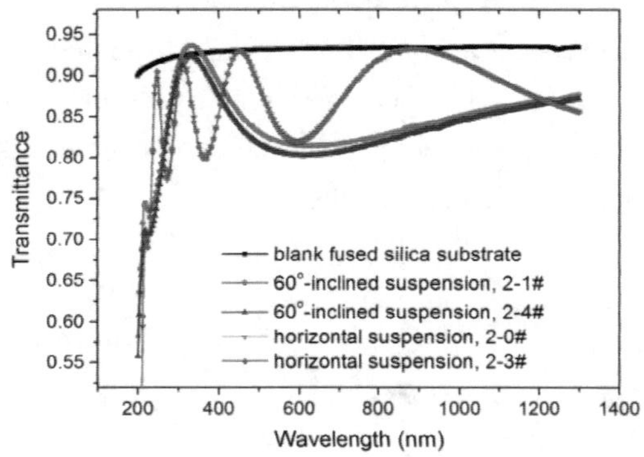

Figure 2. The transmission spectra of HfO_2 film samples for 60°-inclined suspension and horizontal suspension.

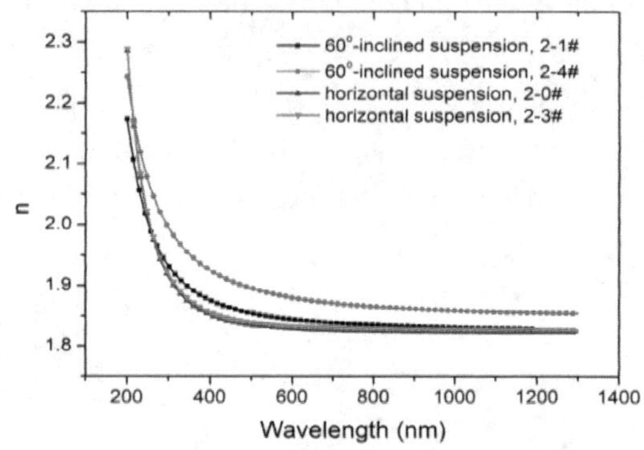

Figure 3. The *n* of HfO_2 film samples for 60°-inclined suspension and horizontal suspension.

Figure 3 depicts the dependence of *n* on light wavelength derived by the OptiLayer software, showing that the *n* of HfO_2 films at the edge of sample holder are larger than those of HfO_2 films at the center regardless of suspension way of sample holder. Moreover, *n* varies a lot between the two samples (sample *2-1#* and sample *2-4#*) of inclined suspension, while the *n* is almost same for samples (sample *2-0#* and sample *2-3#*) of horizontal suspension, coinciding with above analysis on the sample spectra. The obvious difference on *n* of sample *2-1# and sample 2-4#* reveals that the n of HfO_2 film is also relatively inhomogeneous along the radial direction of sample holder for inclined suspension. Table 2 lists some results calculated by the OptiLayer software from the spectra and gives a quantitative analysis on the variation of n. With regard to inclined suspension, the *n* at λ=1053 nm of sample *2-4#* increases about 0.028 compared to sample *2-1#*, and the variation amplitude is as large as 1.52%. However, the *n* at λ=1053nm of sample *2-0#* increases slightly compared to sample *2-3#*, about 0.005, corresponding to a variation amplitude of only 0.25%. Moreover, the *n* inhomogeneity of sample *2-1#* and sample *2-4#* is -4.2% and -1.5%, respectively, where the negative sign means *n* decreases from substrate to film surface. Whereas, the *n* inhomogeneity of sample *2-0#* and sample *2-3#* are same with each other, only -0.3%. The larger *n* inhomogeneity of inclined samples than those of horizontal samples are attributed to the severer crystallization and more columnar growth for films deposited on inclined substrate [13]. Thus, it could be concluded that sample *2-1#* is of severer crystallization and more columnar structure. The more columnar structure the film has, the more pores the film has, subsequently resulting in the decrease of *n*. Thus, the n of sample *2-1#* is smaller than that of sample *2-4#*. Regarding to horizontal suspension, the n inhomogeneity of sample*2-0#* and *2-3#* are almost same, suggesting that they are of similar structure, thus possessing similar n.

A simulation had been performed by the OptiLayer software to quantitatively analyze the effects of *n* variation on a sampling mirror, of which the reflectance was supposed to be required to 0.2%±0.025% at λ=1053nm. The structure of the mirror during simulation was designed to be "fused silica\$HfO_2$\$SiO_2$", in which the thickness of HfO_2 film and SiO_2 film were kept constant, and n values (in table 2) for HfO_2 films and n=1.4 for SiO_2 films were adopted without consideration of the dispersion of *n*. The simulation results were plotted in figure 4. Figure 4a shows that the variation of reflectance of sampling mirror is about 0.1% with *n* variation of 0.028 (in table 2) for inclined suspension, far beyond the required limit of 0.025%. By contrast, the reflectance varies only about 0.015% with n variation of 0.005 (in table 2) for horizontal suspension. In other words, the reflectance inhomogeneity along the radial direction of *Φ*560 mm-aperture sample holder is as large as ~50% for inclined suspension, while it is only ~7.5% for horizontal suspension. From this point of view, horizontal suspension is superior to inclined suspension for fabricating larger aperture optics that require highly accurate transmittance or reflectance with the existing process conditions.

Table 2. Results calculated from the spectra by the OptiLayer software.

Suspension way of sample holder	Sample	Film thickness (nm)	*n* (λ=1053nm)	Inhomogeneity of *n* along the film thickness	Variation of *n* (λ=1053nm)	Variation of the inhomogeneity of *n*
60°-inclined	*2-1#*	87.18	1.8298	-4.2%	0.028	2.66%
	2-4#	84.56	1.8576	-1.5%		
horizontal	*2-0#*	244.80	1.8231	-0.3%	0.005	~0
	2-3#	243.88	1.8276	-0.3%		

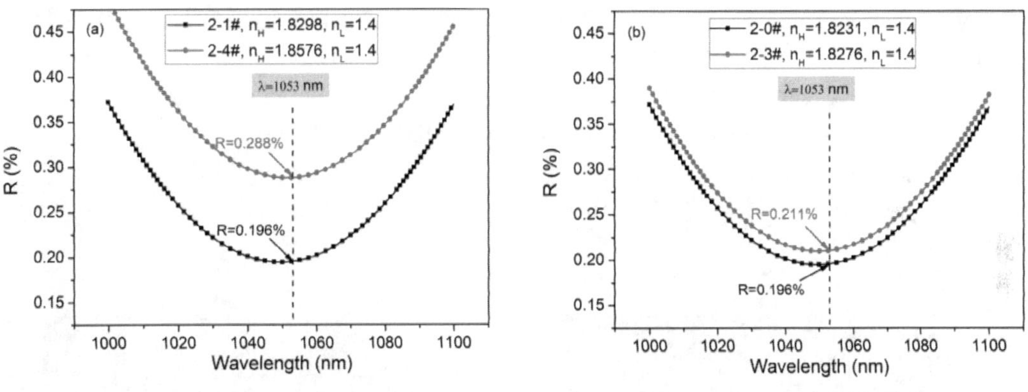

Figure 4. Simulations by the OptiLayer software on the effects of *n* variation on the reflectance of a sampling mirror for 60°-inclined suspension (a) and horizontal suspension (b)

Figure 5 plots the X-ray diffraction (XRD) spectra of HfO_2 film samples. All the curves in figure 5 have been normalized for comparative analysis. It's clear that all the spectra present several strong peaks, demonstrating that all HfO_2 films have crystallized. And the crystalline phase has been proved to be monoclinic by comparing with the powder diffraction file (PDF#06-0318) [13, 14]. Although the film thickness (in table 2) of horizontal samples are almost three times those of inclined samples, the peak intensity of horizontal samples are nearly equal to those of inclined samples, indicating that the horizontal samples have crystallized much more slightly than the inclined samples. Similarly, the peak intensity of sample *2-4#* is weaker than that of sample *2-1#* for inclined suspension, suggesting that sample *2-4#* has crystallized more slightly than sample *2-1#*. And these conclusions from XRD spectra further demonstrate the rationality of above analysis on film growth and *n* inhomogeneity of samples.

IOP Conf. Series: Materials Science and Engineering **733** (2020) 012007 doi:10.1088/1757-899X/733/1/012007

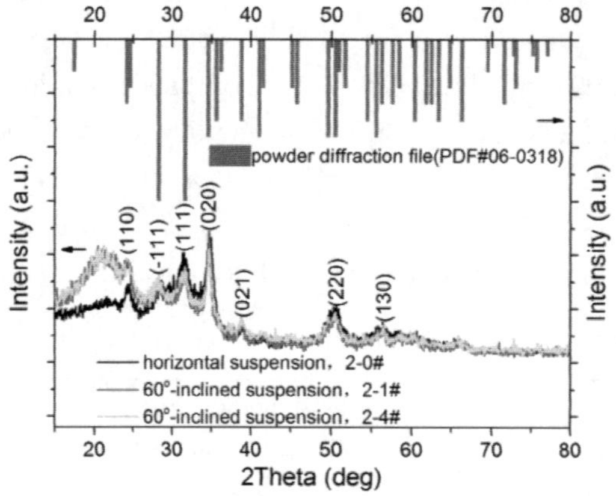

Figure 5. Spectra of X-ray diffraction on the HfO_2 film samples

Figure 6 presents surface morphology graphs of sample *2-1#* and sample *2-4#* for inclined suspension by atomic force microscope (AFM). It's clear that the surface are smooth and full of sphere grains with uniform size, suggesting that both of sample *2-1#* and sample *2-4#* have crystallized. The root-mean-square roughness (R_q) for 5 μm×5 μm is 1.23 nm and 1.41 nm for sample *2-1#* and sample *2-4#*, respectively. Moreover, the sphere grain size of sample *2-4#* is smaller than that of sample *2-1#*, indicating the weaker crystallization of sample *2-4#*, which is consistent with above XRD analysis. Regarding to horizontal samples (sample *2-0#* and sample *2-3#*), AFM graphs in figure 7 show that the sphere grain size are similar with each other, but smaller than those of inclined samples, indicating that the two horizontal samples are of similar growth properties and have crystallized much more slightly than inclined samples. And the R_q is 1.54 nm and 1.36 nm for sample *2-0#* and sample *2-3#*, respectively.

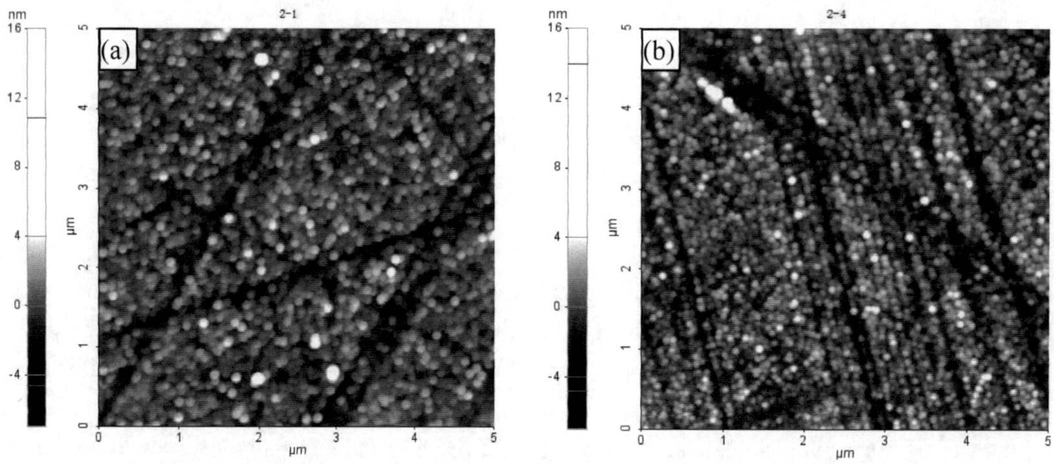

Figure 6. AFM graphs of sample *2-1#* (a) and sample *2-4#* (b) for 60°-inclined suspension

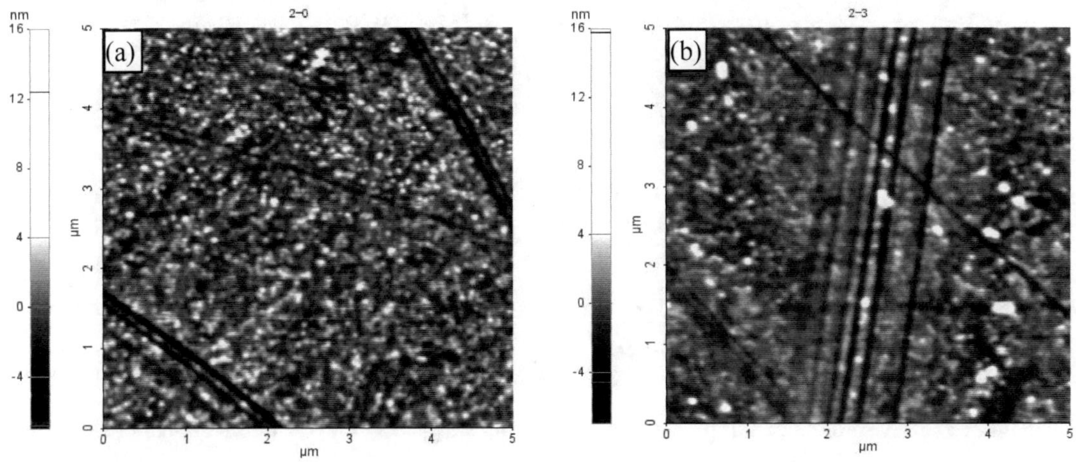

Figure 7. AFM graphs of sample *2-0#* (a) and sample *2-3#* (b) for horizontal suspension

Figure 8 and figure 9 exhibit scanning electron microscope (SEM) photos of the samples. Lots of micro gaps arise due to the aggregation of grains for sample *2-1#* in figure 8a, while no visible aggregation as well as gaps can be found for sample *2-4#* in figure 8b. Thus the emerging of the gaps would be a reason for the smaller n of sample *2-1#* than sample *2-4#*. The optical absorption at λ=355 nm (measured by laser calorimetry and not shown here) is 793.4 ppm and 577.4 ppm for sample *2-1#* and sample *2-4#*, respectively. And the higher absorption for sample *2-1#* would result from the multiple reflections of light among the gaps of columnar in the film [15, 16]. On the contrary to inclined samples, the microstructure of horizontal samples in figure 9 present almost no difference, thus leading to small variation of n between sample *2-0#* and sample *2-3#*.

The above-mentioned differences on crystallization, surface morphology and microstructure among the samples can be discussed in the light of atomic shadowing, reemission and diffusion of ad-atoms [17, 18]. With regard to horizontal suspension, the incidence angle θ_{ic} and θ_{ie} (shown in figure 1b and figure 1c), which are the angles between the evaporating source and the normal of the substrates at the center and edge of sample holder, respectively, are both small, thus the atomic shadowing is depressed, resulting in the smaller grains and dense microstructure. Moreover, the sample holder is far away from the evaporating source and rotates in a planetary way, thus the difference between θ_{ic} and θ_{ie} is small. The small difference of incidence angles leads to the similar co-effects of atomic shadowing, reemission and diffusion of ad-atoms on the film deposition of sample *2-0#* and sample *2-3#*. Therefore, sample *2-0#* and sample *2-3#* exhibit similar film properties and n.

As for 60°-inclined suspension, the incidence angles are relatively larger, thus atomic shadowing is enhanced, resulting in larger grains and more gaps of sample *2-1#* and sample *2-4#* than those of horizontal samples. Moreover, the incidence angle θ_{ie} of sample *2-4#* varies in a much wider range than θ_{ic} of sample *2-1#* during planetary rotation of sample holder. The wider variation range of incidence angle can depress the atomic shadowing, eventually leading to smaller grains, denser microstructure and larger n of sample *2-4#* compared to sample *2-1#*.

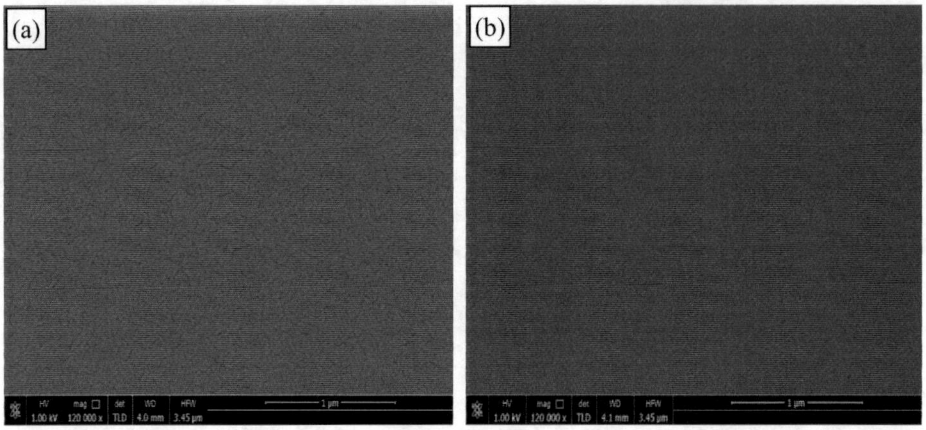

Figure 8. SEM graphs of sample *2-1#* (a) and sample *2-4#* (b) for 60°-inclined suspension

Figure 9. SEM graphs of sample *2-0#* (a) and sample *2-3#* (b) for horizontal suspension

4. Conclusions

HfO_2 films have been deposited by reactive e-beam evaporation and the effects of suspension way of substrates on the large-aperture uniformity of refractive index of HfO_2 films have been comparatively studied. The optical transmittance and OptiLayer software analysis reveal that the refractive index of deposited HfO_2 films for 60°-inclined substrate suspension present inhomogeneity along the normal as well as the radial direction of sample holder, while the refractive index are homogeneous in both directions for horizontal substrate suspension. From this point of view, the horizontal suspension is superior to inclined suspension to fabricate large-aperture optics that require highly accurate transmittance or reflectance under the existing experimental conditions. Regarding to inclined suspension, the film at the edge of sample holder crystallized more slightly and was denser, thus possessed larger refractive index than the film at the center of sample holder. And the inhomogeneity of refractive index for inclined suspension results from the different combined effects of atomic shadowing, reemission and diffusion of ad-atoms during the film deposition.

References

[1] Baisden P A, *et al.* 2016 Large optics for the National Ignition Facility *Fusion Sci. Technol.* **69** 295–351

[2] Stevenson I C and Sadkhin G 2001 Optimum location of the evaporation source: experimental verification. *SVC Technical Conference Proceedings* **44** 306–13

[3] Ramprasad B S and Radha T S 1973 Uniformity of film thickness on rotating planetary planar substrates *Thin Solid Films* **15** 55–64

[4] Musset and Stevenson I C 1988 Obtaining uniformly thick films in coating chambers. *SVC Technical Conference Proceedings* **31** 203–9

[5] Smith D J, *et al.* 1998 Counter-rotating planetary design for large rectangular substrates *SVC Technical Conference Proceedings* **41** 193–6

[6] Oliver J B and Talbot D 2006 Optimization of deposition uniformity for large-aperture National Ignition Facility substrates in a planetary rotation system *Appl. Opt.* **45** 3097–105

[7] Stolz C J, Sheehan L M, Gunten M K, Bevis R P and Smith D 1999 The advantages of evaporation of hafnium in a reactive environment to manufacture high damage threshold multilayer coatings by electron-beam deposition *Proc. SPIE* **3738** 318–24

[8] Alvisi M, Di Giulio M, Marrone S G, Perrone M R, Protopapa M L, Valentini A and Vasanelli L 2000 HfO_2 films with high laser damage threshold *Thin Solid Films* **358** 250–8

[9] Gallais L, Capoulade J, Yves J, Commandré M, Cathelinaud M, Koc C and Lequime M 2008 Laser damage resistance of hafnia thin films deposited by electron beam deposition, reactive low voltage ion plating, and dual ion beam sputtering *Appl. Opt.* **47** C107–13

[10] OptiLayer GmbH https://www.optilayer.com/

[11] Tikhonravov A V, *et al.* 2011 Optical parameters of oxide films typically used in optical coating production *Appl. Opt.* **50** C75–85

[12] Liu H S, et al. 2014 Correlation between properties of HfO_2 films and preparing parameters by ion beam sputtering deposition *Appl. Opt.* **53** A405–11

[13] Franta D, *et al.* 2011 Optical characterization of HfO_2 thin films *Thin Solid Films* **519** 6085–91

[14] Gao J, *et al.* 2016 Microstructure, wettability, optical and electrical properties of HfO_2 thin films: Effect of oxygen partial pressure *J. Alloys Compd.* **662** 339–47

[15] Charles C, *et al.* 2013 Correlation between structural and optical properties of WO_3 thin films sputter deposited by glancing angle deposition *Thin Solid Films* **534** 275–81

[16] Cansizoglu M F, *et al.* 2010 High optical absorption of indium sulfide nanorod arrays formed by glancing angle deposition *ACS Nano* **4** 733–40

[17] Tokas R B, *et al.* 2015 Oblique angle deposition of HfO_2 thin films: quantitative assessment of indentation modulus and micro structural properties *Mater. Res. Express* **2** 035010

[18] Tokas R B, *et al.* 2016 Effect of angle of deposition on micro-roughness parameters and optical properties of HfO_2 thin films deposited by reactive electron beam evaporation *Thin Solid Films* **609** 42–8

IOP Conf. Series: Materials Science and Engineering **733** (2020) 012008 doi:10.1088/1757-899X/733/1/012008

Used of bottom ash waste from power plant and tailing as mine backfill

M Masniyom[1] and W Jaiboon[2]

1 Lecturer, Prince of Songkla University, Songkhla, Thailand
2 Mining Engineer, Patong Tungsong Company, Nakonsritamarat, Thailand
E-mail: manoon.ma@psu.ac.th

Abstract. The using of bottom ash and tailing were by product from Mae Moh coal fired power plant as mine backfill. Laboratory was investigated the chemical, physical and mechanical properties of backfill containing bottom ash, tailing and cement. Using cast of backfill materials to test the ability of compressive strength in box size 5x5x5 cm. Four different types of backfill that portland cement type 1 was replaced by bottom ash, in the proportion of percent 5, 10, 15 and 20 by weight uncured and cured in water for 7 and 28 days. Backfill materials characterization bottom ash have been examined in grain size distribution, chemical composition analyzed by X-ray fluorescence (XRF) of backfill material. For the mechanical property was test by uniaxial compressive strength machine. The result was found that mixing ratio of bottom ash can be used as backfill material ranging from 15-20 % by weight. The mechanical property of backfill mixture suitable for use with compressive strength from 6.4-19.7 MPa and cured in water 28 days was maximum compressive strength.

1. Introduction

Waste from Mae Moh coal fired power plant was fly ash and bottom ash. The bottom ash is the companion to fly ash in process of coal fired with an approximately amount of 20 % by volume of the total ash. It's particle is porous, irregular and coarse grain than fly ash. The bottom ash produced annually in Thailand with an estimated amount of 750,000 tons, has been mostly dumped in landfill site.[1]

Mine backfilling that can solve such puzzles as shortage of backfilling aggregate, contamination of tailing occupied area, construction and maintenance of tailing reservoir, is an operative technology to reduce the backfilling cost, turn waste to reuse eliminate pollution effectively. Tailing cement backfilling body is a kind of low grade concrete which strength depend mainly on bond strength. [2] During the past ten years, significant studies have been completed which focused on the fundamental reasons for the strength variations between the different testing methods, other discoveries were also made which have assisted in maximizing the strength of backfill.

The objective of this paper have done the optimal ratio among cement, bottom ash and tailing based on laboratory investigation. The mechanism of bottom ash backfill will be revealed according to the physical, chemical characteristics, and mechanical properties will be evaluated based on bottom ash and cement ratio.

2. Materials and methods

In this study use the ground bottom ash size was 100 mesh and tailing was 30 mesh collected from Mae Moh power plant, Lampang province and Portland cement type 1 show in figure 1 that mix all materials ratio show in Table 1 which was replaced by bottom ash, in the proportion of percent 5, 10, 15 and 20 by weigh uncured and cured in water for 7 and 28 days. The physical properties (Bulk density, Specific gravity and Water content) and chemical analysis by X-ray fluorescence (XRF) of backfill materials show in Table 2-3, the major component of bottom ash was (SiO_2), (Al_2O_3), (CaO) and (Fe_2O_3) , the major component of tailing was (SiO_2), (Al_2O_3), the major component of cement was (SiO_2), (Al_2O_3) and (CaO). Furthermore uniaxial compressive strength (UCS) test to evaluate mechanical property that can be use as suitable mine backfill.

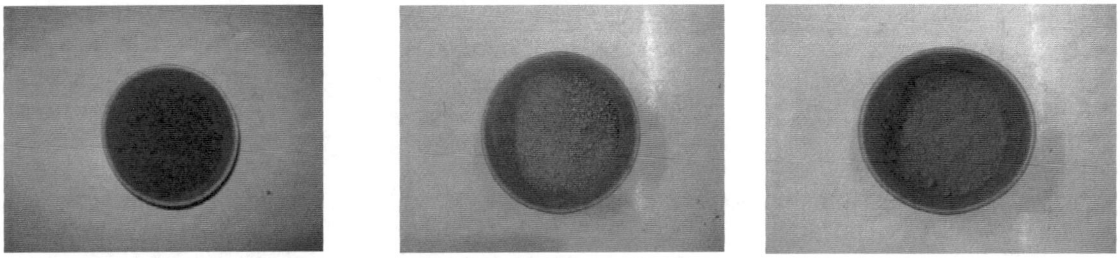

(a) Bottom ash (b) Tailing (c) Cement

Figure 1. Backfill materials

Table 1. Mixture ratio of backfill materials

Mixture No.	Cement %	Cement (g)	Tailing (g)	Bottom Ash (g)	Water (g)
1C5	5	11.5	161	57.5	41
2C10	10	23	161	46.5	41
3C15	15	34.5	161	43.5	41
4C20	20	46	161	23	41

Table 2 Physical characteristic of backfill materials

Sample	Bottom ash	Tailing	Cement
Bulk density (g/cm^3)	1.82	1.67	1.72
Specific gravity (g/cm^3)	2.95	2.53	3.04
Water content (%)	1.43	2.11	0.78
Size (mesh)	100	30	200

Table 3. Chemical composition of backfill materials

Chemical composition (%)	Bottom ash	Tailing	Cement
(Na_2O)	1.10	-	-
(MgO)	1.64	-	1.24
(Al_2O_3)	17.75	11.27	5.22
(SiO_2)	29.89	82.88	20.84
(SO_3)	1.83	-	2.41
(K_2O)	4.37	0.34	0.22
(CaO)	18.68	0.20	66.28
(TiO_2)	0.64	0.04	-
(Fe_2O_3)	20.42	2.72	3.2
(SrO)	0.17	-	-
(ZrO_2)	0.02	0.04	-
(BaO)	0.24	-	-

3. Results and discussion

The experimental result of cemented bottom ash backfill show that the bottom ash power plant waste has small particle, mine tailing waste has large particle and cement binder has fine particle when mixed in water to form slurry, the action mechanism of all materials active gelatinization and filling compaction and hardening are ability all component react which mineral phase in cement and its hydrates product $Ca(OH)_2$.

Using cast of backfill materials to test the ability of compressive strength in box size 5x5x5 cm. Four different types of mortar where Ordinary Portland cement type 1 was replaced by bottom ash, in the proportion of percent 5, 10, 15 and 20 by weight uncured and cured in water for 7 and 28 days. The test result is shown in Table 4. The result was found that mixing ratio of bottom ash can be used as backfill material ranging from 15-20 % by weight. The mechanical property of backfill mixture suitable for use with compressive strength from 6.4-19.7 MPa and cured in water 28 days was maximum compressive strength.

Table 4. The result of backfill mixture uniaxial compressive strength test

Mixture No.	Uncured (MPa)	Cured in water for 7 days (MPa)	Cured in water for 28 days (MPa)
1C5	0.8	1.2	1.3
2C10	3.0	4.1	5.9
3C15	6.4	8.0	9.8
4C20	8.0	11.6	19.7

4. Conclusion

Bottom ash waste from power plant and mine tailing can be used in the cemented backfill and can be used to replace partly cement. Initial strength of bottom ash cemented backfill is low for all mixture, but after curing in numerical control cement standard curing box for 28 days in water was found that mixing ratio of bottom ash can be used as backfill material ranging from 15-20 % by weight maximum compressive strength 19.70 MPa.

5. References

[1] Kasemchaisiri R and Tangtermsiri S, 2008 Properties of Self-Compacting Concrete in Corporating Bottom Ash as a Partial Replacement of Fine Aggregate ScienceAsia 34 pp 88-95
[2] Manca P Massaci G Massida L and Rossi G, 1984 Stabilization of mill tailings for mining backfill with Portland cement and fly ash Trans. Instn. Min. Metall 93 pp 48-54
[3] Zhang R Y ,2003 Mechanism study and experiment of fly ash used for tail sand cement filling, Journal of Southern Institute of Metallurgy 24(2) pp 1-5
[4] Li J Zhuang X Querol X Font O and Moreno N, 2018 A review on the applications of coal combustion products in China. International Geology Review, 680(6), 821-828
[5] Ramme B and Tharaniyil M, 2013 Coal Combustion Products Utilization Handbook (3rd Ed.) We Energies Publication: Wisconsin
[6] H K Kim H K Lee, 2011 Use of power plant bottom ash as fine and coarse aggregates inhigh-strength concrete Construction and Building Materials 25 pp 1115–1122
[7] H K Lee H K Kim E A Hwang, 2010 Utilization of power plant bottom ash as aggregates in fiber-reinforced cellular concrete Waste Management 30 pp 274–284

IOP Conf. Series: Materials Science and Engineering **733** (2020) 012009 doi:10.1088/1757-899X/733/1/012009

Experimental research on sound absorption properties of impedance gradient composite with multiphase

R X Liu [1,2], **D L Pei** [1,2] **and Y R Wang** [1,2]

1 Institute of Mechanics, Chinese Academy of Sciences, Beijing, China
2 School of Engineering Science, University of Chinese Academy of Sciences, Beijing, China
E-mail: yurenwang@imech.ac.cn

Abstract. In order to improve the sound absorption performance of the sound absorbing material in water, impedance gradient underwater sound absorbing composite with multiphase (IGCM) is prepared in this study. By superposing several layer materials of different impedance gradients, the mixture is poured in the mold layer by layer so that three kinds of IGCM with different mass density gradients of iron powder, tungsten powder and mixture of iron powder and tungsten powder are prepared. Under two operating conditions of the incident angle of 0°and 20°, the results show that the three IGCMs have good broadband absorption characteristics. The sound absorption frequency ranges of the three samples of IGCM under normal incident conditions are concentrated between 6kHz and 10kHz. The sound absorption performance of IGCM under oblique incidence is better than that under normal incidence, especially the IGCM with iron-tungsten powder, which could achieve wide-band strong sound absorption from 2kHz to 10kHz under the condition of oblique incident. The IGCM with iron-tungsten phase under sound wave's incident angle of 20° exhibits better sound absorption performance than the other two, which provides a theoretical reference for the design and manufacture of gradient sound absorbing materials.

1. Introduction

It is well known that sound waves propagate in water in the form of longitudinal waves that are not easily attenuated, which becomes the main means of underwater detection information[1]. In order to improve the stealth ability of underwater detectors, underwater sound absorbing materials are used[2]. The impedance gradual sound absorbing material is a sound absorbing material whose impedance changes according to certain gradients in the thickness direction of the material. It is strongly designable and easy to realize the multi-physics coupling requirement [3], and expected to realize the ultra-wide frequency sound absorption effect.

As for studies published, Pedersen et al. found that the impedance-transition-type acoustic structure can obtain a wider sound absorption band[4]. Emery discovered that impedance could reduce the reflection coefficient at high frequency, and the problem of the peak value of the reflection coefficient could be solved by employing multi-layer structures with different thickness and impedance[5]. Beretti et al. found that the four-layer silencing structure had the functions of sound absorption and vibration reduction when the thickness was constant[6]. Forest et al. chose SiO_2 aerogel particles with particle size of 80um and 3.5mm to make double-layer sound absorption structure, including impedance matching layer and sound absorbing layer. It was found that its sound absorption performance was higher than that of the uniform structure of the same thickness in the low frequency range[7]. Spivack et al. studied the problem of acoustic scattering of irregular surfaces of impedance changes, and derived the acoustic scattering field formula under random impedance changes[8]. Andrew studied the problem of acoustic propagation in air layers of different heights and performed simulation analysis using the finite element method[9]. Shang Erchang studied the approximation of the reflectivity of the gradient absorption layer, and theoretically explained the effectiveness of the impedance gradient sound absorption layer[10]. The method of approximate solution layer by layer to transfer matrix was proposed, and the acoustic characteristics of the sound absorbing structure were discussed by He Zuoyong[11]. Yang Xue et al. found the optimized design of the material in the low frequency range of 1kHz-4kHz improved the sound absorption coefficient [12]. Li Bo found that the sound absorption coefficient of the composite polymer particles was higher in the range of 1 kHz to 6 kHz than that of the single polymer particles[13]. Tang Huiping et al. found that the arrangement of different porosity had a significant effect on the sound absorption performance of the gradient structure, and to some extent, the sound absorption coefficient in the frequency band of 0.8 kHz-1.6 kHz was improved to Between 0.8-1[14]. In summary, the acoustic wave frequency band of the impedance gradient sound absorbing material needs to be broadened, and the research on the sound absorption characteristics of the material under oblique incidence is less, so its application is easily limited.

Aiming at the problems above, this paper develops a new type of impedance gradient underwater sound absorbing composite with multiphase（IGCM）, and tests the sound absorption performance of the material under normal incidence (0°)and oblique incidence (20°). The results of the wide-band strong sound absorption experiment with oblique incidence are obtained, which broadens the sound absorption band of the underwater impedance gradient sound absorbing material, and provides a theoretical reference for the broadband design of the underwater impedance gradient sound absorbing material.

2. Experimental

2.1. Preparation of IGCM

In the experiment, PTMG/TDI produced by Wanhua Chemical Co. Ltd. was used to prepare the polyether polyurethane for the experiment as the matrix material, because the polyether polyurethane has strong water resistance[15][16],strong adhesion and the same acoustic impedance as that of water so that it could be permanently and firmly paste with the coated object. According to the fact that the thickness of the viscoelastic substrate sound absorbing material is no less than 50mm[3], three kinds of IGCM with the thickness of 55mm were prepared by respectively blending iron powder, tungsten powder and iron-tungsten powder. The acoustic impedance of each layer of materials was close to that of water medium by one side, whereas gradually increased to that of steel body surface by the other side. And there was a superposition of the sound absorbing layers whose impedance gradient changes due to different mass densities inside so that sound waves could be absorbed and attenuated. Iron powder, tungsten powder were used as the additional phase (the average particle diameter is 25um), three sets of IGCM were prepared by blending method. The composition is shown in Table 1.

Table 1. Components of three kinds of IGCM

Layers	1	2	3	4	5	6	7	8	9	10	11
Matrix material	1	1	1	1	1	1	1	1	1	1	1
Iron	0	0.2	0.4	0.8	1.0	1.5	2.0	2.5	3.0	3.5	4.0
Tungsten	0	0.2	0.4	0.8	1.0	1.5	2.0	2.5	3.0	3.5	4.0
Iron / Tungsten	0	0.2	0.4	0.8	1.0	1.5	2.0	2.5	3.0	3.5	4.0
$\rho(g/cm^3)$	1	1.2	1.4	1.8	2.0	2.5	3.0	3.5	4.0	4.5	5.0

The physical maps of the three sets of IGCM prepared in the experiment are shown in Fig.1.After testing and calculation, it is verified that the acoustic impedance of the three groups of materials increases by gradients in the thickness direction, shape rules, and size also meet the sound absorption coefficient test requirements as is shown in Fig. 2.

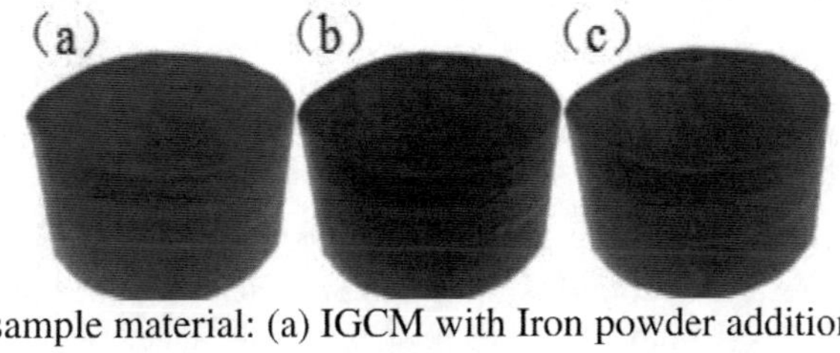

Figure 1. Experimental sample material: (a) IGCM with Iron powder additional phase; (b) IGCM with Tungsten powder additional phase ; (c) IGCM with Iron-tungsten powder additional two-phase.

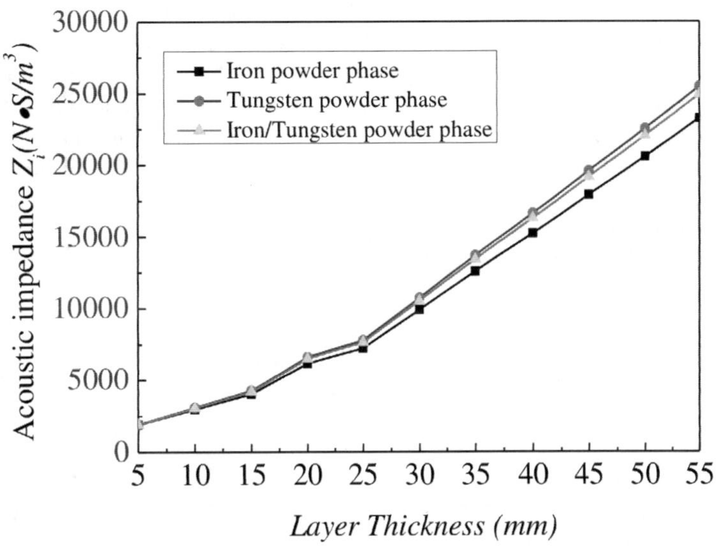

Figure 2. Impedance curves of three IGCM.

2.2. Test Methods

The underwater acoustic impedance sound tube is used to test the sound absorption coefficient of IGCM under the condition of normal and oblique incidence sound waves. Fig. 3 (a) is a schematic diagram of the underwater acoustic impedance sound tube, in which the incidence of sound wave comes from the side which has the same acoustic impedance with water. Fig. 3(b) is an enlarged schematic view of a sample of IGCM placed in the sound tube. According to the method provided by literature [17], a wedge-shaped metal tube structure with the oblique angle of 20° is made for the incidence of sound wave. Since the sound wave is a longitudinal wave, there is no cutoff frequency for the sound wave propagating in the impedance tube. When the wavelength is much larger than the transverse dimension of the tube, the sound waves propagating in the tube are always plane waves, and the wave vector is parallel to the axial direction of the tube. The frequency range and the wavelength of the sound tested in this study satisfies this condition, so it is still plane wave parallel to the axial direction of the tube. Under the conditions of normal temperature, normal pressure and frequency range of 0.2 kHz to 10 kHz, the underwater sound absorption coefficients of the material under normal incidence of 0° and oblique incidence of 20° were tested.

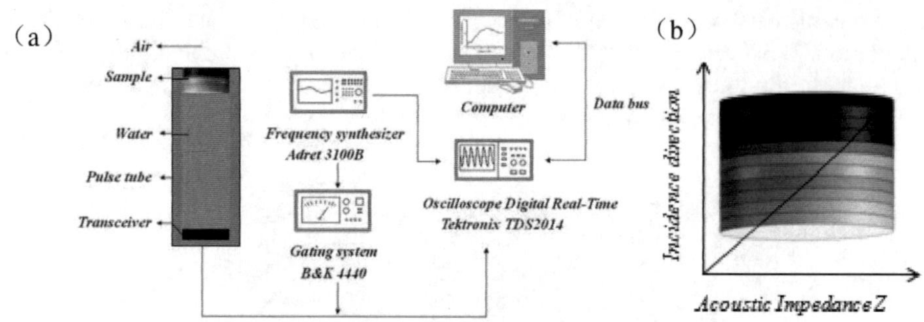

Figure 3. Schematic diagram for (a) sound absorption coefficient of acoustic impedance tube test and (b)sample and its acoustic impedance.

3. Results and discussion

The underwater sound absorption coefficients of different samples into which sound wave transmit are shown in Fig. 4. The sound absorption coefficient of IGCM with tungsten powder is larger than that with iron powder when the frequency ranges from 7kHz to 8kHz; The sound absorption coefficient of IGCM with iron-tungsten mixed powder is higher than that with iron or tungsten powder when it comes to frequencies except the range from 6.5 kHz to 7.5 kHz, among which the sound absorption coefficient reaches 1.0 with all the sound waves were absorbed when the frequency is around 10kHz.The reason why the IGCM with tungsten or iron-tungsten powder has higher sound absorption coefficient may be that tungsten has higher density than that of iron, and tungsten particles therefore vibrate stronger than iron particles under lower frequency, which led to intensified fraction at inter-phase and transferring sound energy to heat energy for better sound absorption. The reason why IGCM with iron-tungsten powder has higher sound absorption coefficient may be that some of iron particles vibrate stronger than tungsten particles at higher frequencies, and therefore compensated more sound energy than IGCM with tungsten powder, making better sound absorption performance of IGCM with iron-tungsten powder.

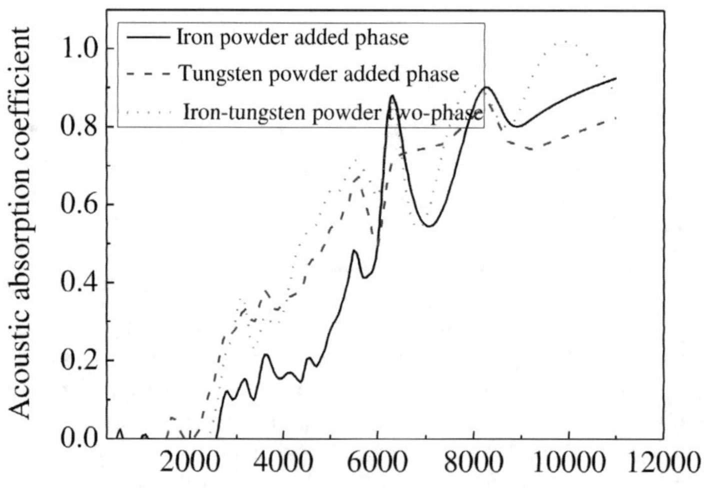

Figure 4. Acoustic absorption coefficient of three sets of IGCM at normal incidence

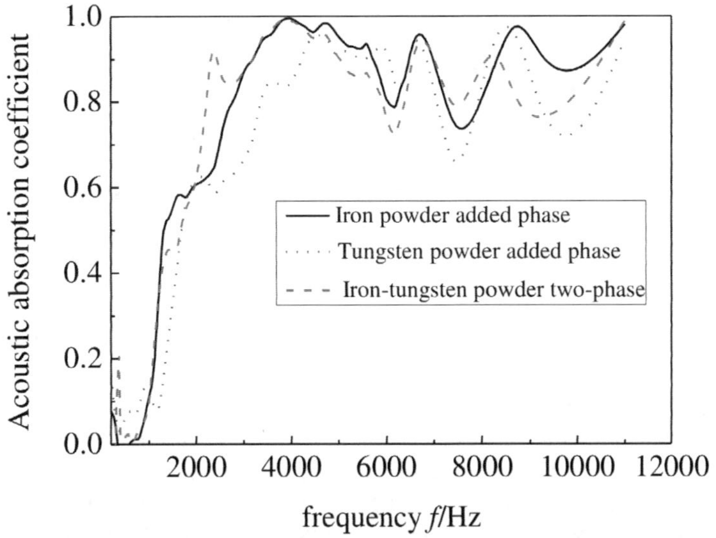

Figure 5. Acoustic absorption coefficient of three sets of IGCM at normal incidence.

Fig.5 shows the sound absorption coefficients of IGCM which the incident angle of the sound wave is 20°.The IGCM with iron-tungsten powder has a relatively high sound absorption coefficient over the entire test frequency range, and the sound absorption performance is relatively optimal. Maybe due to the added iron powder's low density, particles in the sample with iron powder additional phase and that with iron-tungsten powder mixed phase tungsten powder vibrates more strongly in higher frequency range under oblique incidence, leading to a better sound absorption energy consumption effect compared with that with tungsten powder phase. The phase particles vibrate more strongly, so the sound absorption energy consumption effect is better than that of the sample of IGCM with tungsten powder additional phase. At low frequency, the sound absorption coefficients of the sample with tungsten powder additional phase and that with iron-tungsten powder mixed phase are higher than that with the iron powder additional phase when the frequency ranged from 5 kHz to 7 kHz, maybe because the tungsten powder additional phase has particles of a higher density. Meanwhile, it has stronger vibration than iron powder particles, and the phase friction energy consumption is larger. However, in the frequency range lower than the above case between 2 kHz and 4 kHz, maybe due to the heavier phase of the tungsten powder particles, the distance between some tungsten powder particles and the tungsten powder particles is close or closely connected, which is not conducive to the energy dissipation mechanism of scattering and thermal energy conversion in the material in the lower frequency range. As is shown in Fig. 6, after a beam of oblique incidence of acoustic waves, the sound waves are shown in the schematic diagram of the refraction and reflection between the layers in the material.

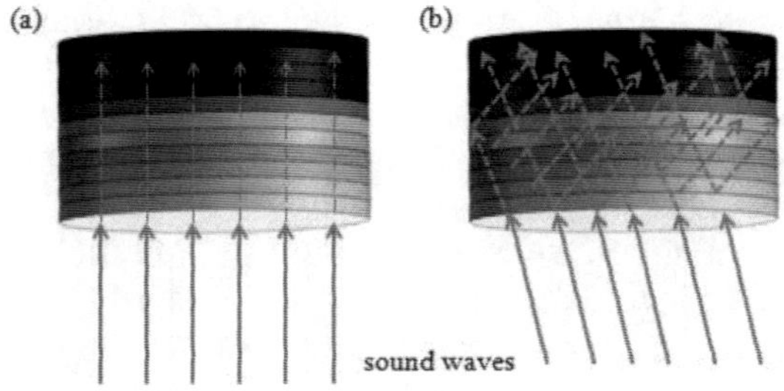

Figure 6. Schematic diagram of acoustic wave propagation in a IGCM at (a) normal incidence and (b) oblique incidence

It is obvious that the sound absorption coefficients measured by the three groups of materials when the sound wave is oblique (20°) were higher than normal incidence, and the frequency range with a good sound absorption coefficient in the case of oblique incidence is wider. Analyzing the reason, it may be that when the sound wave incidence is at an angle of (20°), after the sound waves enter the material, the longitudinal wave in the interlayer of the impedance gradient changes into a transverse wave that is more easily attenuated, causing a large shear deformation and causing the incident wave, making the energy consumed in large quantities. In addition, after the incidence comes obliquely into the material, multiple reflections and refractions occur between the two layers with the impedance gradient changing accompanied by multiple waveform conversions, so that the acoustic waves are attenuated more when oblique incidence occur, thus having better sound absorption effect.

4. Conclusions

In this paper, the underwater sound absorption performance of IGCM is studied under normal and oblique incidence conditions. The experimental materials are prepared by mature techniques, and the sound absorption coefficients of the material samples of IGCM under the above two conditions are measured in the underwater acoustic impedance tube test system. The results showed that the sound waves are effectively absorbed in the IGCM, and the sound absorption frequency range under oblique incidence is wider, the sound absorption coefficient is higher, the effect is better than that under normal incidence. In addition, the sound absorption performance of IGCM with iron-tungsten powder mixed phase is better than that of the other two single-phase samples. Therefore, the broadband sound absorption performance under oblique incidence is obtained by the study of this paper. The impedance gradient underwater sound absorbing composite material can provide theoretical support for the structural design of acoustic impedance gradient materials.

Acknowledgements

Thanks to the National Natural Science Foundation of China (Grant No11602269), the Strategic Priority Research Program of the Chinese Academy of Sciences (Approval No XDB22040301) and the Beijing Science and Technology Plan (Approval No Z161100002616034 Z171100000817010).

References

[1] Wang Y R, Miao X H, Jiang H et al. 2017 *Advances In Mechanics* **00** 96–125(in Chinese).
[2] Shi Y X, Xi Z P, Tang H P et al. 2010 *Materials Reports* **1**(in Chinese）.
[3] Zhang Y B, Ren C Y and Zhu X 2013 *Materials Reports* **7** 59–62(in Chinese).
[4] Pedersen P C, Tretiak O and He P 1982 *Acoust. Soc. Am* **2** 327–336
[5] Emery P A 1995 Conference Proceedings of *UDT (London)* pp 527–531
[6] Beretti S and Vassas M 1996 Conference Proceedings of *UDT(London)* pp 300–303.
[7] Forest L, Gibiat V and Hooley A 2001 *Journal of Non-Crystalline Solids*, **285**(1–3):0–235.
[8] Spivack M, Spivack O 2006 *International Congress on Sound and Vibration* **5** 3982–3988.
[9] Peplow, Andrew 2009 *8th European Conference on Noise Control*, v**31**, part3.
[10] E C Shang 1965 *Acta Acustica* **2** 193–197(in Chinese).
[11] He Z Y and Wang M 1996 *Journal of Applied Acoustics* **5** 12–19(in Chinese).
[12] Wang Y S, Yang X, Zhu J H et al. 2006 *Acta Acustica*, **1** 14–18(in Chinese).
[13] Li B, Zhou H and Huang G S 2006 *Polymer Materials Science and Engineering* **3** (in Chinese).
[14] Tang H P, Zhu J L, Ge Y et al. 2007 *Rare Metal Materials and Engineering* **12** 2220–2223 (in Chinese).
[15] Liu H J 2012 *Polyurethane Elastomer Hand-book 2nd Edition* (Beijing:Chamical Industry Press) (in Chinese).
[16] Shi Z G 2011 *Acoustics and Electronics Engineering* **4** 1–3(in Chinese).
[17] Mei J, Ma G C, Yang W et al. 2012 *Physics* **7** 425–433(in Chinese).

IOP Conf. Series: Materials Science and Engineering **733** (2020) 012010 doi:10.1088/1757-899X/733/1/012010

Influence of welding groove on residual stress and distortion in T-joint weld

ChongYi WEI and WenGuang JIANG*

1 School of Mechanical Engineering, Yanshan University, Qinhuangdao 066004, China and
2 Hebei Innovation Center for Equipment Lightweight Design and Manufacturing 066004, China
E-mail: wgj@ysu.edu.cn

Abstract. T-joint welding is the most common type used in heavy equipment industry. Localized heating and subsequent cooling induce residual stress in the weld process. Service loading overlapped with residual stress influences the strength of the welded structure and leads to premature failure of structure. In this study, finite element analysis is used to study the influence of welding groove on residual stress and distortion in T-joint welding. The simulation consists of sequentially coupled thermal and elastic-plastic analysis using an element birth and death method to model the addition of weld metal to the workpiece. The results of numerical simulation are verified by comparing with experiment from published literature. The results show that welding groove has a significant influence on welding residual stress.

1. Introduction

In heavy equipment industry, T-joint fillet weld is widely used in the manufacturing of complicated structures. Localized heating from welding and subsequent cooling process induce residual stress near the weld bead and produce distortion, both of them have negative effects on fatigue performance of welded structures [1]. Hence, the predication of residual stress and distortion in manufacturing process is of great importance.

Under the development of numerical simulation, finite element analysis validated with a few experiments has become a popular technique for the prediction of residual stress and distortion. In addition, finite element analysis is an effective tool to provide insight on welding residual stress that are difficult to measure experimentally.

A substantial number of numerical simulations and experimental works focusing on T-joint fillet weld with emphasis on predication of residual stress on flange plate is available in the literature. Gannon [2] studied the influence of welding sequences on residual stress and distortion in flat-bar stiffened plates by finite element analysis. The results show that welding sequence do not have a significant influence on the distribution pattern of the residual stress, however it affects the peak values. Fu [3] investigated the welding residual stress and distortion in T-joint weld under different mechanical boundary conditions. To improve the computational efficiency, a combined shell and solid FE model was proposed by Perić [4] to calculate

residual stress distribution. The FE simulation results have proved that the proposed shell and solid element technique is an effective method to predict welding distortions and residual stresses for large structures.
For the convenience of component assembly and welding penetration, the existence of groove is necessary in the component to be welded. Many researchers have developed numerical and experimental methods to investigate the effect of groove on welding residual stress.
Sattari-Far [5] used finite element techniques to study the effects of weld groove shape and weld pass number on welding residual stresses in butt-weld pipes. Akbari [6] investigated the effects of pipe wall-thickness, weld groove angle and root opening distance on welding residual stresses in dissimilar butt-welded pipe. Ye [7] discussed the influences of V, K and X groove joints on residual stress and angular distortion in a SUS304 butt-welded joint numerically and experimentally, the simulation results show that groove type has a significant influence on welding residual stress distribution, angular distortion and width of sensitization region. Teng [8] discussed the effects of welding penetration depth on residual stress and distortion.
Despite a number of researches focusing on prediction of residual stress in T-joint weld and effects of groove type in butt welded plate and pipe on residual stress, few researchers had paid attention to the counterpart in T-joint welds. The present study aims to figure out the effect of weld groove on welding residual stress and distortion in T-joint fillet weld. The welding simulation was carried out by a 3-D thermal elastic plastic finite element method and was validated with experimental measured data published by Fu [9] to prove the effectiveness of proposed computational approach. Welding residual stresses and distortions of welded plates with or without groove are discussed.

2. Finite element modelling

Thermal and structural finite element analysis were performed by using ANSYS parametric design language (APDL) code. For the weak structural to thermal field couplings, the problem is treated as a sequentially coupled thermal elastic-plastic analysis. The simulation consisted of two types of analysis. The first was a transient thermal analysis where the temperature distribution caused by a moving heat source was determined. Next, the thermal element was converted into structural element, the temperature distribution of all nodes from the thermal analysis were applied as loads in the structural analysis.
The same finite element is used in both thermal and mechanical analysis. In the thermal analysis, eight-node thermal brick element (SOLID70) is used, eight-node brick structural element (SOLID185) is used to proceed the elastic plastic analysis. A fine mesh is used in the vicinity of the welding seam and heat affected zone. The element size increases with the increase of distance from welding seam. The smallest element size is 2.5×1×1mm on the basic of mesh sensitivity analysis. The temperature dependent material property is taken from Wang [10] as shown is Fig. 1.
To take the influence of groove into account in numerical simulation, two different finite element models (i.e. without groove, or with two grooves in web plate) were prepared. The welding heat input is the same for the same welding velocity. A single-side metal gas (MIG) welding was employed. Welding sequence and direction are adopted from reference [9]. The geometry of plates to be welded is the same. The flange is 500mm×500mm, and the web is 500mm×300mm, thicknesses of both plates are 12mm. Fig.2 shows the finite element models with K groove and no groove.

2.1 Thermal analysis

Thermal analysis was carried out to determine the temperature distributions of welded plates. The governing equation of the transient heat transfer during welding is given by

$$\rho c \frac{\partial \mathrm{T}}{\partial \mathrm{t}}(x,y,z,t) = -\nabla . \vec{q}(x,y,z,t) + Q(x,y,z,t) \tag{1}$$

where T, t, $\vec{q}$ and Q are the temperature, time, the heat flux vector and internal heat generation rate respectively, ρ is the density, c is the specific heat capacity, ∇ is the spatial gradient operator and x, y and z are coordinated in the reference system.

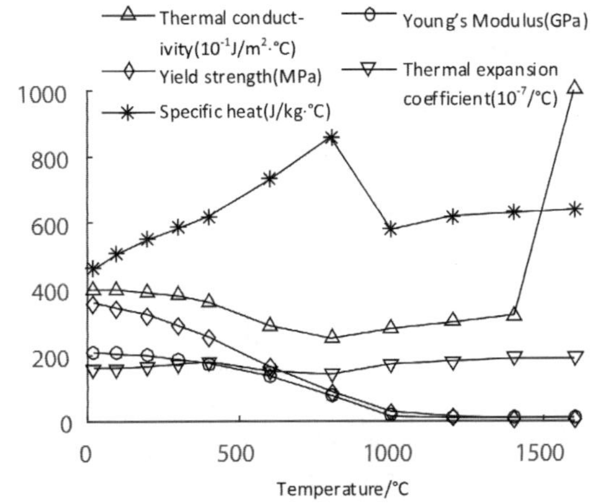

Figure.1 DH36 material properties dependent with temperature

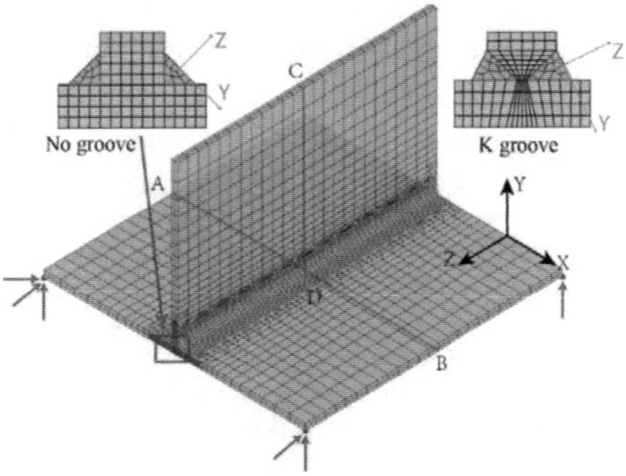

Figure.2 Finite element model, groove type and restraint conditions

A double semi-ellipsoidal heat source proposed by Goldak [11] was used to model the volumetric heat distribution of the moving welding arc. A combined two half ellipsoidal heat source distribution was adopted to describe the heat distribution, which is expressed by the following equations:

$$q_f(x,y,z) = \frac{6\sqrt{3}(f_f Q)}{abc_f\pi\sqrt{\pi}}\exp\left(-\frac{3x^2}{a^2}-\frac{3y^2}{b^2}-\frac{3z^2}{{c_f}^2}\right) \quad (2)$$

$$q_r(x,y,z) = \frac{6\sqrt{3}(f_r Q)}{abc_r\pi\sqrt{\pi}}\exp\left(-\frac{3x^2}{a^2}-\frac{3y^2}{b^2}-\frac{3z^2}{{c_r}^2}\right) \quad (3)$$

These two equations represent the front part and the rear part of the heat source respectively. In these equations, x, y and z are the local coordinates of the heat source, f_f and f_r are the parameters which give the fraction of the heat deposited in the front and the rear parts. Q is the heat input given by $Q = \eta VI$, where I is the current supplied, V is the voltage across the arc and η is the arc efficiency which is assumed to be 0.85, a, b, c_r and c_f refer to the geometry of heat source for the welding being modeled. The heat flux distribution and geometry of heat source is shown is Fig.3. This model suggested that the temperature gradient in the front of the heat source is steeper than the rear one according to experimental experience, an idealized welding pool consistent with experiment could be realized by adjusting the parameters in the equations.

For thermal boundary conditions, convective heat transfer with heat transfer coefficient of 15×10^{-6}W/mm^2°C and radiation heat transfer with emissivity of 0.51 are applied to all free surfaces. In order to account for heat transfer by convection due to fluid flow in the weld pool, an artificially increased thermal conductivity is assumed for temperature above the melting point [12]. The latent heat of fusion is considered due to the solidification of welding pool.

A conventional technique named element birth and death was used to model the addition of molten weld metal as the welding torch progressed along the workpiece. A full FE model was generated from the beginning, all elements representing filler metal were deactivated by assigning them extremely low stiffness. The deactivated elements were reactivated when they came under the influence of the welding torch.

2.2 Mechanical analysis

For the mechanical analysis, the finite element model used in thermal analysis was replaced by the mechanical one. The temperature history calculated in thermal analysis was imposed as a temperature load

IOP Conf. Series: Materials Science and Engineering **733** (2020) 012010 doi:10.1088/1757-899X/733/1/012010

at each time step in the mechanical analysis, which caused the expansion and contraction of the material and plastic deformation occurred where the yield criterion was exceed. The mechanical boundary conditions of the finite element were just to prevent the rigid body motion as shown in Fig. 2.

Mechanical analysis was performed using thermo-elastic-plastic material model with Von Mises yielding criterion and bilinear hardening rule. The mechanical model is based on the solution of three governing partial differential equations of force equilibrium. In tensor notation, these are written as

$$\sigma_{ij,i} + p_j = 0 \tag{4}$$

where p_j is the body force at any point within the volume, σ_{ij} is stress tensor. The total strain increment at the integration point as a function of elastic, plastic strain and thermal strain is governed by

$$d\varepsilon = d\varepsilon^e + d\varepsilon^p + d\varepsilon^{th} \tag{5}$$

where $d\varepsilon^e$, $d\varepsilon^p$ and $d\varepsilon^{th}$ are the elastic, plastic and thermal strain increments respectively.

A user-defined subroutine by using element birth and death technique was utilized to control non-convergence problem in the structural analysis. The temperature of every element in welding bead was checked, an element was deactivated if the temperature exceeded the melting point, it was reactivated in the subsequent cooling process.

3. Verification of the finite element model

The finite element model was validated by comparison of predicted residual stress with experimental results from Fu [9]. The measured points are alone path AB, which is the middle line in the flange and perpendicular with welding line as shown in Fig. 2. The comparison of results from finite element analysis and experiment are given in Fig.4, it can be seen that, for both longitudinal and transverse residual stresses, the finite element predictions show good agreements with experiment measurements. The welding effects on the finial residual stress becomes weaker as the distance to the welding line increases. In addition, it can be seen that the transverse residual stress on the right is higher than the left. This is due to that the right welding seam was welded earlier than left one. The temperature distribution of first welding seam has an effect of preheating on the later one. It functions as thermal tension to the later one to be welded and mitigates the transverse residual stress.

The results of numerical simulation show that an extremely large tensile residual stress occurs near the weld bead and it turns into compressive stress away from the welding bead in longitudinal stress. The maximum longitudinal residual stress reaches yield strength, and the distribution in the path AB shows a M pattern. The transverse residual stress is influenced by welding sequence, which conforms with the proposition of Fu and Teng [9, 13].

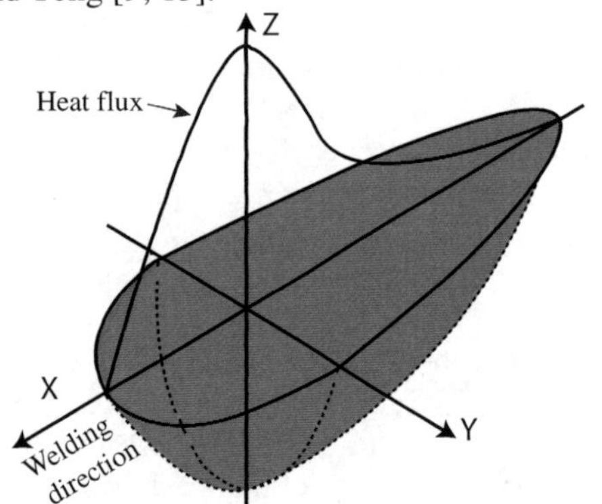

Figure.3 Double ellipsoid heat source model

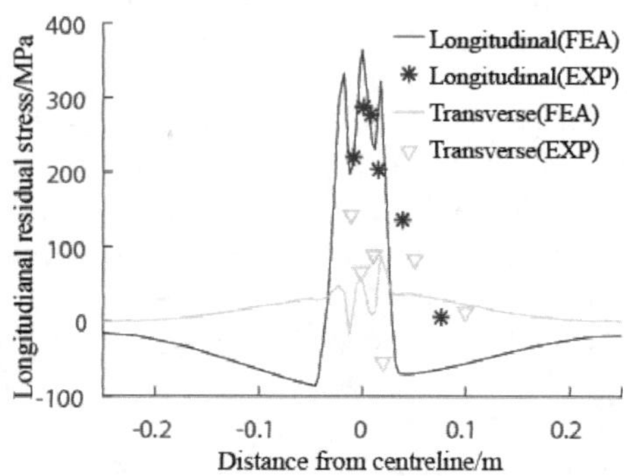

Figure.4 Results comparison of finite element analysis and experiment

4. Results and discussion

4.1 Results of thermal analysis

The welding pool is the fusion zone where temperature exceeds 1450°C in the numerical simulation, the welding pool size from numerical simulation approach welding groove geometry size, which is the prerequisite to predict reasonable residual stresses and distortions. The welding pool of no groove and K groove are shown in Fig.5. The geometry size of welding pool without groove predicted matches with the experiment measured by Fu [9]. The fusion zone of K groove welding indicates that the welding penetration quality is guaranteed by enough penetration depth and overlap with groove shape.

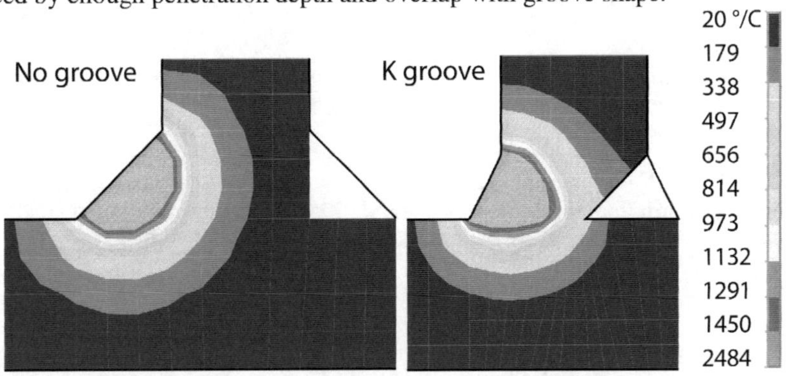

Fig.5 The welding pool in numerical simulation

4.2 Results of structural analysis

The distribution of longitudinal residual stress along path AB is shown in Fig.6, it can be seen that the tensile residual stress of K groove is larger than without groove welding. The tensile residual stress region for without groove is wider than that of K groove welding. However, the compressive residual stress for the case of without groove welding is larger.

Fig.7 shows the distribution of transverse residual stress along path AB for the different cases. It shows that both the magnitude and distribution pattern of transverse residual stress are more sensitive to the existence of groove than longitudinal residual stress. The width of high tensile transverse residual stress is opposite with counterpart in longitudinal residual stress, tensile residual stress region for without groove is narrower than K groove welding. Besides, the location of peak value in transverse residual stress is different from K groove.

The three-dimensional distribution of longitudinal residual stress on the right lower surface of the flange in K groove and without groove welding are shown in Fig.8. It can be seen that longitudinal residual stresses in K groove welding are larger and show less fluctuation, and tensile residual stresses are concentrated in welding bead. The residual stresses in no groove welding shows more volatility, and the change of residual stress from tensile to compressive is less sharp.

The existence of groove in welding influences the magnitude of heat input, the K groove welding represent larger penetration depth, which corresponds to a concentration in heat input or an increase of heat input in local zone. The consequences of heat input increase leads to the reduction of temperature variation of upper and lower surface of the flange, and the concentration of heat affected zone. Therefore, the K groove welding shows larger longitudinal tensile residual stress and narrower tensile stress zone. In addition, the concentration of heat input contributes to higher temperature zone, such as the heat input of the first welding seam strongly affects the second seam, which results in accumulation of temperature and larger transverse tensile residual stress.

The longitudinal and transverse residual stresses in web plate alone path CD are shown in Figs.9-10, it can be seen that longitudinal residual stress in the web is similar with the flange, the high tensile residual stress region is narrower in K groove welding. Both of K groove and no groove welding show compressive transverse residual stress near the welding bead, the compressive residual stress is larger in K groove welding for more heat input on web plate.

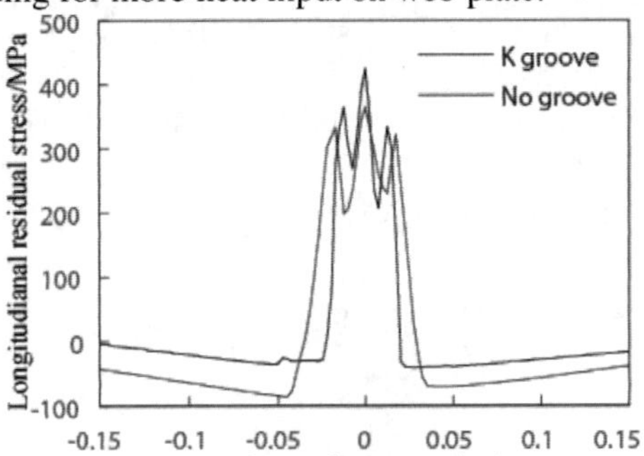

Figure.6 Longitudinal residual stress distribution for different groove

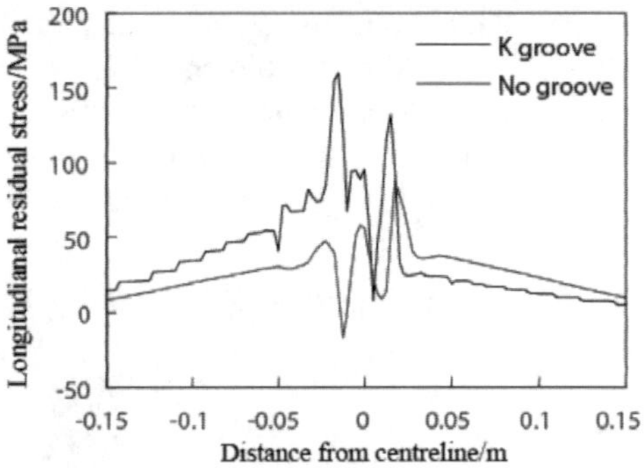

Figure.7 Transverse residual stress distribution for different groove

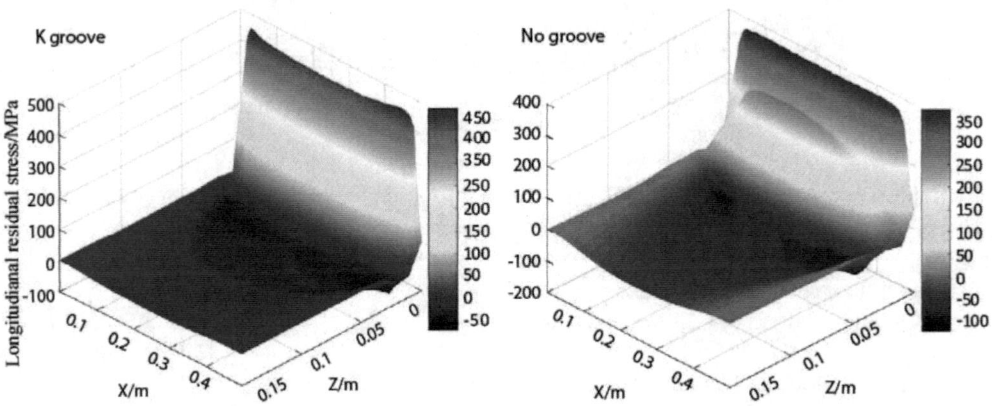

Figure.8 3D longitudinal residual stress distribution

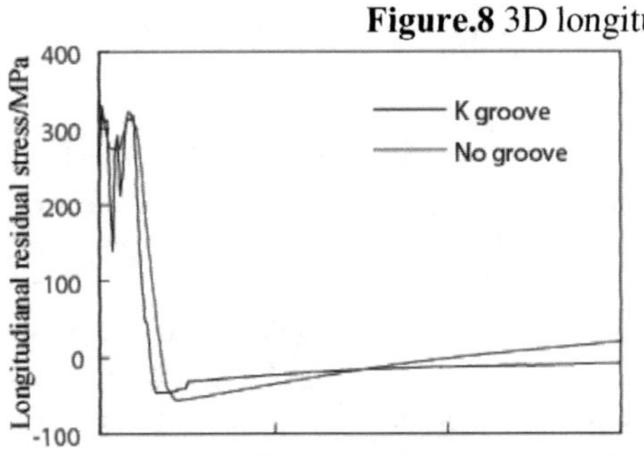

Figure.9 Longitudinal residual stress in web

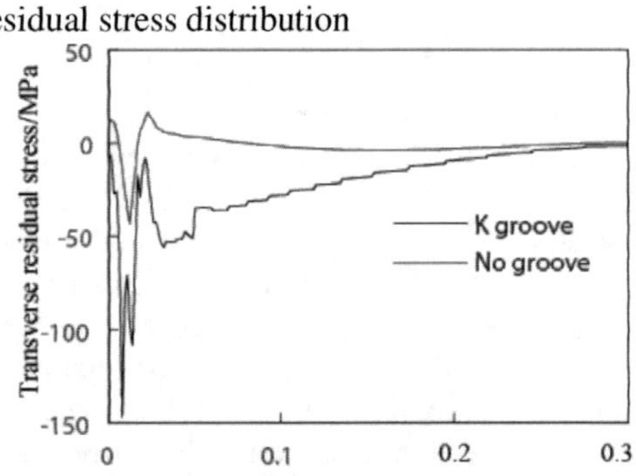

Figure.10 Transverse residual stress in web

4.3 Effects of groove on distortion

Fig.11 compares the vertical deflection of the flange along path AB, the results of finite element analysis in no groove welding are compared with measured experiment results of literature [9], it shows that the simulated results match the corresponding experiment well. The discrepancy between K groove and no groove welding is not significant, the vertical deflection of no groove is slightly larger than K groove welding, which comes from the different location of heat source on the flange.

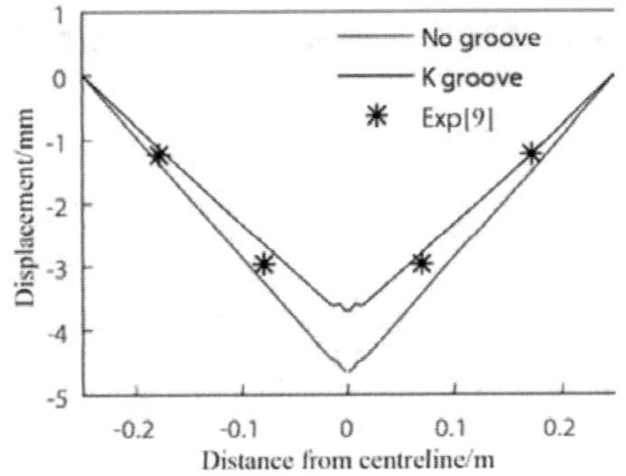

Figure.11 Comparison of vertical deflection of the flange

5. Conclusion

This research employs the finite element analysis to evaluate the effects of groove on residual stress and distortions in T-joint welding. Thermal elastic-plastic analysis to predict the temperature distribution, residual stress and vertical deflection distribution, the numerical results are in good agreement with experiment measurement results from published literature, which indicates that the current numerical method is suitable to the prediction of welding residual stress and distortion. The existence of groove in T-joint welding has a significant influence on residual stress and distortion. The peak value of longitudinal and transverse residual stress in K groove welding is higher than those in no groove welding, the higher region of residual stress is narrower in K groove welding. The influence of groove on vertical deflection in T-joint welding is not significant.

Reference

[1] M.N.J.E.F.A. James, **18** (2011) 1909-1920.
[2] L. Gannon, L. Yi, N. Pegg, M.J.M.S. Smith, **23** (2010) 385-404.
[3] G. Fu, M.I. Lourenco, M. Duan, S.F.J.J.o.C.S.R. Estefen, **102** (2014) 121-135.
[4] M. Perić, Z. Tonković, A. Rodić, M. Surjak, I. Garašić, I. Boras, S.J.M. Švaić, *Design*, **53** (2014) 1052-1063.
[5] I. Sattari-Far, M.R.F.J.I.*J.o.P. Vessels, Piping*, **86** (2009) 723-731.
[6] D. Akbari, M. Farahani, N. Soltani, *Journal of Strain Analysis for Engineering Design* **47** (2012) 73-82.
[7] Y. Ye, J. Cai, X. Jiang, D. Dai, D.J.A.i.E.S. Deng, **86** (2015) 39-48.
[8] T.L. Teng, C.P. Fung, P.H. Chang, W.C.J.I.J.o.P.V. Yang, *Piping*, **78** (2001) 523-538.
[9] G. Fu, M.I. Lourenço, M. Duan, S.F.J.M.S. Estefen, **46** (2016) 30-55.
[10] H. Wang, Z. Wang, P. Bai, S. Hu, Y. Zhang, W.J.A.i.E.S. Ren, **117** (2018) 1-7.
[11] J.A. Goldak, M. Akhlaghi, *Computational welding mechanics*, 2005.
[12] D. Deng, H.J.C.M.S. Murakawa, **37** (2006) 0-277.
[13] T.L. Teng, P.H. Chang, W.C.J.C. Tseng, *Structures,* **81** (2003) 273-286.

Effect of Ca/K Modified ZSM-5 on Catalytic Cracking of Low Carbon Olefin

Xiaohua Liu, Wenqing Huang*,Yi Man, Dingbo Wang ,Jianzhun Jiang
1 Sinopec Beijing Research Institute of Chemical Industry, Beijing 100013, PR China
*Corresponding author E-mail: huangwq.bjhy@sinopec.com

Abstract. In this paper, the structural and acidic changes of ZSM-5 zeolite modified by K and Ca were studied. Meanwhile we use in-situ FTIR to study the reaction of isobutylene on ZSM-5 zeolite before and after modification, investigating reaction mechanism and optimal reaction time. The results shows that the amount of weak acid increases in the modified zeolite. The Ca-ZSM-5 can not only effectively control the hydrogen transfer reaction to inhibit the formation of aromatic hydrocarbons, but also promote the occurrence of dehydrogenation cracking reaction; while the K-ZSM-5 is not effective in the cracking reaction.

1.Introduction

Low-carbon olefins such as ethylene and propylene are important basic raw materials in the petrochemical industry, their demand is increasing with the development of the chemical industry. The traditional method of ethylene production uses a non-catalytic steam pyrolysis process, and the propylene is produced by co-production of ethylene which has limited capacity. In recent years, the low-carbon olefin catalytic cracking process has become the most promising new production technology for its advantages of low reaction temperature, high product yield and flexibility distribution. The key to the cracking process is the catalyst. ZSM-5 is one of the common catalysts used in this process due to its special pore structure, acid properties and good anti-coking ability [1-4], It has been widely reported that the amount and strength of acid sites are adjusted by element modification to improve the performance [5-8]. The suitable modification method is one of the important factors that directly affect the yield of low-carbon olefins. Therefore, it is necessary to study the effect of different element modified ZSM-5 on catalytic cracking of low carbon olefin.

In this work, ZSM-5 zeolite was modified to investigate the effect of element modification on the pore structure and surface acidity by Ca and K. After that we found the relationship between catalysts and cracking reaction after using the in-situ FTIR.

2.Experimental

2.1 Catalyst preparation

The unmodified ZSM-5 was obtained with ultrafine SiO_2 as carrier and high silica-alumina ratio ZSM-5 as active component, followed by adding the amount of the phthalocyanine powder and the dilute nitric acid. After drying and calcination the gel, all the catalysts were crushed to particles.

Preparation of Ca and K modified ZSM-5 zeolite: The pre-treated ZSM-5 were prepared by a wet incorporation technique with 1% aqueous solution of $Ca(NO_3)_2{\cdot}4H_2O$, KNO_3 of appropriate concentration. After drying and calcination, modified catalysts loaded with different elements can be obtained, respectively labelled as Ca-ZSM-5, K-ZSM-5.

2.2. Catalyst characterization

Nitrogen adsorptions were performed with a Micromeritics ASAP 2020 automated gas adsorption analyzer. Prior to analysis, the samples were outgassed at 350℃ for 12 h. The specific surface area, pore size and pore volume were calculated according to the isotherm.

The acidities of the catalysts were determined by the NH_3-TPD using a Micromeritics Autochem II 2920 automatic chemical adsorption instrument. The sample (0.10 g) was loaded into a U-shaped quartz tube and pretreated at 600℃ for 1h in a flowing He. After the retreatment, the sample was cooled to 40℃ and saturated with NH_3 for 1h. Subsequently, the physically adsorbed NH_3 was removed in a flow of dry He for 1 h. When the baseline was stable, NH_3-TPD was carried out from ambient temperature to 600℃ at a heating rate of 10℃ /min,and analyzed by a TC detector.

The acid types of the samples were determined by adsorption of pyridine on a Nicolet AVATAR-360 FTIR. Prior to the measurements, the samples were pressed into self-supporting discs and activated in the IR cell attached to a vacuum line at 500℃ for 1.5 h under 10^{-6} Torr, then cooled to room temperature where pyridine vapor was introduced into the cell for 1h. After the adsorption is saturated, the sample was heated to the corresponding temperature, and a spectrum was recorded at each step.

2.3. Evaluation of Catalytic performance

In-situ FTIR transmission spectroscopy was used to characterize the surface information on ZSM-5 in the isobutylene cracking reaction. Prior to the measurements, the samples were pressed into self-supporting discs and activated in the IR cell attached to a vacuum line at 500℃ for 1.5 h under 10^{-6} Torr. After the sample was pretreated, the isobutylene gas was introduced in situ. When the gas flow rate was stable, the temperature began to heat to 520℃ at a heating rate of 12℃/min, and a spectrum was recorded at each step.

TPSR experiment used Micromeritics Autochem II 2920 automatic chemical adsorption instrument with one on-line mass spectrometer(QIC 20, HIDEN) to detect the fragment ion peaks ,isobutylene(m/z=56) and propylene (m/z=48) in real time.

3.Results and discussion

3.1 Structure of ZSM-5 zeolites

IOP Conf. Series: Materials Science and Engineering **733** (2020) 012011 doi:10.1088/1757-899X/733/1/012011

Fig. 1 and Table 1 highlights the adsorption/desorption isotherms of ZSM-5 zeolite before and after modification. Fig.1 indicates the catalysts have similar pore structures. The volume and specific surface area are fairly close for all three samples in Table 1. The pore size of the Ca-ZSM-5 and K-ZSM-5 decreased slightly, but it was not enough to cause a large change in the pore structure of the zeolite .The BET and microspore surface areas are slightly decreased after Ca modification, attributed the dealuminization phenomenon occurs during the modification process. The structure collapses after the skeleton aluminum falls off, resulting in a decrease in specific surface area and pore volume [7].

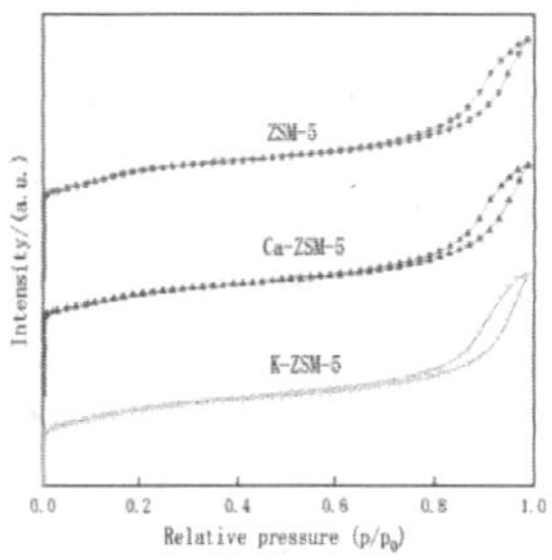

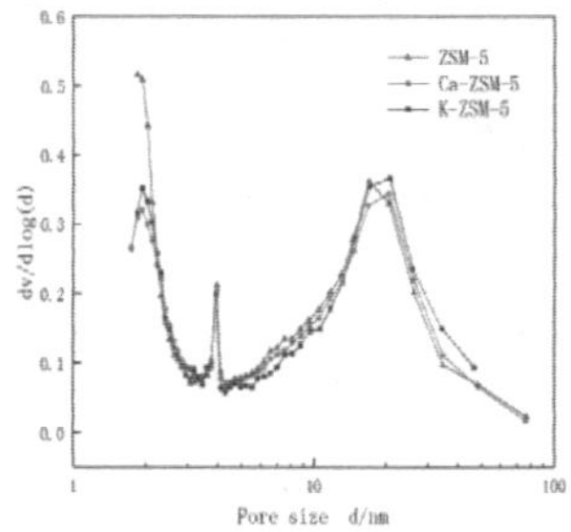

Figure 1. N_2-adsorption/desorption isotherm and pore distribution of zeolites

Table 1. Physical properties of zeolites

Sample	S_{BET}/ ($m^2 \cdot g^{-1}$)	S_{micro}/ ($m^2 \cdot g^{-1}$)	S_{meso}/ ($m^2 \cdot g^{-1}$)	V_{total} ($cm^3 \cdot g^{-1}$)
ZSM-5	362.52	285.54	76.98	0.13
Ca-ZSM-5	338.84	261.01	77.83	0.12
K-ZSM-5	364.77	287.94	76.83	0.13

3.2 Acidic property of ZSM-5 zeolites

The acidity and acid strength of the zeolites were determined by NH_3-TPD, and the results are shown in fig 2. We sorted acidic properties of zeolites by total acid amounts/(mmol • g^{-1}): Ca-ZSM-5(0.50)>ZSM-5(0.45)>K-ZSM-5(0.38). Fig 2 illustrates the weak acid sites (low temperature desorption) of Ca-ZSM-5 were significantly higher than ZSM-5, but the strong acid sites (high temperature desorption) were lower than ZSM-5.Meanwhile the weak acid sites (low temperature desorption) of K-ZSM-5 are higher than ZSM-5, but the strong acid sites (high temperature desorption) are almost absent

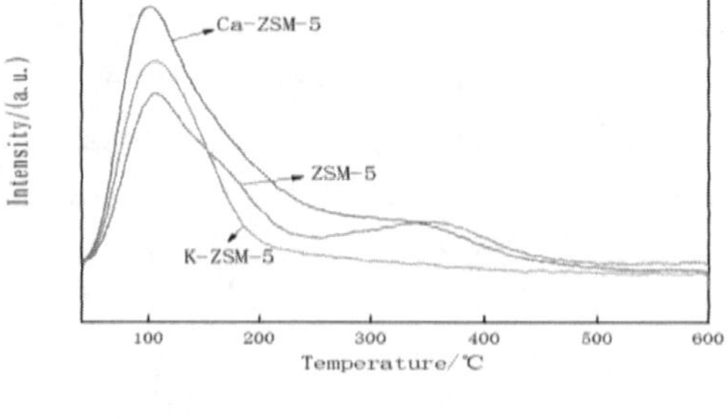

Figure 2. TPD of NH_3 profiles of ZSM-5 zeolites

In order to further detect the acid type of ZSM-5 zeolites, pyridine adsorption FTIR was used to illustrate the acid type and strength. Fig 3. represents the variation of IR bands in 1400-1600 cm^{-1} region as the increase of the temperature. There are three bands in the region. The one at 1450 cm^{-1} indicates the presence of Lewis acid sites. The1540 cm^{-1} one is attributed to the Brønsted acid sites. The other at 1490 cm^{-1} is attributed to the pyridine species interacting with both the two kinds of acid sites[9] Ca-ZSM-5 (Fig 3B) has an increased amount of Lewis acid and a reduced amount of Brønsted acid compared to ZSM-5 (Fig 3A). It is indicated that Ca-ZSM-5 increases the weak acid and decreases the strong acid. The amount of acid in K-ZSM-5 (Fig 3C) was reduced compared to ZSM-5 (Fig 3A) and the amount of Brønsted acid was significantly less than Lewis acid. This corresponds to the NH_3-TPD characterization data. The results suggest the introduction of K and Ca can weaken the acid strength and reduce the acidity, and the original strong acid sites of ZSM-5 are converted into weakly acidic sites. The dehydrogenation cracking reaction mainly occurs in the acidic center of alkaline earth metal modification. The reduction of acidity can inhibit the hydrogen transfer reaction and reduce the side-reaction of the olefin product, thereby promoting the dehydrogenation cracking reaction, which is the main reason for increasing the yield of olefin [7].

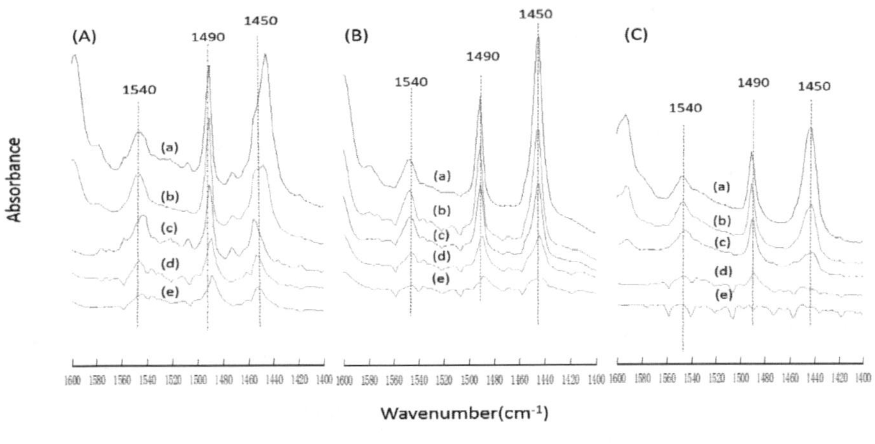

Figure 3. IR spectra of pyridine adsorption on ZSM-5 zeolites (A)ZSM-5(B)Ca-ZSM-5(C)K-ZSM-5 at different temperature (a)RT(b)100 (c)150(d)200(e)300℃

3.3 Operando experiment study on Catalytic Cracking of Isobutylene

FTIR can give structural information on the surface-adsorbed species, especially the structure information of the adsorbed species under the reaction conditions. High-quality and time-resolved spectra of the catalyst (Fig.4) were recorded by the FTIR spectrometer under real reaction conditions, which provided abundant information about the surface adsorbed species and their evolutions. According to literature [10], the bands at 2965 and 2860 cm^{-1} are attributed to the asymmetric and symmetric C-H stretching vibrations of the $-CH_3$ group, respectively. The band at 2934 cm^{-1} is associated with the asymmetric C-H stretching vibration of the $-(CH_2)-$ group, The band at 1467 cm^{-1} is attributed to the C-H bending vibration of the $-(CH_2)-$ group, and that at 1458 cm^{-1} to the asymmetric C-H bending vibration of the -CH_3 group. Fig. 4A illustrates that the absorption peak 2860 cm which attributed to the methylation vibration gradually increases, from the start of the reaction to 150 min, and the methylene absorption peak (2934 cm^{-1}) gradually decreases. The characteristic peaks of isobutylene, 1396 and 1386 cm^{-1}, also immediately and continuously decreased in intensity. The reaction temperature increases at a rate of 2℃/ min. These spectra suggest as the reaction temperature rises to 300℃, the increase of 2860 cm^{-1} peak and the decrease of the 1396 and 1386 cm^{-1} mean that the molecular chain length decrease, It is an important indicator of the progress in the cracking reaction. In addition, after 150 min of reaction, that is, when the reaction temperature was higher than 300℃, the band at 1505 cm^{-1} related to the C=C vibration of an aromatic ring appeared [10]. The reason is the acid strength of ZSM-5 is high, and the cracked product is re-adsorbed, resulting in the occurrence of side reactions such as aromatization reaction [11]. Isobutylene cracking reaction over Ca-ZSM-5 and K-ZSM-5 were also studied by in-situ FTIR, shown in Fig. 4B and 4C. The cracking reaction occurred on the Ca-ZSM-5 from room temperature to about 300℃, but no characteristic peak associated with the aromatic ring was observed at about 1505 cm^{-1}. This indicates that the acidity of the zeolites modified by Ca can be reduced, and the hydrogen transfer reaction can be inhibited to form aromatic hydrocarbons, while the dehydrogenation cracking reaction can be promoted [7]. K-ZSM-5 hardly undergoes a cracking reaction which is because the amount of Brønsted acid modified by K is much lower than that of Lewis acid, and the acidity is mainly weak acid. It is reported in the literature [12] that the Brønsted acid on ZSM-5 is the active center of the catalytic cracking reaction of olefins. The Lewis acid does not contribute to the activity of the cleavage. At the same time, some scholars do not recommend alkali metals such as K to modify zeolites because in most catalytic reactions, the effect is not obvious [13].

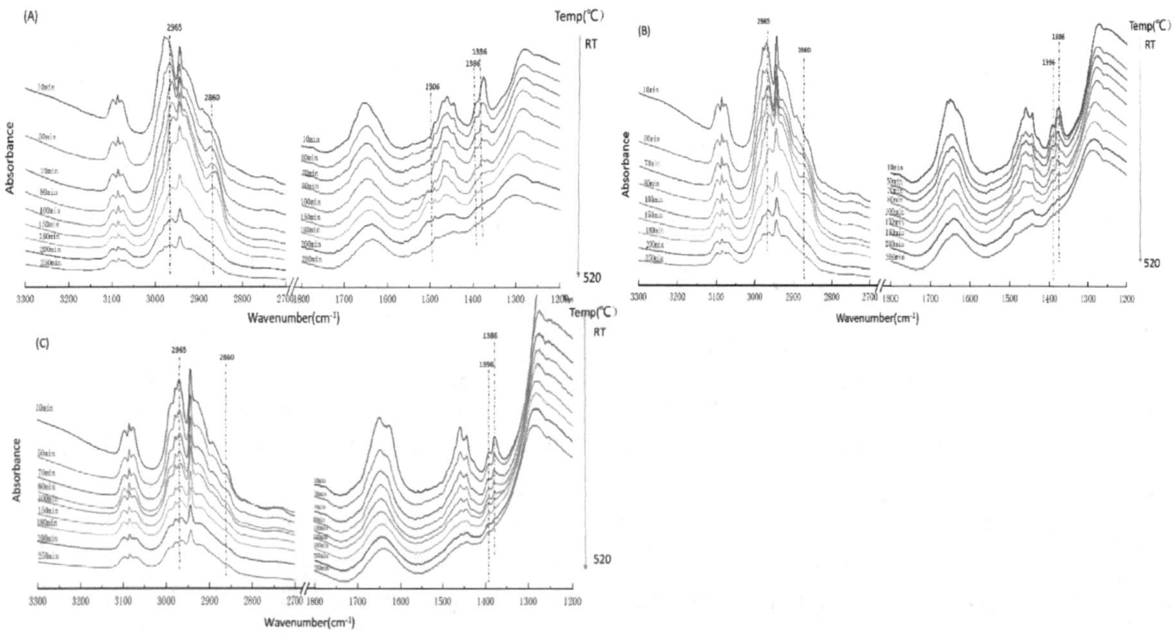

Figure 4. In-situ FT-IR spectra of isobutylene cracking on the ZSM-5 zeolite surface (A)ZSM-5 (B)Ca-ZSM-5(C)K-ZSM-5

In order to further verify the reaction conditions of isobutylene on the zeolite, the change of isobutylene(m/z=56) and propylene (m/z=48) fragment ion peaks was detected by TPSR experimental. Fig. 5 shows that when the catalytic cracking reaction is carried out in ZSM-5 and Ca-ZSM-5 zeolites, isobutylene decreases and propylene increases significantly as the temperature rises from room temperature to 300 °C. However, these peaks on K-ZSM-5 did not change, indicating that no catalytic cracking reaction occurred. This is consistent with in-situ FTIR. Thus, we can conclude that the optimal reaction temperature for the catalytic cracking reaction is about 300 °C.

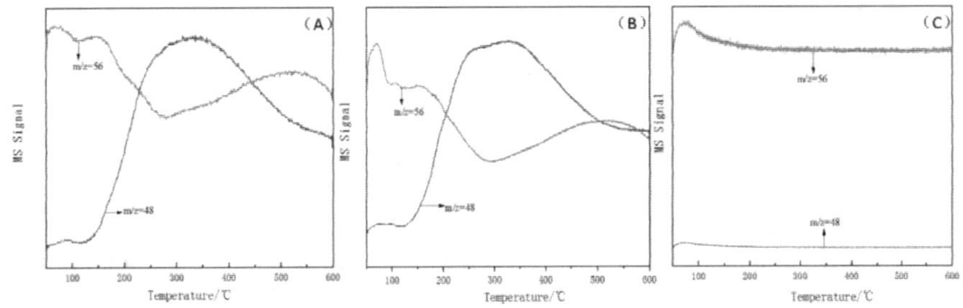

Figure 5. TPSR of isobutylene cracking on the ZSM-5 zeolite surface(A)ZSM-5

(B)Ca-ZSM-5(C)K-ZSM-5

4.Conclusions

The shaped ZSM-5 was modified by Ca and K. The characterization results shows that the pore volume and specific surface area of zeolites are not much different, but the introduction of K and Ca can weaken the acid strength and reduce the acidity. Meanwhile, in-situ FTIR shows that ZSM-5 had a side reaction such as aromatization, while the Ca-ZSM-5 did not appear related peak and K-ZSM-5 hardly undergoes a cracking reaction. The Ca-ZSM-5 can not only effectively control the hydrogen transfer reaction to inhibit the formation of aromatic hydrocarbons, but also promote the occurrence of dehydrogenation cracking reaction; Furthermore the optimum temperature of the reaction are about 300 °C.

References

[1] Zhu, X., Li, X., Xie, S. et al. *Catal Surv Asia.* 2009 **13**: 1-8.
[2]Liu D , Choi W C , Lee C W , et al. *Catal Today*. 2011, **164**(1):154-157.
[3]Tzoulaki D, Jentys A, Pérez-Ramírez J, et al. *Catal Today*. 2012, **198**(1): 3-11.
[4]Xue N, Chen X, Nie L, et al. *J.Catal*. 2007, **248**(1): 20-28.
[5] Lv J, Hua Z, Ge T, et al. *Micropor Mesopor Mat.* 2017, **247**: 31-37.
[6] Li J, Li T, Ma H, et al. *Ind Eng Chem Res*. 2015, **54**(6): 1796-1805.
[7] Wang L, Ozawa K, Komatsu T, et al. *Appl Catal A-Gen.* 2011, **407**(1-2): 127-133.
[8] Varzaneh A Z, Towfighi J, Mohamadalizadeh A. *J Anal App Pyrol*. 2014, **107**: 165-173.
[9] G. Poncelet and M.L. Dubru, *J. Catal*. 1978, **52**: 321.
[10] S. Rousseau, O. Marie, P. Bazin, M. Daturi, S. Verdier, V. Harlé, *J. Am. Chem. Soc.* 2010, **132**: 10832-10841
[11] Chen C J, Rangarajan S, Hill I M, et al. *Acs Catal*. 2014, **4**(7): 2319-2327.
[12] Coelho A, Costa L, Marques M M, et al. *Appl Catal, A-Gen.* 2012, **413**: 183-191.
[13] Zhu X, Liu S, Song Y, et al. *Catal Lett*. 2005, **103**(3-4): 201-210.

Effect of C/C preform density on oxidation properties of C/C-SiC composites

X Y Yao, W Li and G H Feng

State Key Laboratory of Solidification Processing, Northwestern Polytechnical University, Xi'an Shaanxi 710072, PR China
E-mail: yaoxiyuan@nwpu.edu.cn

Abstract. In order to study the effect of C/C preform on the oxidation properties of C/C-SiC composites, four kinds CVD-SiC coated C/C-SiC composites with different initial densities of C/C preform were prepared. The thermogravimetric oxidation behavior from 25 to 1450℃ of prepared C/C-SiC composites and the isothermal oxidation of CVD-SiC coated C/C-SiC composites under static air at 1500℃ were studied. And the bending properties of SiC coated C/C-SiC composites before and after oxidation were investigated. The results show that the initial oxidation temperature of C/C-SiC composites increases first and then decreases with the increase of initial C/C density, which is determined by the porosity and SiC content of the composites. When the C/C density is 1.0 g/cm^3, the initial oxidation temperature is 633℃. The CVD coated C/C-SiC composites with initial C/C density of 1.0 g/cm^3 exhibited the best static oxidation resistance at 1500℃. The weight loss is 4.23% after oxidation for 10 h at 1500℃, and the bending strength retention rate is 71.2%.

1. Introduction

C/C-SiC composites combine the advantages of carbon fiber, SiC matrix and fiber reinforced ceramic matrix composites. The most prominent feature is that some carbon matrix in C/C composites is replaced by SiC matrix with better oxidation resistance and high hardness [1]. It is a new functional integrated composite material which combines thermal protection, structural load-bearing and oxidation resistance. Since its emergence as a high temperature thermal structural material, it has been widely used in aerospace, mechanical engineering and braking systems [2-4].

The preparation of C/C-SiC is mainly divided into two steps. Firstly, porous C/C composites are prepared by chemical vapor infiltration (CVI) or polymer infiltration and pyrolysis (PIP), and then SiC matrix is prepared by CVI, PIP or reactive melt infiltration (RMI) [5-10]. As the key intermediate in the preparation of C/C-SiC composites, the basic structural parameters (such as density, porosity and structure, etc.) of C/C porous materials have a significant effect on the structure and mechanical properties of C/C-SiC composites.

CVI and PIP method has attracted much attention due to its advantages of low preparation temperature, designable matrix structure and easy preparation of complex shape components. Du[6] studied the effects of infiltration solution concentration and preform density on the density and mechanical properties of C/C-SiC composite. Yang [7] et al. prepared C/C-SiC composites with density of 1.34, 1.52 and 1.62 g/cm^3 by liquid phase densification using PCS as precursor. The flexural strength and oxidation resistance at 1500℃ of C/C-SiC composites were compared. At present, the research work of C/C-SiC composites mainly focuses on the study of rapid densification and mechanical properties by CVI+RMI process. There are few reports on the effects of different C/C initial densities on the oxidation resistance C/C-SiC by PIP.

In this paper, the thermogravimetric oxidation behavior of C/C-SiC composites prepared with different C/C preform densities was studied. And then the isothermal oxidation resistance and bending properties of C/C-SiC composites coated with CVD-SiC coating were investigated under static air at 1500℃ for long time.

2. Experimental

2.1. Preparation and properties of CVD coated C/C-SiC composites

The C/C-SiC composites were fabricated by CVI and PIP process. Porous 2D needle C/C preform with densities of 0.8, 1.0, 1.2 and 1.4 g/cm_3 by CVI were infiltrated using PCS (Yellow-brown solid with softening point of 210℃) as precursor, and then pyrolyzed at 1100℃. C/C-SiC composites were prepared by repeating PIP process for eight times. The density and porosity of C/C-SiC composites are 1.78, 1.84, 1.96, 1.70g/cm^3 and 13.30, 9.52, 5.51, 7.17% respectively.

SiC coatings were prepared by low-pressure chemical vapor deposition (LPCVD) using MTS as precursor, hydrogen as reaction gas and argon as dilution gas. The deposition temperature was 1150-1250℃, and the pressure was 4-8 kPa.

2.2. Measurement

Thermogravimetric test was carried out in STA 409 PC/PG synchronous thermal analyzer of NETZSCH Company with the size of 4×4×4 mm. The starting temperature was 25℃, the ending temperature was 1450℃, and the heating rate was 10℃/min.

The composite material was cut into 55×10×4 mm dimensions for oxidation and bending mechanical properties testing. The isothermal oxidation tests of the coated specimens were carried out in air at 1500℃ in an electrical furnace. The specimens were weighted by an electronic balance with a sensitivity of ± 0.1 mg. Flexural strengths of the coated specimens before and after oxidation were measured by three-point bending test on the CMT-5304 universal machine with the span of 40 mm and the strain rate of 0.5 mm/min.

The crystalline structure and microstructure of C/C-SiC composites were structurally analyzed by X-ray diffraction (XRD, X' Pert PRO) and scanning electron microscopy (SEM, JSM-6460).

3. Discussion

3.1. Oxidation behavior of C/C-SiC composites

The oxidation behavior of different C/C-SiC composites in air was compared by thermogravimetric analysis, as shown in Figure 1. The oxidation process can be divided into the following four stages: no weightlessness stage, rapid weightlessness stage, weightlessness reduction stage and weightlessness balance stage. The rapid weight loss is mainly due to the rapid oxidation of carbon fiber and matrix carbon. In weight loss reduction stage, the weight loss rate of the composite decreases with the increase of temperature because of the oxidation of SiC matrix. On the one hand, the oxidation of SiC shows oxidation weight gain, and its oxidation rate increases with the increase of temperature. On the other hand, its oxidation product SiO_2 can effectively reduce the permeability of oxygen and slow down the oxidation of composites. Therefore, the oxidation rate of composites decreases with the increase of temperature at this stage.

The weight loss rate of C/C-SiC composites increases sharply with the increase of oxidation temperature, reaching the maximum value at 900-1000℃, and then decreases with the increase of temperature, and then tends to zero. The final residual mass decreases with the increase of C/C density. Combined with the corresponding TG-DTG data (as shown in table 1), the oxidation behavior of the composite was further analyzed. The initial oxidation temperature of samples increases first and then decreases with the increase of C/C density. The composites prepared with a density of 0.8 and 1.4 g/cm^3 had a low oxidation starting temperature, and the composites began to lose weight at

544°C.When the C/C density is 1.0g /cm^3, the initial oxidation temperature is 633°C, which infers that the C/C-SiC composite material prepared by C/C with a density of 1.0g /cm^3 has best oxidation resistance. SiC can improve the initial oxidation temperature of the material by occupying the surface active point[11]. However, when the initial density of C/C is 0.8 g/cm^3, the prepared C/C-SiC composite material has a high porosity, and the pores provide channels for oxygen infiltration, so the initial oxidation temperature of the composite material is low. So the initial oxidation temperature of C/C-SiC composite is determined by the porosity and SiC content.

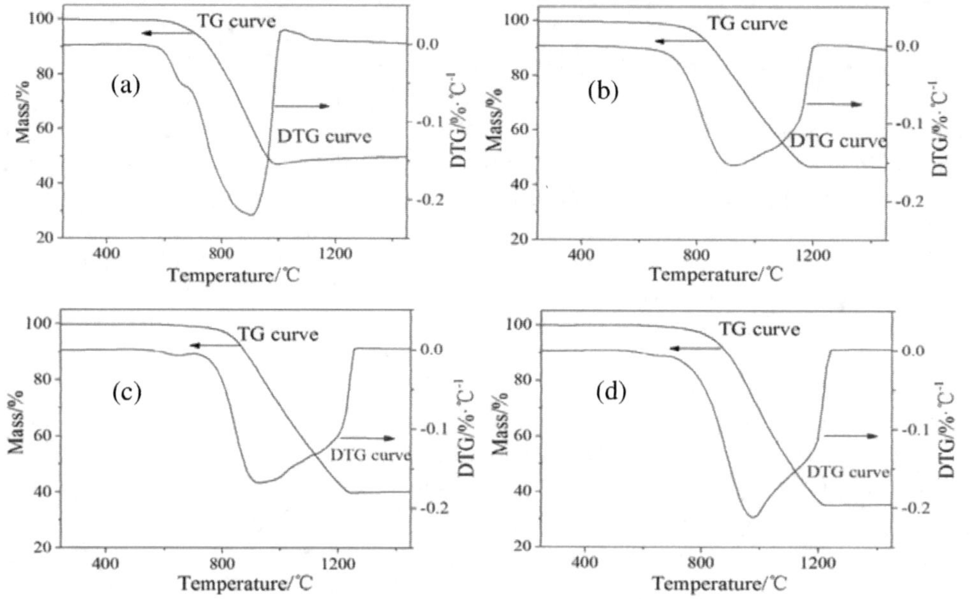

Figure 1. The TG and DTG curves of C/C-SiC composites
(a) 0.80g/cm^3; (b) 1.00 g/cm^3; (c) 1.20 g/cm^3; (d) 1.40 g/cm^3

Table 1. The TG-DTG analysis results of C/C-SiC composites by C/C with different density

C/C density (g/cm^3)	T_i (°C)	Tmax (°C)	R (wt.%)
0.8	544	900	49.8
1.0	633	929	46.4
1.2	573	929	40.2
1.4	544	974	35.4

The temperature of maximum oxidation rate increases with the increase of C/C density. The temperature point of the maximum oxidation rate is mainly affected by the SiC content. Because SiC began to slowly oxidize at 800℃, and its oxidation performance was oxidation weight gain. At the same time, the slow oxidation of SiO_2 can partially slow down the diffusion of oxygen into the material, thus reducing the oxidation rate. The higher the SiC content is, the more obvious the above phenomenon is. The oxidation weight loss rate of the material starts to decrease at a lower temperature. Therefore, The temperature of maximum oxidation rate of C/C-SiC composites prepared by low density C/C preform is lower. However, the porosity of the C/C-SiC composite prepared with low density C/C is relatively large, which leads to poor oxidation resistance of the composite at the initial oxidation stage. In conclusion, the oxidation properties of C/C-SiC composites prepared in this study are mainly due to the interaction of SiC content and sample defects, and the composites with a density of 1.0 g/cm^3 have good oxidation property.

3.2. High temperature oxidation resistance of SiC coated C/C-SiC composites

In order to further investigate the oxidation resistance of C/C-SiC composite at high temperature for a long time, the SiC coating was prepared by CVD on the surface of C/C-SiC composite. Figure 2 shows the XRD spectrum of SiC coating prepared by CVD process. The diffraction peaks of SiC (111), (220) and (311) planes were corresponded to 35.7 °, 60.3 ° and 72.1 °respectively, which shows that the SiC coating prepared by CVD is mainly composed of β-SiC.

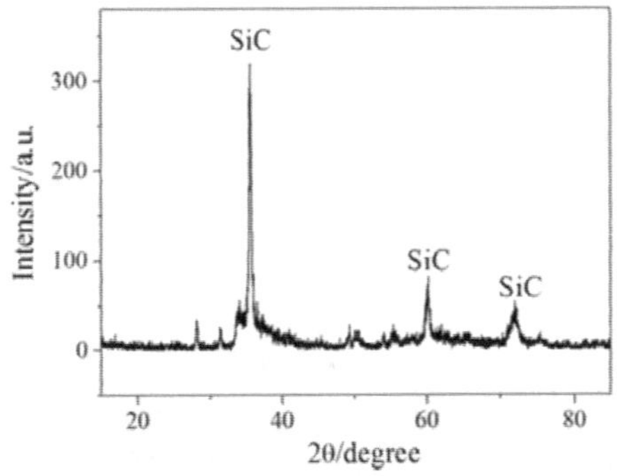

Figure 2. XRD pattern of the surface of the coating prepared by CVD

Figure 3 shows surface and cross section SEM photos of SiC coating prepared by CVD method. The coating is cauliflower like, mainly composed of SiC pellets, which are fully fused between balls. The section morphology of CVD-SiC coating shown that the coating thickness was relatively uniform, about 50 μm. The coating appears in two distinct layers, mainly due to two cycle of SiC deposition. At the same time, obvious deposition defects, i. e. penetrating cracks, can be found in the coating.

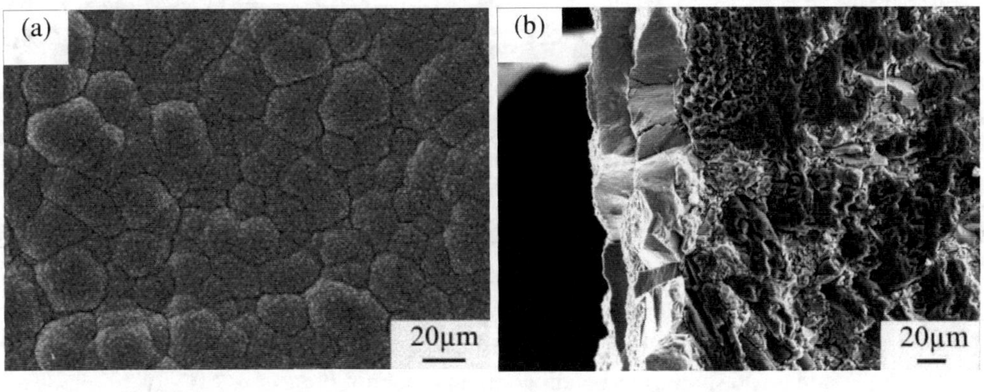

Figure 3. SEM micrographs of the SiC coating prepared by CVD (a) surface; (b) cross-section

In order to investigate the effect of oxidation weight loss on the mechanical properties of coating composites, the oxidation test was stopped when the oxidation weight loss was 4.0-5.0 %. The coating composites all exhibit oxidation weight loss, and the oxidation weight loss increases with the increase of oxidation time (as shown in figure 4). When the C/C density is 0.8g /cm^3, the weight loss is 4.58% after oxidation for 5 h. While the C/C density is 1.0g /cm^3, the weight loss reaches 4.23% after 10 h oxidation. However, as the density of C/C preform continue to increase, the oxidation performance of the composites decreases again. Especially when the C/C density is 1.4g /cm^3, there will be obvious oxidation holes in C/C-SiC composite due to the failure of oxidation protection of local SiC coating. In conclusion, C/C-SiC composites with initial C/C density of 1.0 g/cm^3 exhibit good high temperature oxidation resistance.

Figure 5 shows the SEM photos of surface and section of coated C/C-SiC samples after static air oxidation for 10 h at 1500℃. It can be seen that a layer of oxide film is formed on the surface. Due to the mismatch of thermal expansion coefficient between coating and C/C-SiC composite during rapid cooling from 1500℃ to room temperature, the tensile stress of the coating results in cracking or peeling of the coating. The appearance of coating defects provides diffusion channels for oxygen, which leads to oxidation of carbon matrix and carbon fibers. Figure 5(b) shows that the carbon matrix is obviously oxidized and the carbon fibers are acicular due to oxidation.

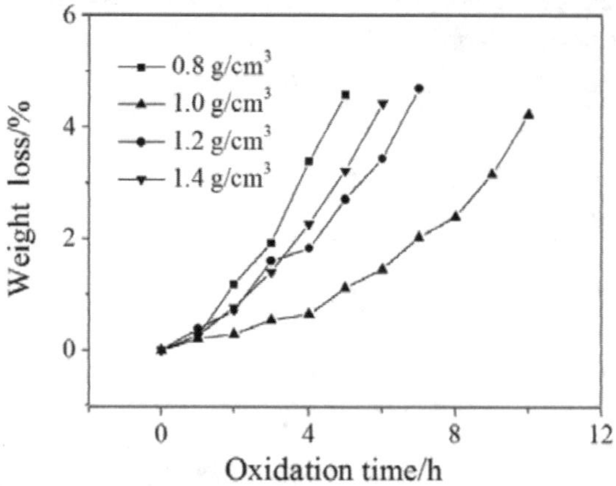

Figure 4. Oxidation curves of CVD-SiC coated C/C-SiC composites in air at 1500℃

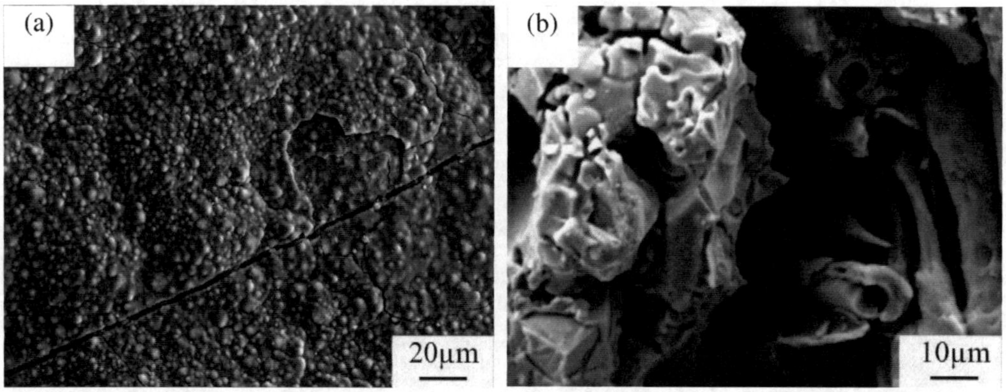

Figure 5. SEM micrographs C/C-SiC after oxidation at 1500 °C for 10 h (a) surface; (b) cross-section

The residual bending strength of coating C/C-SiC composite after oxidation is shown in table 2. When the C/C density is 1.0g /cm^3, the bending strength retention rate of the composite is 71.2% after oxidized for 10 h at 1500℃. And the C/C density is 1.4 g/cm^3, large oxidation holes appear locally due to oxidation of the composites, and the bending strength cannot be tested. C/C-SiC composites with initial C/C density of 0.8 g/cm^3 have low final density (1.78 g/cm^3) and high content of defects such as voids in the composites, which provide a channel for oxygen permeation into the composites, so their oxidation resistance is not ideal. However, the content of SiC phase infiltrated into C/C-SiC composites with initial C/C density of 1.4 g/cm^3 is relatively low, so it also shows poor oxidation resistance. According to the weight loss and the residual bending strength after oxidation at 1500℃, the composites with initial C/C density of 1.0 g/cm^3 have the best oxidation resistance.

Table 2. Flexural strength of coated C/C-SiC composites after oxidation

C/C density (g/cm^3)	Flexural strength of C/C-SiC (MPa)	Flexural strength after oxidation (MPa)	Retention rate (%)
0.8	207.9	159.4	76.7%
1.0	170.7	121.6	71.2%
1.2	161.9	110.6	68.3%
1.4	147.5	/	/

4. Conclusion

The initial oxidation temperature of C/C-SiC composites prepared by PIP process increases first and then decreases with the increase of initial C/C density. When the preform density is 1.0 g/cm^3, the initial oxidation temperature of C/C-SiC composites reaches 633℃. And the composite exhibits good oxidation resistance at 1500℃ in static air after the preparation of CVD-SiC coating on C/C-SiC composites. After oxidation for 10 h, the weight loss was 4.23% and the bending strength retention rate was 71.2%. The oxidation resistance of C/C-SiC composites is mainly affected by SiC content and material defects. The C/C-SiC composites with C/C preform density of 1.0 g/cm^3 have relatively excellent oxidation resistance. This study can provide theoretical support for the performance optimization of C/C-SiC composites.

5. References

[1] Hu L, Li C J and Zhang Y H 2018 Carbon 1 18-23.
[2] Krenkel W and Berndt F 2005 Materials Science and Engineering A 412 177-81.
[3] Zhang Z, Hao Z B and Yan L S 2008 Carbon 2 29-35.
[4] Yang F H, Wang Y Y, Liu X R, Zhou C L, Yang L P and Jiang K 2018 Solid State Phenomena 28 402-7.
[5] Yan L S, Li H J, Cui H and Wang T 2006 Journal of Inorganic Materials 21(3) 664-70.
[6] Wu S, Chen L, Qian L, Zhang J B, Pan J F and Zhou C J 2008 Journal of the Chinese ceramic society 36(7) 973-7
[7] Du H N, Du H L, Zhou W C and Zhu D M 2006 Carbon Techniques 25(6) 19-22.
[8] Yang X, Cui H and Yan L S 2007 Carbon 3 21-25.
[9] Zhang Y H, Xiao Z C, Wang J P, Yang J F and Jin Z H 2009 Materials Science and Engineering A 502(1-2) 64-69.
[10] Liu Y, Fu Q G, Li H J and Li G M 2016 Materials China 35(2) 128-135.
[11] Deng J Y, Liu W C, Du H F, Cheng H M and Li Y Y 1999 Journal of the Chinese society 27(3): 357-61.

Study of the Dielectric Properties of Condenser Material Used in SF6 Insulated RIP Bushing

Zhang Shiling[1]

1 State Grid Chongqing Electric Power Company Chongqing Electric Power Research Institute, Chongqing, China
E-mail: 526793305@qq.com

Abstract. All In the frequency range of 10^{-1}~10^{6}Hz and the temperature range of -100~200°C, the experimental dielectric spectrum of the condenser material used in UHVDC RIP bushing has been measured by Novocontrol broadband dielectric spectroscopy. The measured data has been fitted applied to nonlinear numerical calculation of the Havriliak-Negami (HN) equation, by which the characteristic parameters of HN mathematical model have been got. The studies show that the genetic algorithm combining with the least-square curve fitting method can ensure the uniqueness and accuracy of the fitting parameters. In the frequency range of 10^{-1}~10^{6}Hz and the temperature range of -40~60°C, β relaxation process mainly takes place in the condenser material. In temperature range of 70~150°C, α and β relaxation processes can be observed. In the temperature range of 160~200°C, the α relaxation and DC conduction processes are obvious. The DC conductivity and relaxation time have significant temperature dependence which can be fitted well with Arrhenius equation.

1. Introduction

At present, a large number of dry-type DC bushing have been used in the construction of UHVDC transmission lines in China. Its operation status directly determines the safety and the stability of transmission lines[1]. Therefore, it is of great significance to monitor and evaluate the performance of dry-type bushing in actual operation. The most commonly used nondestructive monitoring method for dry bushing is the dielectric spectrum measurement, which mainly depends on the accurate measurement of dielectric properties of epoxy impregnated paper composites for dry bushing cores, and the post-processing of the measured results to extract the characteristic parameters characterizing the properties of materials.

The breakdown characteristics and space charge distribution of epoxy/crepe paper composites were studied in literature [2]. The space charge distribution characteristics of the pure epoxy near glass transition temperature T_g were analyzed in literature [3]. However, few literatures specifically measured and fitted the dielectric properties of epoxy impregnated paper for UHVDC dry bushing cores. Based on this, samples of epoxy impregnated paper composite for dry bushing were prepared and observed by polarization and scanning electron microscopy. It was found that there were obvious interfaces between the epoxy resin and the wrinkle paper, which could cause the polarization effect of Maxwell-Wagner interface and lead to the complexity of dielectric properties of sub-composites [4]. The glass transition temperature (T_g) of core composites was obtained by DSC measurement, which can provide the basis for determining the test temperature range in subsequent dielectric spectroscopy measurements. At the same time, non-linear numerical fitting method combining genetic algorithm and least squares is proposed. The approximate values of fitting parameters are obtained by genetic algorithm, and then the fitting parameters are solved accurately by least squares method. This method can ensure the uniqueness and accuracy of fitting parameters. Furthermore, the frequency spectrum and the temperature spectrum characteristics of the core material were measured by Novocontrol broadband dielectric spectrometer in the range of 10^{-1}-10^{6} Hz and -100-200°C. The results show that in the frequency range of 10^{-1}-10^{6} Hz, the relaxation process is the main one in the temperature range of -40-60°C, and the relaxation process is also found in the temperature range of 70-150°C. The relaxation and the DC conductivity processes are remarkable in the temperature range of 160-200°C. Based on improved Havriliak-Negami (HN) equation and the non-linear numerical calculation, the measured data are fitted to obtain characteristic parameters of the dielectric spectrum mathematical model. The results show that both the DC conductivity and the relaxation polarization time have remarkable temperature dependence, which conform to the Arrhenius formula. In the laboratory environment, the dielectric spectrum fitting method of UHV dry DC bushing core material is studied and its characteristic parameters are obtained, which provides the effective method for understanding the dielectric properties of dry bushing core, and can guide the condition monitoring and performance evaluation of dry bushing in actual operation[5-7].

2. Formatting Modified HN Model and Characteristic Parameters

The complex permittivity of the epoxy impregnated paper composites under Havriliak-Negami (HN) model function can be expressed as formula (1):

$$\varepsilon_{HN}^{*}(\omega) = \varepsilon_{\infty} + \frac{\Delta\varepsilon}{(1+(i\omega\tau_{HN})^{\beta})^{\gamma}} \tag{1}$$

Formula (1) ε_{∞} is the optical frequency relative permittivity $\Delta\varepsilon = \varepsilon_{s} - \varepsilon_{\infty}$, and ε_{s} is the static relative permittivity, τ_{HN} is the relaxation polarization time under HN model, and β 、 γ is the distribution parameter of the relaxation time. In addition, $\varepsilon_{HN}^{'}$ and $\varepsilon_{HN}^{''}$ the real part and the imaginary part of the complex permittivity [8]:

$$\varepsilon_{HN}^{'} = r(\omega)\cos[\gamma\psi(\omega)]\Delta\varepsilon + \varepsilon_{\infty} \tag{2}$$

$$\varepsilon_{HN}^{''} = r(\omega)\sin[\gamma\psi(\omega)]\Delta\varepsilon \tag{3}$$

Among them:

$$r(\omega) = [1 + 2(\omega\tau_{HN})^{\beta}\cos(\beta\pi/2) + (\omega\tau_{HN})^{2\beta}]^{-\gamma/2} \tag{4}$$

$$\psi(\omega) = \arctan\{\sin(\beta\pi/2)/[(\omega\tau_{HN})^{-\beta} + \cos(\beta\pi/2)]\} \tag{5}$$

Therefore, there are five characteristic parameters in single HN model function ε_{∞}, $\Delta\varepsilon$, τ_{HN}, β, γ. Based on the experimental data of broadband dielectric spectroscopy, the specific values of the above five characteristic parameters can be fitted, and the frequency point ω_{p} of the peak value $\varepsilon_{HN}^{''}$ can be calculated by formula (6):

$$\omega_{p} = (1/\tau_{HN})[\sin((\beta\pi)/(2+2\gamma))]^{1/\beta}[\sin((\beta\gamma\pi)/(2+2\gamma))]^{-1/\beta} \tag{6}$$

When the testing temperature is lower than the glass transition temperature T_{g} of the dry bushing core material, the dielectric properties of the material can be generally characterized by (1) ~ (6) formula. When the test temperature is higher than T_{g}, the DC conductivity process of the material is remarkable, and the formula (7) should be introduced into the formula (3):

$$\varepsilon^{''} = \sigma_{0}/(2\pi f\varepsilon_{0}) \tag{7}$$

Among them, the characteristic parameters of DC conductivity are vacuum dielectric constant and f is test frequency. For the DC dry bushing core composites, the HN model mentioned above needs to be further revised to (8)~(9) form due to combined effect of electrode polarization and Maxwell-Wagner interface polarization at temperatures higher than T_{g}.

$$\varepsilon_{HN}^{'} = r(\omega)\cos[\gamma\psi(\omega)]\Delta\varepsilon + \varepsilon_{\infty} + A\omega^{-n} \tag{8}$$

$$\varepsilon_{HN}^{''} = r(\omega)\sin[\gamma\psi(\omega)]\Delta\varepsilon + \sigma_{0}/(i\varepsilon_{0}\omega^{s}) \tag{9}$$

IOP Conf. Series: Materials Science and Engineering **733** (2020) 012013 doi:10.1088/1757-899X/733/1/012013

A, N and s are the characteristic parameters to characterize the high temperature polarization process. For composites with complex dielectric properties, the polynomial HN equation can be introduced to characterize several independent relaxation polarization processes. As shown in Formula (10):

$$\varepsilon_{HN}^{*}(\omega)=\varepsilon_{\infty}+\sum_{i=1}^{m}\frac{\Delta\varepsilon_i}{(1+(i\omega\tau_{HN_i})^{\beta_i})^{\gamma_i}}+\frac{\sigma_0}{i\varepsilon_0\omega^{s}} \tag{10}$$

These two relaxation processes are commonly referred to as relaxations. The relaxation time constants and distribution parameters of two processes are sum, respectively. Formula (11) can be transformed to obtain the real and imaginary expressions of the dielectric polarizability, taking into account that:

$$(i)^{\alpha}=(e^{i\pi/2})^{\alpha}=\cos((\alpha\pi)/2)+i\sin((\alpha\pi)/2) \tag{11}$$

Substitute the above formula to get:

$$\chi^{*}(\omega)=[\varepsilon_{\infty}+\frac{\chi_s[1+(\omega\tau)^{n}\cos\frac{n\pi}{2}]}{1+2(\omega\tau)^{n}\cos\frac{n\pi}{2}+(\omega\tau)^{2n}}+\frac{\xi\cos\frac{\gamma\pi}{2}}{\varepsilon_0\omega^{\gamma}}]$$
$$-i[\frac{\chi_s(\omega\tau)^{n}\sin\frac{n\pi}{2}}{1+2(\omega\tau)^{n}\cos\frac{n\pi}{2}+(\omega\tau)^{2n}}+\frac{\sigma_{dc}}{\varepsilon_0\omega}+\frac{\xi\sin\frac{\gamma\pi}{2}}{\varepsilon_0\omega^{\gamma}}] \tag{12}$$

Within the test range of 10^{-4}-10^{4}Hz, the dielectric response characteristics of pure epoxy and epoxy impregnated paper can usually be described by the modified Cole-Cole model with the single or two relaxation processes. If the single relaxation process is used, the seven parameters of P_1~P_7 can be obtained. For the model of double relaxation time constant, 10 parameters of P_1~P_{10} can be obtained, of which P_8~P_{10} represents the relaxation process.

3. Test Results and Analysis

Figure 1 (a) is microscopic observation result of wrinkle paper in bushing core. Cellulose is evenly distributed in the same direction, and coarse cellulose and fine cellulose are alternately arranged in wrinkle paper. Figure 1 (b) shows that for epoxy impregnated paper composites, wrinkle paper is tightly adhered to pure epoxy, and the structure can ensure mechanical toughness of dry busing core.

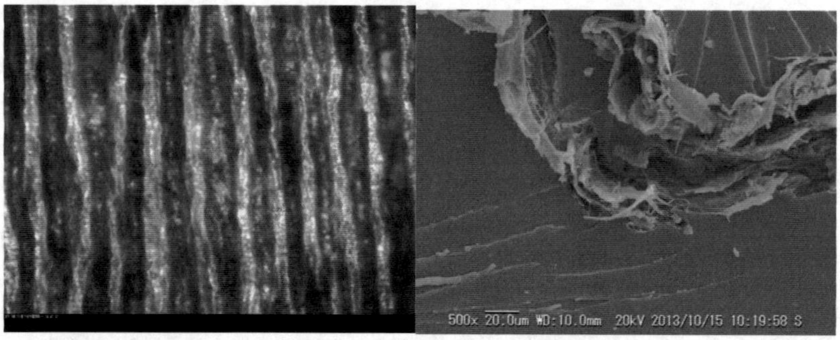

(a) Crepe paper (b) Section

Figure 1 Micro-structure of specimen

The electrodes used in the breakdown test are copper, and the surface of the electrodes is polished. The thickness of the prepared sample includes 0.25mm, 1.08mm and 2.69mm. The step-up speed in the breakdown test is 4kV.s^{-1}. The test results are shown in Figure 2. The breakdown test data are processed and analyzed by the Wei-bull distribution function with two parameters [8]. According to Wei-bull probability and statistics theory, the breakdown probability of the epoxy impregnated paper under electric field E is as follows:

$$F(E;\alpha,\beta)=1-\exp\left\{-\left(\frac{E}{\alpha}\right)^{\beta}\right\}=0.632 \quad (13)$$

In the formula *E*, the breakdown field strength/kV.mm^{-1}, breakdown field strength when breakdown probability is 63.2%, i.e. the scale parameter/kV.mm^{-1}, and the shape parameter are used.

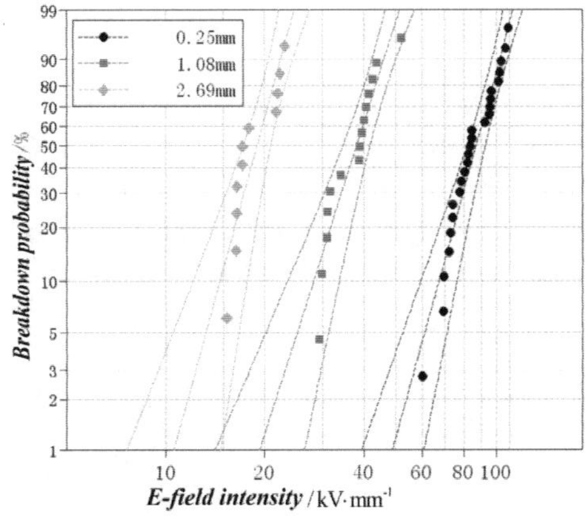

Figure2 Relationship between breakdown field strength and thickness of the sample

Using the laser thermal conductivity analyzer and the thermal expansion analyzer to measure thermal performance parameters of epoxy impregnated paper materials for failure bushing is shown in Figure 3.

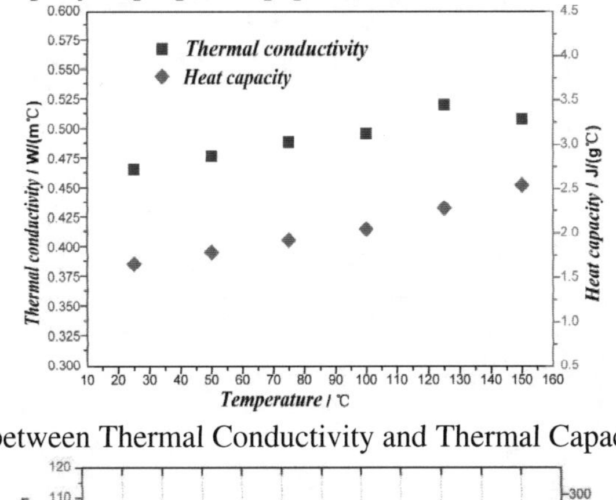

(a) The Relation between Thermal Conductivity and Thermal Capacity with Temperature

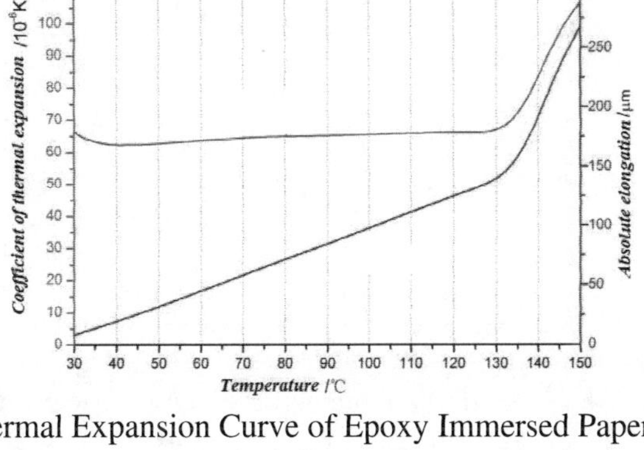

(b) Thermal Expansion Curve of Epoxy Immersed Paper Material

Fig.3 Relationship between properties of epoxy impregnated paper and temperature

Figure3 (a) shows that thermal conductivity of epoxy impregnated paper material is about 0.463W/(m °C) at room temperature of 25°C. The orderly arrangement of cellulose in wrinkle paper is beneficial to the heat flux transmission. The thermal conductivity of epoxy impregnated paper material increases first and then decreases in the temperature range of 25-150°C, and the maximum value appears near 125°C. This indicates that the temperature of the converter bushing core should be controlled in the range of 25-125°C in actual operation. At this time, the thermal conductivity of the material has the "negative feedback" effect. With the slight increase in degree, the thermal conductivity increases, the thermal conductivity of material increases, and the temperature decreases. If the operating temperature of material exceeds 125°C, the heat inside the core is difficult to derive effectively due to the decrease of thermal conductivity, which is not conducive to the safe operation of the bushing. The thermal capacity of epoxy impregnated paper materials is kept between 1.5 and 2.5 J/(g°C) in the temperature range of 25-150°C. Figure 3 (b) shows that thermal expansion coefficient of epoxy impregnated paper increases slowly in the temperature range of 30-150°C, then increases abruptly. The inflection point of this change is near 130°C, which indicates that the cellulose structure of corrugated paper inside the epoxy impregnated paper material has the buffer effect on the thermal stress. In actual operation, the deformation of the core caused by thermal stress is relatively small in the temperature range of 30-120°C. If the temperature exceeds 130°C, the insulation defects such as cracking of epoxy impregnated paper material will occur in the core due to the long-term effect of thermal stress.

Routine test of bushing core is usually carried out at room temperature (20°C). For the actual bushing operation, the hottest temperature inside the core is about 90°C. To illustrate the change of dielectric properties of the insulating material of bushing core at 150°C, the dielectric spectrum test results in frequency domain at 20°C, 90°C and 150°C are listed as shown in Figure 4.

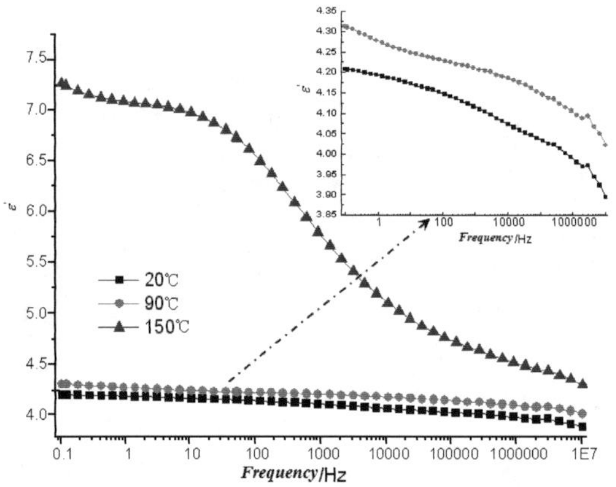

(a) Real Frequency Domain Dielectric Spectra of Dielectric Constants

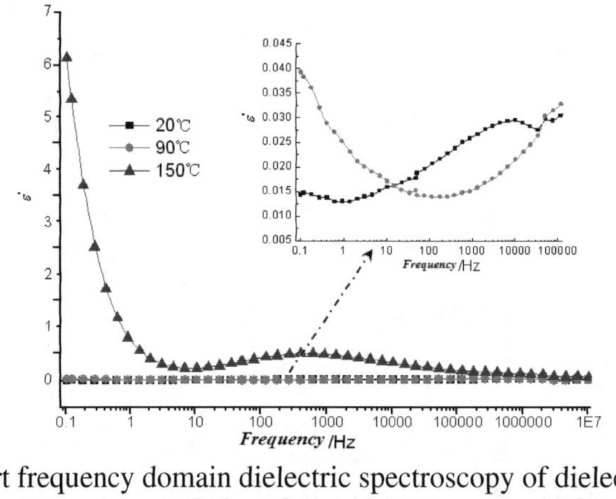

(b) Imaginary part frequency domain dielectric spectroscopy of dielectric constant

Fig.4 Dielectric spectrum characteristics of material parameters in frequency domain

Figure 4 shows that the dielectric spectrum in frequency domain varies significantly with temperature. The real part of dielectric constant decreases with the increase of frequency *f* at 20°C and 90°C. The reason is that the real part of dielectric constant of main insulating material of SF6 gas insulating bushing can be expressed as follows:

$$\varepsilon_r = \varepsilon_\infty + \frac{\varepsilon_s - \varepsilon_\infty}{1+\omega^2\tau^2} \tag{14}$$

It can be seen from formula (15) that displacement polarization and relaxation polarization can be established in low frequency region (1), so the real part of dielectric constant tends to be equal. The real part of the number shows the decreasing trend. For the epoxy sample, the sum value decreases first and then increases with the increase of frequency, because the tangent value of loss angle can be expressed as follows:

$$\tan\delta = [\frac{\gamma}{\omega\varepsilon_0} + (\varepsilon_s - \varepsilon_\infty)\frac{\omega\tau}{1+\omega^2\tau^2}]/(\varepsilon_\infty + \frac{\varepsilon_s - \varepsilon_\infty}{1+\omega^2\tau^2}) \tag{15}$$

The formula is dielectric conductivity, relaxation polarization time constant, static dielectric constant and optical frequency dielectric constant. It can be seen from formula (16) that with the increase of the frequency of alternating electric field, the polarization lag electric field changes, and the relaxation polarization can not be fully established, so the loss caused within the week decreases, making the sum value of the epoxy insulation sample decrease with increase of the frequency. In the high frequency region, the relaxation polarization is also less than that. The sum value increases with the increase of frequency because of the increase of the number of cycles per-second, which increases the dielectric loss. It can also be seen from Figure 4 that the values of each parameter are higher than those of the other two temperatures at 150°C, and the imaginary part of dielectric constant and the tangent value of loss angle increase sharply at low frequencies (<1Hz). The main reasons for this phenomenon are as follows: (1) glass transition of epoxy resin has exceeded that of epoxy resin at 150°C. At temperature T_g, the dielectric is in the high elastic state, the polar groups and chains in macro-molecules can be oriented along electric field direction, and the dipole polarizes in the high elastic state, so the dielectric constant and the tangent value of the loss angle increase significantly; (2) the dielectric conductivity increases significantly at high temperature, and the Formula(15) shows that the term will be in the low frequency condition.

4. Temperature Translation Characteristics

The frequency dependence of dielectric spectra of epoxy impregnated paper composites was fitted at the specific temperature. At the same time, temperature is closely related to the dielectric spectra: the dielectric spectra of materials have characteristics of temperature translation during the continuous change of temperature. The dielectric spectrum data at five temperatures are fitted and plotted in Figure 5 with temperature range of 160 ~200°C as an example and the temperature interval of 10°C. At high temperatures and low frequencies, the real part of dielectric constant shows an upward trend, which can not be fitted with theoretical curve, mainly due to existence of electrode polarization.

All of them exhibit the characteristics of temperature translation, that is they tend to move to high frequency region. At the same time, with the increase of temperature, the conductivity polarization intensifies and the loss peak of polarization moves to the high frequency region. The main reason is that the relaxation time is approximately exponential to the temperature T. With the increase of T, the characteristic frequency f_c moves to high frequency direction. Temperature translation characteristics of Cole-Cole diagram are shown in Figure 5. It can be seen that if the temperature is further increased, the relaxation time is very small and polarization is established quickly, the relaxation polarization loss can be kept up with the change of electric field, so that the relaxation polarization loss is small and negligible, and conductivity loss increases rapidly with the increase of temperature.

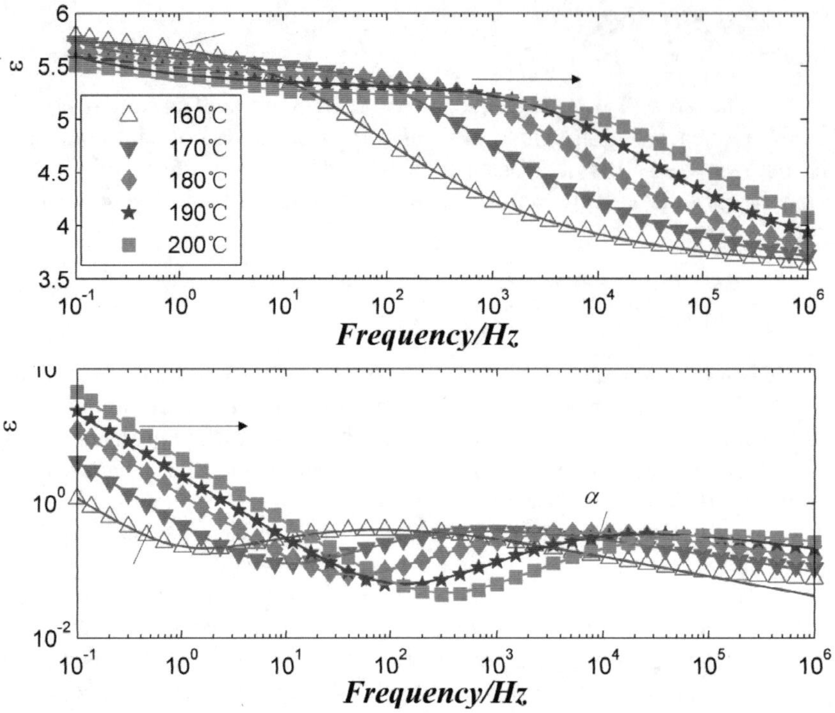

Figure 5 Temperature translation characteristics of dielectric spectra

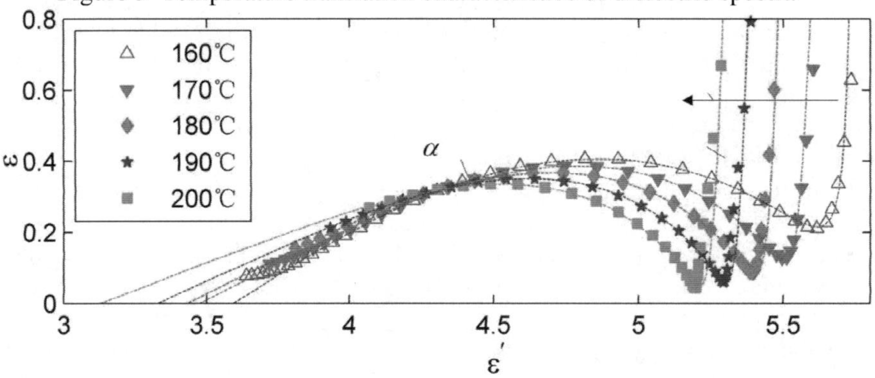

Figure 6 Complex Plane Map

Temperature dependence of dielectric spectra of epoxy impregnated paper composites was analyzed. The temperature spectra of the epoxy impregnated paper composites were plotted in Figure 6 at the frequency of 50 Hz in the temperature range of -10 ~200°C. The figure shows that the real part of the dielectric constant is basically unchanged in the temperature range of -100~130°C. It rises sharply after 130°C, and the peak D appears near 188 °C.

The imaginary part has two peaks A and B in the temperature range of -100 to 200°C, near -25°C and 170°C, respectively, and the lowest value C near 110°C. The reason why the peak B of imaginary part lags behind the peak D of real part is that although the polarization is established rapidly in high temperature region, it does not mean that the polarization has been fully established. Only when the temperature rises to make the polarization fully established can the maximum be reached.

5. Summary

In this paper, the spectrum and temperature spectrum curves of UHVDC dry bushing core material are measured and fitted. The conclusions are as follows:

(1) The uniqueness and accuracy of the fitting parameters can be guaranteed by the genetic algorithm combined with least squares curve fitting method, and the binomial Havriliak-Negami (HN) equation with 11 parameters can be well fitted.
(2) The measured data are fitted by the improved HN equation. The measured and fitted data are in good agreement with each other at 20℃and 170℃. From the low temperature to high temperature, the polarization peaks appear in turn, and after 160℃, the remarkable process of DC conductivity polarization is observed.
(3) The dielectric spectra of the core material are strongly dependent on frequency and temperature, and obvious temperature translation characteristics are found in the high temperature range of 160~200℃. Meanwhile, the DC conductivity and relaxation polarization time of the core material have significant temperature dependence, which basically conform to Arrhenius formula.
(4) The dielectric properties of single sample of the bushing core material are studied in this paper. The next step is to measure the spectrum and temperature spectrum curve of the actual bushing core, summarize difference and relationship between the two, and apply the conclusion to the evaluation and monitoring of the insulation state of the dry bushing.

References

[1] Shen Wei, Cao Wen, Wu Kai. Study of characteristics of breakdown strength and space charge in epoxy/paper composite[J]. Transactions of China Electrotechnical Society, 2013,28(1): 1-6.
[2] Chen Shaoqing, Peng Zongren, Wang Xia. Effects of glass transition temperature on space charge in epoxy resin[J]. Transactions of China Electrotechnical Society, 2011,26(8): 227-230.
[3] Liu Jun, Zhou Lijun, Wu Guangning. Dielectric frequency response of oil-paper composite insulation modified by nanoparticles[J]. Proceedings of the CSEE, 2011,31(28):144-153.
[4] Antun Mikulecky. Influence of temperature, moisture content and aging on oil impregnated paper bushings insulation[J]. IEEE Transactions on Dielectrics and Electrical Insulation, 2013, 20(4): 1421-1427.
[5] Stefan Wolny, Artur Adamowicz, Marek Lepich. Influence of temperature and moisture level in paper-oil insulation on the parameters of the Cole-Cole model[J]. IEEE Transactions on Power Delivery, 2014, 29(1): 246-250.
[6] S.Wolny, M.Zdanowski. Analysis of recovery voltage parameters of paper-oil insulation obtained from simulation investigations using the Cole-Cole model[J]. IEEE Transactions on Dielectrics and Electrical Insulation, 2009, 16(6): 1676-1680.
[7] C. Vanga Bouanga, H. Couderc, M.F.Frechette, S. Savoie. Dielectric study of low glass transition temperature cycloaliphatic UV-curable epoxy networks[J]. IEEE Transactions on Dielectrics and Electrical Insulation, 2012, 19(4): 1269-1282.
[8] Xu Hao, Shao Dan, Yang Honghui, Yan Wei. Surface analysis of Ti/Sb-SnO2 electrode after electrolysis[J]. Journal of Xi'an Jiaotong University, 2014,48(2): 77-82.

The design of swirl burner in a pulverized coal gasifier by numerical analysis

Qin Dachuan[1],Li Bing[1],Wang Mingquan[2], Yang Chao[3]

1 Huadian electric power research institute Co.,LTD., Hangzhou
2 Guizhou huadian bijie thermal power electric Co. LTD, Bijie
3. Guizhou huadian tangzhai power generation Co. LTD, Guiyang

Abstract. In this paper, the eddy-dissipation-concept model is used to simulate the influence of burners of different structures on the performance of gasifier. The effects of the annular inner and outer passage, annular air passage and annular pulverized coal passage on the spatial distribution of flow field, temperature field and grouping field in gasifier were investigated. The simulation results show that pulverized coal can obtain higher carbon conversion rate by going through the annular outer passage of the burner. As the number of air swirl S increases, the high temperature flame in the gasifier becomes more dispersive and the temperature gradient becomes smaller. The high temperature flame zone in the gasifier is more diffuse when pulverized coal goes through the annular outer channel than the annular inner channel, which can reduce the temperature of the high-temperature flame zone.

1. Introduction

Coal gasification technology is an important way to convert coal into syngas (CO, H_2) to realize clean and efficient utilization of coal[1-6]. The current domestic market operation of the gasifier are mainly GSP gasifier developed by Siemens company[7], Texaco gasifier developed by Texaco petroleum corporation in America[8], Shell gas furnace developed by Shell company[9], Two-stage colliding gasifier developed by East China University of Science and Technology[10], Space furnace developed by China aerospace science and industry corporation, MHI gasifier developed by mitsubishi heavy industries, etc., there are many reports on numerical simulation of gas solid flow field and temperature field in these commercial gasifiers. Wu yuxin [11] carried out numerical simulation of Texaco gasifier, used simplified PDF method to describe turbulent gas phase chemical reaction, and wrote UDF description for coke heterogeneous phase reaction to obtain the distribution characteristics of temperature, CO and H2 in Texaco gasifier. Zhou junhu [12] conducted three-dimensional numerical simulation of Shell gasifier by using euler-lagrange method, and analyzed the distribution characteristics of flow field in cold state of gasifier, particle running track, temperature field and fixed carbon burnout in hot state of gasifier. Kumar [13] analyzed the effects of different gas-phase turbulence models and coke reaction models on the flow field and temperature field in MHI gasifier by describing the interaction between turbulence and chemical reaction used the finite rate/vortex dissipation model. There are few reports on the influence of burner structure on the flow field characteristics in the gasifier, and the burner structure directly affects the gas flow field inside the gasifier, the residence time of pulverized coal particles, the gas-solid mixing characteristics and the heat and mass transfer and heat properties, so it is the core of the gasifier design.

The paper would use the Eddy Dissipation Concept（EDC）model to consider the 12-step total package equation in the component transport framework, the reliability of the mathematical model used in this paper was verified by the BYU [14] experimental gasifier running at simulated atmospheric pressure firstly, and then the structure of preheating burner designed by the research group would be optimized, this paper investigated the pulverized coal to go inside the annular channel and channel, circular air channel swirl number, number of annular passage vortex pulverized coal gasifier gasification performance impact respectively. The purpose of this paper is to provide a reasonable scheme for the burner structure of preheating gasifier which will be used in the experimental bench set up by the research group, and ensure that the pulverized coal gasifier in high carbon conversion and particle residence time, and provide quality fuel for experiment system downstream of the combustion chamber in order to ultimately through the technology realization of low rank coal combustion stability, high efficiency and low NOx emissions synergies.

2. Numerical calculation model and research object

2.1. The research object

The physical model. There is a great difference between the heterogeneous reaction kinetic parameters of coke under high pressure and normal pressure, while the operating pressure of preheating gasifier is normal pressure. Therefore, the BYU experimental gasifier operating at normal pressure was selected to verify the reliability of the mathematical model used in this paper. The diameter of the nozzle inlet center channel is 13mm, and the temperature is 367 K. The outer diameter of the annular channel is 28.6mm, and 450 K of water vapor enters. Pulverized coal is carried into the furnace by mixed gas from the central channel; The simulated coal is Utah coal. The gasifier air equivalent ratio is only 0.25, so a single nozzle top gasifier is chosen. In order to achieve high carbon conversion rate, the design of burner nozzle of gasifier should ensure the residence time of pulverized coal as long as possible. On the other hand, strong turbulent mixing between pulverized coal and air in gasifier is needed to ensure that pulverized coal is quickly dispersed to participate in the reaction, so as to achieve a higher carbon conversion rate, which is also conducive to the precipitation of coal nitride. According to the calculation requirements of process design, the total height of preheating gasifier body is 1950mm, and the inner diameter of furnace is 400mm. The burner is composed of three channels. The central channel is the ignition device of gasifier, and the outer part is the annular air circulation channel and annular pulverized coal circulation channel, with an included Angle of 15°. Pulverized coal is carried into the furnace by a small amount of air, the designed flow rate of pulverized coal channel is 5m/s, and the designed flow rate of air channel is 15m/s. The coal for simulation is selected as shenmu coal. Detailed numerical simulation coal parameters are given in table 1. Firstly, the influence of ring inner and outer channels on gasification performance of gasifier was investigated. On this basis, the influences of cyclones in annular air passage and annular pulverized coal passage on gasification performance of gasifier are investigated. In this paper, the cyclones S is defined as the ratio of tangential component and axial component of air inlet.

Table 1 coal industry analysis

Name	Unit	Utah Coal	Shenmu coal
Industry analysis			
Fixed carbon	%	43.70	57.61
Volatile	%	45.60	30.23
Moisture	%	2.40	5.85
Ash	%	8.30	6.31
HHV_{ar}	MJ/kg	29.80	27.11
Elemental analysis			
C	%	71.00	80.43
H	%	6.00	4.47
O	%	12.70	13.72
N	%	1.30	1.15
S	%	0.50	0.24

IOP Conf. Series: Materials Science and Engineering **733** (2020) 012014 doi:10.1088/1757-899X/733/1/012014

3. Results and discussion

3.1. The difference of pulverized coal channel between inner annulus area and outer annulus area

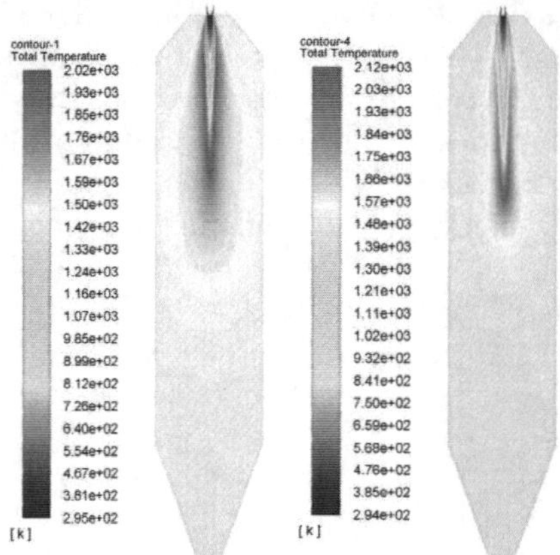

（a）operating condition 1 （b）operating condition 2

FIG.1 temperature distribution in gasifier in inner and outer annular gap zones

FIG. 1 shows the simulation results of the temperature field distribution in the gasifier in the inner and outer annular gaps. It can be seen that under the same air swirl number, the high temperature flame zone in the gasifier in which pulverized coal goes through inner and outer annular space channels is mainly concentrated in the central area of the upper half of the gasifier, jet coal is heated rapidly and the process of volatilization occurs in this region, at the same time Volatiles, coke residue in coal and syngas (CO, H_2) in gasifier are rapidly mixed with swirling air, so that Exothermic oxidation occurs; The temperature of both sides of the upper half and the lower half of the gasifier gradually decreases, mainly resulting in heterogeneous endothermic reactions of coke with CO_2, H_2O and H_2. By comparing FIG. 1a and b, it can be seen that the high temperature area in the furnace presents a more diffuse flame pattern when coal powder goes through the outer annular passage. However, the high temperature area in the furnace chamber where pulverized coal goes through the inner passage is concentrated in a small narrow and long area with a large temperature gradient in the upper half centre, and the peak temperature in the furnace chamber is 110K higher than that in the outer passage. This is because the high-speed swirling air enters the furnace and rapidly expands the rotating radius under the action of centrifugal force, while the coal powder channel located outside the air channel can make the coal powder more directly dispersed by the high-speed swirling air in the inner channel.

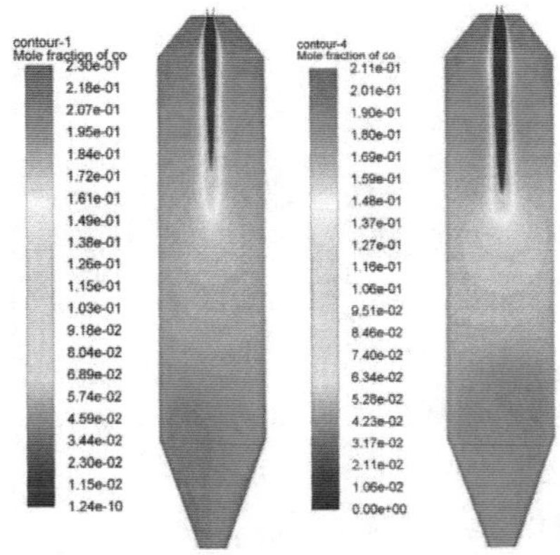

(a) distribution of CO in operating condition 1 (b) distribution of CO in operating condition 2

FIG. 2 distribution of CO and H_2 in gasifier in inner and outer annular gaps

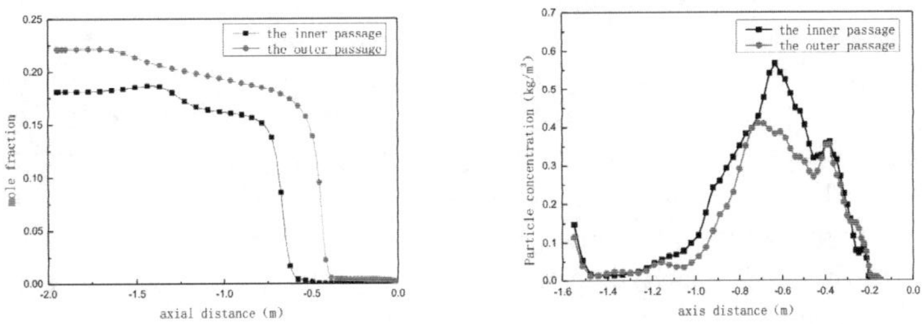

(a) distribution of CO along the central axis

(b) distribution of particle concentration near the edge wall

FIG. 3 axial CO distribution in the inner annular space and outer annular space and particle concentration distribution near the wall surface

FIG. 2 shows the distribution simulation results of CO and H_2 in the gasifier under working conditions 1 and 2. It can be seen that the regions with higher CO and H_2 concentrations in the gasifier are mainly concentrated in the gasification area, that is, on both sides of the central flame area of the lower half and upper half of the gasifier, and the end of the high-temperature flame area shows an increasing trend upward and downward. The higher concentration in the upper half of the gasifier is caused by gasification reaction because of pulverized coal flows back into this area carried by the backflow upward air on both sides air due to high-speed jet air. As the distribution of CO and H_2 is relatively close, FIG. 3a only shows the distribution of CO on the central axis. In FIG. 3a, the connection between the burner and the furnace is the coordinate origin, and downward is negative. It can be seen that the mole fraction of CO at the gasifier outlet when pulverized coal goes through the outer channel is 0.22, while the mole fraction of CO at the gasifier outlet when pulverized coal goes through the inner channel is 0.16 It indicates that coal can obtain higher carbon conversion rate by using the outer channel. Figure 3b shows the particle concentration distribution at the straight section of the gasifier with a length of 1400 mm and 2 mm from the side wall (the same goes for the rest), two kinds of working condition of particle concentration distribution near the side wall are first increasing then decreasing, and there is a maximum points, which value is smaller when pulverized coal walk outside channels, this is because the sea coal go outside when the channel is more likely to be return air suction roll entrained gasifier to the upper section of the gasification zone. To sum up, pulverized coal was selected as the basis of the following simulation.

3.2. The difference of swirl number in air passage

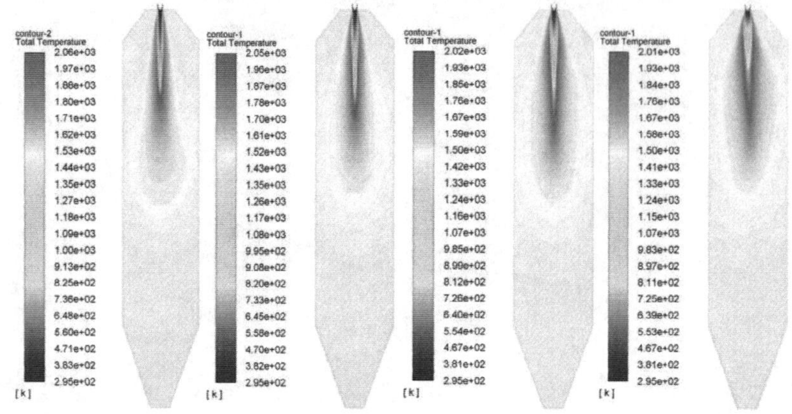

（a）S=0 （b）S=0.58 （c）S=1 （d）S=1.73

FIG. 4 temperature distribution in gasifier under different air swirl number

Figure 4 shows the gasifier temperature distribution under different air swirl number, it can be seen that temperature field distribution is similar in the gasifier under different swirl number S, the difference is that the scope of the gasifier in high temperature increases gradually present a more diffuse and full flame shape with the increase of swirl number S, flame temperature gradient decreases in high temperature at the same time. When S=1, the diameter of the upper half of the gasifier in the high temperature region can reach half of the furnace diameter. This shows that with the increase of the number of air cyclones, pulverized coal can be quickly dispersed in the high temperature area, and then participate in the reaction, which is also conducive to the precipitation of nitrogen compounds in the coal.

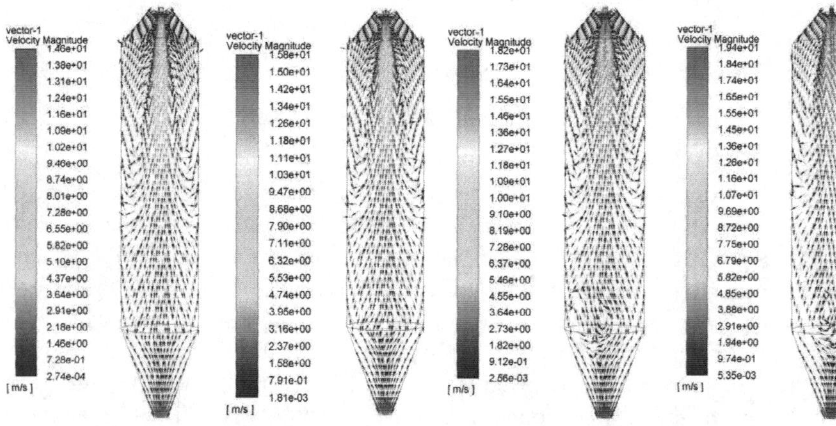

（a）S=0 （b）S=0.58 （c）S=1 （d）S=1.73

FIG. 5 velocity distribution in gasifier under different air swirl number

FIG. 5 is the vector diagram of gas velocity in gasifier under different air swirl Numbers. Similarly, the velocity of high-speed jet gas decreases gradually after entering the gasifier and divided into two parts in the middle part of the furnace: One part continues down to the exit, and the other part backflows upward under the action of high speed jet gas inverse pressure gradient. Under the condition that the air inlet flow rate remains unchanged, the maximum velocity of gas near the burner increases with the increase of the number of swirls, which will generate a greater adverse pressure gradient, it is conducive to sucking more pulverized coal into the upper half of the gasification area. In the shrinking section of the gasifier outlet, two large vortex groups appear in the two working conditions S=1 and S=1.73, and a small vortex group appears in the working condition S=0.58, while no obvious vortex group appears in the working condition S=0.

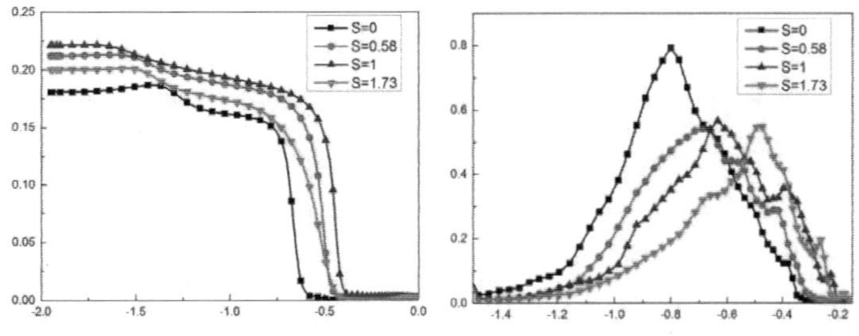

(a) distribution of CO along the central axis

(b) distribution of particle concentration near the edge wall

FIG. 6 swirl number distribution and particle concentration distribution in different air annular channels

FIG. 6 shows the axial CO mole number distribution along the central axis of gasifier and particle concentration distribution near the wall under different air swirl number. The analysis of figure 6a shows that the distribution trend of CO concentration on the central axis of the gasifier under different S is relatively similar, that is, there is almost no CO concentration within 0.5m from the inlet, and then it suddenly increases around 0.5m from the inlet, and then slowly increases until the outlet. When S=0, the concentration of CO at the gasifier outlet is about 0.16; when the swirl number S increases from 0.58 to 1, the CO at the gasifier outlet increases from 0.21 to 0.22; when the swirl number S continues to increase to 1.73, the CO concentration at the gasifier outlet decreases to 0.2. Figure 6b shows the particle concentration distribution near the straight segment hearth side wall, it can be seen that when the number of air swirls S increases gradually, the concentration of pulverized coal in the wall of the maximum distance between point and gasifier burner gradually decreases, and pulverized coal will be carried faster to the gasifier wall with the increase of S, this may be the reason why CO concentration in gasifier exit is lower instead when the swirl number increases to 1.73. In conclusion, when the number of air swirl is 1, it is selected as the basis of the following simulation.

4. Conclusion

Through numerical simulation of BYU experimental gasifier, this paper verifies that the numerical model used in this paper can be used to reasonably predict the distribution of temperature field and grouping field in preheating gasifier. In the design conditions selected in this paper, the influences of the swirl number of the annular inside and outside channels, the annular air channel and the swirl number of the annular coal channel on the flow field and grouping and dividing characteristics of the gasifier are investigated aiming at the design and optimization of the burner structure of the preheating gasifier. The following conclusions are drawn: (1) the numerical calculation results show that the high temperature flame zone in the gasifier is more diffuse when pulverized coal goes through the annular outer channel than the annular inner channel, which can reduce the temperature of the high-temperature flame zone by about 110K, and obtain higher CO mole fraction at the outlet of the gasifier. (2) as the number of cyclones in the annular air passage increases from 0 to 0.58, 1 and 1.73, the high temperature area of the upper half of the gasifier gradually becomes larger, showing a more diffuse flame form, but at the same time, coal powder will be carried to the wall of the gasifier faster. When the swirl number reaches S=0.58, 1 and 1.73, a vortex group favorable to gas-solid secondary mixing can be formed at the lower end of the gasifier. When the swirl number S=1, the CO concentration at the gasifier outlet is the highest. (3) when the swirl number in the pulverized coal channel is about 0.58, there is no swirl number in the pulverized coal channel, and a low temperature zone is formed near the burner nozzle of the gasifier, which is conducive to reducing the nozzle head temperature, but will reduce the carbon conversion rate.

5. References

[1]. J L Xu and H F Liu and J Wang 2011 Simulation study of multiphase turbulent reaction flow in GSP gasifier *Chemical engineering*. 39(11):56-60.

[2]. D Yang, X L Ye and J Guo 2008 Numerical simulation and model experimental study on flow field optimization of wet flue gas desulfurization *J. Power technology and environmental protection*. 34(2):9-13.

[3]. W P Shan, A Y Liu and D H Li 2008 Application research of large particle ash interception technology in SCR flue gas denitrification equipment of coal-fired power plants *J. Power technology and environmental protection*. 34(2):20-23.

[4]. F H Zhu, J S Li and X Y Ma 2018 Emission and control of unconventional pollutants in clean coal power flue gas *J. Power technology and environmental protection*. 34(1):23-26.

[5]. B Bryan, J Rabovitser, S Nester and S Wohadlo 2005 Methane de-NO_X for Utility PC Boilers *Gas Technology Institute*.

[6]. J Rabovitser, B Bryan, R Knight, S Nester, S Wohadlo, A G Tumanovsky, E N Tolchinsky, E H Verbovetsky, R Lisauskas and R Beittel 2003 Development and testing of a novel coal preheating technology for NOx reduction from pulverized coal-fired boilers *in: Mega Symposium*

[7]. C Liu, S Hui, S Pan, D Wang, T Shang and L Liang J 2015 The influence of air distribution on gas-fired coal preheating method for NO emissions reduction *Fuel* 139 (2015) 206-212.

[8]. C Liu, S Hui, X Zhang, D Wang, H Zhuang and X Wang 2015 Influence of type of burner on NO emissions for pulverized coal preheating method *Appl. Therm. Eng.* 85 (2015) 278-286.

[9]. C Liu, S Hui and S Pan 2014 Experimental Investigation on NOx Reduction Potential of Gas-Fired Coal Preheating Technology *ENERGY & FUELS*. 28(9):6089-6097.

[10].Q G Lv, J G Zhu and T L Niu 2007 High temperature preheating method for pulverized coal *China*. 200710175526.3 , 2007.

[11].Y X Wu, J S Zhang and M M Wang 2007 3d numerical simulation of Texaco gasifier with simplified PDF model Chinese journal of chemical engineering. 58(9): 2369-2374.

[12].J H Zhou, J P Kuang and Z J. Zhou 2007 Numerical simulation of hot and cold flow field distribution characteristics, nozzle heating and slag layer model of pulverized coal gasifier Chinese journal of electrical engineering. 27(20):30-35.

[13].Kumar M and Ghoniem A F 2012 Multiphysics Simulations of Entrained Flow Gasification. Part II: Constructing and Validating the Overall Model Energy & Fuels. 26(1):464-479.

[14].Brown B W, Smoot L D and Smith P J 1988 Measurement and prediction of entrained-flow gasification processes Aiche Journal. 34(3):435-446.

IOP Conf. Series: Materials Science and Engineering **733** (2020) 012015 doi:10.1088/1757-899X/733/1/012015

Study on the Effect of Fabric Parameters on the properties of Wave-Transparent Composites

Tian Lan[1], Yin Liang[1], Zhongyuan Lu[1], Nan Li[1], Yuchen Pei[1]

1. Aerospace Institute of Advanced Materials & Processing Technology, 100081, Beijing, CHINA

E-mail: lantiansky2016@163.com

Abstract: Novel fabric structures of wave-transparent composites having excellent dielectric property were successfully prepared using autoclave moulding. The influence of fabric parameters (such as weaving patterns, hollow ratio, mix structure) on the properties of quartz fiber-reinforced modified cyanate ester composite were systematically investigated. The results show that satin weave fabric-reinforced composites have better mechanical performance than plain weave fabric--reinforced composites. With hollow ratio of fabric increasing the dielectric properties of composites becomes better, except for the mechanical performance. The prepared hollow/solid quartz fabric-reinforced composites in this study display lower dielectric constant and better mechanical properties than quartz/Aramid III and quartz/Poly-p-phenylene benzobisoxazole(PBO).

1. Introduction

Radar antenna technology is developing toward higher frequency, wider bandwidth, lower power consumption and higher integration which bring higher requirements on high performance wave-transparent composites. To meet the demands of modern science and technology information industry, these wave-transparent materials has not only good processing characteristics and high mechanical property, but also lower dielectric constant, lower dielectric loss and outstanding heat resistance. Quartz fiber, as a representative of excellent performance reinforced phase, has outstanding thermal and mechanical properties with extremely low and stable dielectric loss over wide temperature and frequency ranges. However, Quartz fiber also has several performance disadvantages including higher dielectric constant, higher density etc,. which can not fully meet the requirements of new generation radar system. Here, in order to overcome the drawbacks of quartz fiber, we present a novel fabric structure for fabricating a series of wave-transparent composites. The influence of fabric parameters, such as weaving patterns, hollow ratio, mix structure, on the properties of quartz fiber-reinforced modified cyanate ester composite were systematically investigated. It is supposed that the weaving parameters play an important role in determining properties of wave-transparent composites.

2. Experimental

2.1 Materials

The modified cyanate ester resin was synthesized by our laboratory. A series of different quartz fibers were purchased from Hubei Feilihua (Jing Zhou) co. Aramid III fibers and PBO fibers were supplied by China Bluestar Chengrand (Cheng Du) co. Acetone was purchased from Beijing chemical works (Bei Jing).

2.2 Preparation

The modified cyanate ester resin (MCE) solution was prepared at a concentration of 50 wt% in acetone. A range of different fabric prepregs (weight of resin/fiber = 1:1) were obtained by impregnation process. Several layers of as-received fabric layer were piled into a metal plate which had been coated releasing agent. After that, the metal plate was placed into the autoclave and the samples were cured according to the following schedule: 100 ^{o}C for 2h, 120 ^{o}C for 3.5h, 130 ^{o}C for 4.5h at 850 kPa as the applied pressure.

2.3 Characterization

The tensile properties of the composites were determined using dumbbell specimens with size 180 mm in length, 10 mm in width and 2.2 mm in thickness at 1 mm/min crosshead speed on Instron 5569 instruments, USA. The compression performance of the composite were determined using rectangular specimens with size 110 mm in length, 10mm in width and 2mm in thickness in the shear loading mode. The flexural properties of the composites were determined using rectangular specimens with size 50mm in length, 15mm in width and 2.2mm in thickness in the three point bending mode. The interlaminar shear properties of the composites were determined using ladder-like specimens with size 35mm in length, 15mm in width and 15mm in thickness at 7 mm/min speed on hydraulic universal testing machine. An average value of 5 tests for the single code was reported as the result. The dielectric properties of the composites were measured using circular specimens having the diameter of 50.5mm and 3.3mm as a thickness according to the ASTM D150-11 standard.

3. Results and discussion

3.1 Effect of weaving patterns on mechanical properties of composite

The patterns of quartz fabrics mainly divided into plain weave and satin weave as to fiber reinforcement. Two kinds of fabric composites are prepared and their mechanical performances are shown in Figure.1. Generally, these results indicate that the mechanical property of satin weave composite sample was lower than that of plain weave composite. However, the interlaminar shear strength of composite may not vary much between satin weave and plain weave.

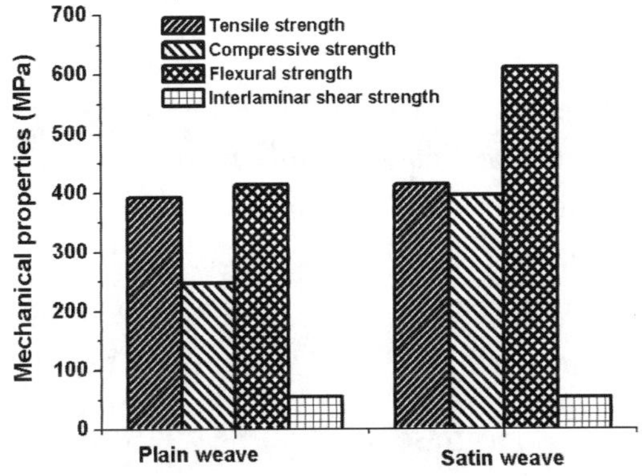

Figure.1 The mechanical properties of plain weave composite and satin weave composite.

3.2 Influence of hollow ratio on mechanical and dielectric properties of composite

Hollow ratio is one of important factor in quartz fibers, which represents the ratio of inner diameter and outer diameter of hollow quartz fiber. Dielectric performance of 32% and 50% hollow ratio composite were analyzed and shown in Table1. It can be seen that both their dielectric constant and loss factor are less than 2.65 and 0.004, respectively.

Table 1. Influence of different fabric hollow ratio on the dielectric properties of composite

Hollow ratio	Dielectric constant（10GHz）	Loss factor（10GHz）
32%	2.61	0.003
50%	2.50	0.003

Mechanical properties of 32% hollow ratio of composite was measured and compared with that of 50% hollow ratio of composite as shown in Fig. 2. Compare to the 32% hollow ratio of composite, the tensile strength (268 MPa), compressive strength (239 MPa), flexural strength (351MPa) and interlaminar shear strength (41.3 MPa) approximately decreased by 16.3%, 4.4% , 4.8% and 18.5%, respectively, when the hollow ratio of quartz fabric reach the maximum (50%). All of tensile strength, compressive strength, flexural strength and interlaminar shear strength decreased, which could be caused by hollow structure.

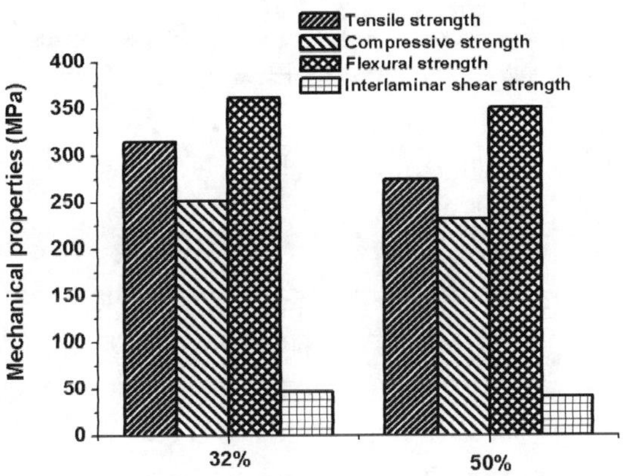

Figure.2 The mechanical properties of different fabric hollow ratio of composite.

3.3 Effect of mix structure on mechanical and dielectric properties of composite

In order to improve the properties of the wave-transparent composites, the fiber reinforcements is employed with hollow/solid quartz fiber, quartz/aramid III fiber and quartz/PBO fiber, respectively. Dielectric properties of different mix structure of composites were performed and results are shown in Table 2. Quartz/aramid III fiber-reinforced composite and quartz/PBO fiber-reinforced composite have a dielectric constant of 3.37 and 3.42, respectively. However, it can be seen that the dielectric constant and loss factor of hollow/solid quartz fiber-reinforced composite is less than 2.8 and 0.004, respectively.

Table.2 Influence of different mix structure of fibers on the dielectric properties of composite

Mix structure	Dielectric constant（10GHz）	Loss factor（10GHz）
Hollow/solid quartz fiber	2.73	0.003
Quartz/aramid III fiber	3.37	0.007
Quartz/PBO fiber	3.42	0.003

Fig.3 shows the evolution of tensile strength, compressive strength, flexural strength and interlaminar shear strength collected during the mechanic test on the composite materials. It can be seen that the highest tensile strength of composite is quartz/aramid III fiber-reinforced composite (634 MPa). However, hollow/solid quartz fiber-reinforced composite possess fine mechanical performance, which included tensile strength (421MPa), compressive strength (393MPa), flexural strength (614MPa) and interlaminar shear strength (55MPa).

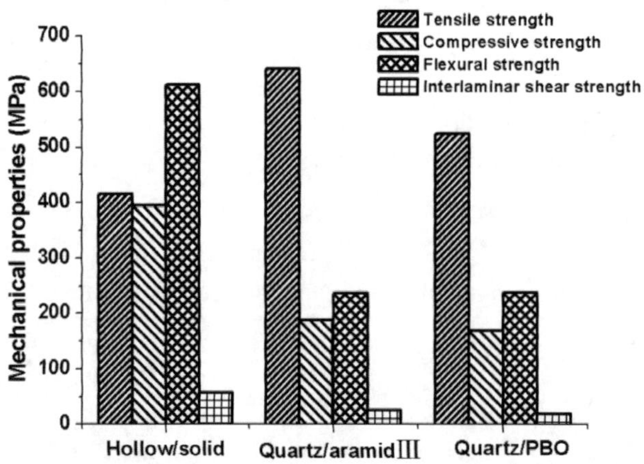

Figure.3 The mechanical properties of different mix structure of composite.

4. Conclusion

Novel fabric structures of wave-transparent composites having excellent dielectric property and mechanical performance were successfully prepared using autoclave moulding. The influence of weaving parameters on the dielectric constant, loss factor, tensile strength, compressive strength, flexural strength and interlaminar shear strength of mixed fabric/MCE composites was evaluated. In generally, hollow/solid fabric-reinforced MCE composites have a better dielectric performance and mechanical properties than others.

5. References

[1] K.S. Santhosh Kumar, C.P. Reghunadhan Nair, K.N. Ninan, Investigations on the cure chemistry and polymer properties of benzoxazine-cyanate ester blends, *Eur. Polym. J.* 45 (2009) 494-502.

[2] C.H. Lin, S.J. Huang, P.J. Wang, H.T. Lin, S.A. Dai, Miscibility, microstructure, and thermal and dielectric properties of reactive blends of dicyanate ester and diamine-based benzoxazine, Macromolecules. 45 (2012) 7461-7466.

[3] González C, Llorca J. Mechanical behaviour of unidirectional fiber-reinforced polymers under transverse compression: microscopic mechanisms and modeling. Compos Sci Technol. 2007, 67(13): 2795–806.

[4] Gojny FH, Wichmann MH, Fiedler B, Bauhofer W, Schulte K. Influence of nanomodification on the mechanical and electrical properties of conventional fibrereinforced composites. Compos Part A[J]. 2005;36(11):1525–35.

[5] Bazhenov S L, Kuperman A M, Zelenskii E S, et al. Compression failure of unidirectional glass-fiber-reinforced plastics. Compos Sci Technol, 1992, 45: 201–208.

[6] Hamerton I. Chemistry and Technology of Cyanate Ester Resins. London: Blackie Academic & Professional, 1994: 59.

Fabrication and ablation resistance research of C/ZrC-SiC composites

Yi YU[1], Bai-hong JIANG[1], Wei LIU[1], Xin JIN[1], Bao-peng ZHANG[1], Jun-peng LIU[1]

1 Engineer, Advanced Materials & Processing Technology Beijing 100074, P. R. China
E-mail: ccinyuyi@163.com

Abstract. C/ZrC-SiC composites were prepared by precursor impregnation pyrolysis (PIP) process. The ablation resistance of the composites was studied. The residual mechanical property of the composites after ablation was compared and analyzed. The results show that the prepared C/ZrC-SiC composites have good ablation resistance under high temperature oxidation environment. The ablation amount in large area is less than 0.1 mm. The ablation products on the surface of the composites are mainly composed of SiO_2 and ZrO_2, forming a denser oxide film, which has a good protective effect on the structure of the composites. The data of residual mechanical properties indicate that the synergistic effect of high temperature and thermal stress under ablation environment may lead to the decrease of mechanical strength of composites.

1. Introduction

With the increase of flight speed, high reliability and reusability of aircraft, more and more requirements are put forward for materials. Therefore, it is urgent to carry out research on ultra-high temperature and ablation resistant materials. Ultra-high temperature ceramic matrix composites are a new kind of ultra-high temperature materials, which have important applications in the leading edge, wing rudder, engine combustion chamber, engine tail nozzle and other components of the new generation hypersonic vehicle [1-3].

C/ZrC-SiC composite is a widely used ultra-high temperature ceramic matrix composite [4-8], which has the advantages of low density, ablation resistance, oxidation resistance and good mechanical properties. The thermal structure parts of aircraft are exposed to high temperature and high speed gas, and they work under oxidative atmosphere. The phenomenon of oxidative ablation is very serious. In addition, the mechanical properties of materials are also very important during high speed flight. Therefore, it is very urgent to study the ablation mechanism of high temperature ceramic matrix composites and its residual mechanical properties after service for improving the performance of materials.

In this paper, C/ZrC-SiC composites were prepared using zirconium-silicon integrated ceramic precursor. The ablation resistance of C/ZrC-SiC composites was tested. The ablation mechanism and mechanical properties of C/ZrC-SiC composites under high temperature oxidation environment were studied, which laid a foundation for further improving the ablation resistance of C/ZrC-SiC composites in the future.

2. Experiment

2.1. Preparation of C/ZrC-SiC Composites

Carbon fiber preform was prepared by Jiangsu Tianniao Science and Technology Co., Ltd. The braiding method is puncture. The initial volume density of the preform is 0.5 g/cm3. Polycarbosilane (PCS) is produced by Suzhou Sailifei Company. Its softening point is 170-220 °C, and the yield of ceramics is 58%. Zirconium-silicon integrated ceramic precursor (ZS) is synthesized by Institute of Chemistry Chinese Academy of Sciences.

Carbon fiber preforms deposit pyrolytic carbon interfacial layer in the depositing furnace. When the density of the preforms reaches 1.30 g/cm3 or higher, and then, they are impregnated repeatedly with ZS precursor and Polycarbosilane solution until the density exceeds 2.10 g/cm3 When the weight increment is less than 1.0%. C/ZrC-SiC composites are prepared.

2.2. Ablation Performance Test of C/ZrC-SiC Composites

The surface of the sample was ablated by oxyacetylene flame. Considering the long-term oxidation resistance of C/ZrC-SiC composites, the ablation conditions were as follows: the diameter of the nozzle was 2 mm, the ablation angle was 90 degrees, and the distance from the muzzle of oxyacetylene to the center of the sample surface was 20 mm. Oxygen pressure is 0.4 MPa, acetylene pressure is 0.095 MPa, the ratio of oxygen flow rate to acetylene flow rate is 2:1, ablation time is 600 s, sample size is 30 mm *10 mm.

2.3. Analysis And Characterization

According to Archimedes principle, the volume density of C/ZrC-SiC composites was measured by drainage method.The bending strength of the composites was measured by three-point bending method using INSTRON 4505 electronic universal testing machine. The specimen was straight strip with a size of 55 mm *10 mm *4 mm and a loading rate of 0.5 mm min-1.

Microscopic morphology of the samples was observed by Hitachi S4800 scanning electron microscopy in Japan. Elemental composition (EDS) of the material surface and ablated surface was analyzed by SYSTEM SIX energy spectrometer. The shear strength of the specimens was measured by short beam shearing method. The size of the specimens was 18 mm x 6 mm x 43 mm.The oxyacetylene test was carried out at Dalian University of Technology in 2300K oxidation environment for 600 seconds.

3. Results And Discussions

3.1. Morphology analysis

Sampling and analysis of ablated composites are carried out as shown in Fig. 1. According to the internal surface morphology of the ablated material, it can be found that the ablation depth of the composite material is small. From the figure, it can be seen that under the action of airflow scouring, the ablated oxide layer has a large number of granular bulges. The thickness of sample area is 10 mm before test, and the average thickness of sample area after ablation (including oxide layer) is 10.48 mm, which may be due to the increase of macro-thickness of material due to the protrusion of oxide layer. In order to measure the actual ablation amount of sample area, the oxide layer is removed by grinding the oxide layer. After the white oxide layer is completely removed, the average thickness of sample area is measured again at 9.92 mm. Based on this, it is estimated that the ablation amount in large area is less than 0.1 mm.

After analysis in many places on the surface of the sampling area, the bare morphology of the fibers was not found, which indicated that the oxide layer was continuous and the protective effect of the material was good.

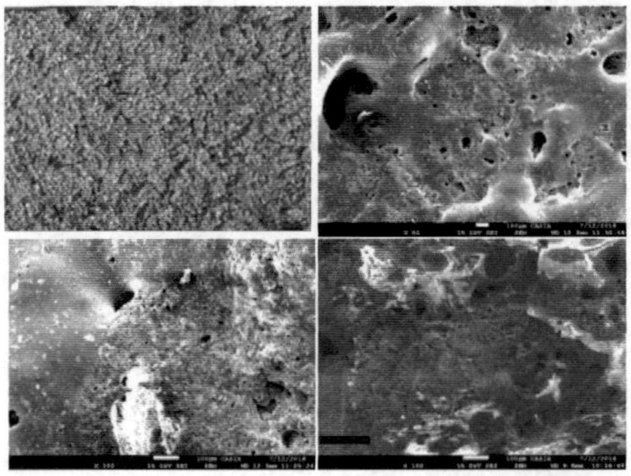

Figure 1 Microscopic Morphology Photographs at Different Positions in Sampling Area

Fig. 2 is a picture of element distribution of ablation products. It shows that the ablation products are mainly composed of SiO_2 and ZrO_2, and most of them are SiO_2. Figure 3 shows the micro-morphology of ablation products on the surface of the sample area. Table 1 shows the EDS analysis results of ablation products in Figure 3. The oxidation products are mainly composed of O, Si, Zr, Al, Mg and other elements. The atomic ratio of Si: Zr is about 9:1, (Si + Zr): O is 1:2, indicating that the oxidation products are mainly SiO_2 and ZrO_2; the oxide films on the surface are continuous, and no further oxidation reaction occurs inside the material.

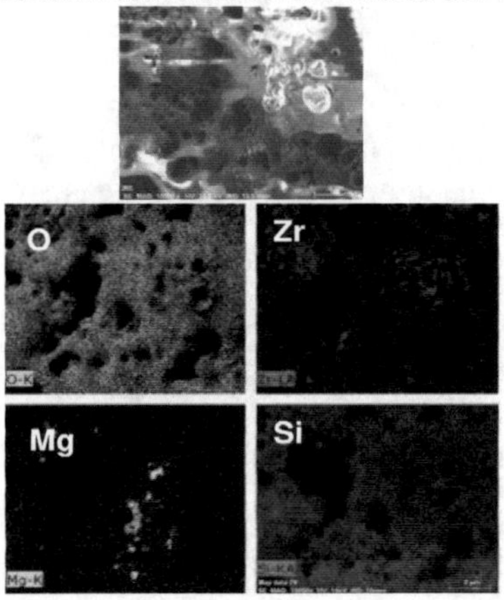

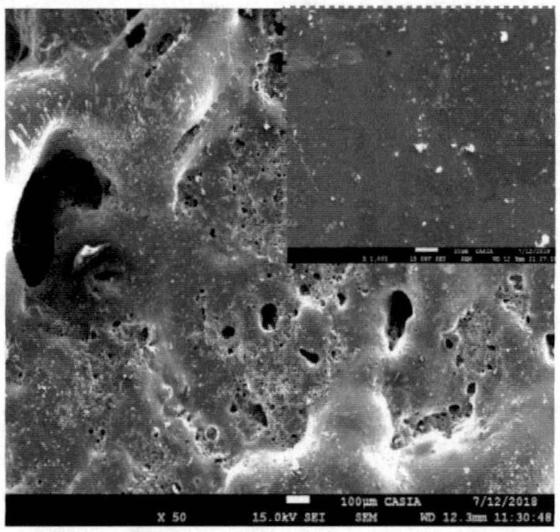

Figure 2 Elemental surface distribution of ablation products **Figure 3** SEM photograph of ablation product surface morphology

Table 1. Analysis of EDS elements in ablation products.

Element	Mass Ratio(%)	Atomic Ratio(%)
O	42.58	60.44
Si	33.46	27.06
Zr	13.25	3.30
Al	8.71	7.33
Mg	2.00	1.87

3.2. Mechanical Properties Analysis

In order to compare the mechanical properties of composites before and after ablation, the ablated materials were sampled and processed to bending strength and shear strength test splines. Because the thickness of ablated parts of the material was different, the actual sampling thickness was based on the residual thickness of the material. By comparing and analyzing the mechanical properties of the samples with the corresponding thickness of the material body, the failure behavior of the materials under the examination conditions was explored. From the data in Fig. 4, it can be seen that the residual shear strength of the debris is about 80% of the body. From the data in Fig. 5, it can be seen that the residual bending strength data of debris are discrete. On the one hand, it may be due to the harsh assessment environment, which results in the different oxidation and ablation conditions at different parts and the uneven residual thickness, resulting in the large deviation of the calculated mechanical properties data. On the other hand, it is possible that the thermal stress in high temperature environment will damage the material to a certain extent, resulting in a decline in mechanical properties. The bending residual strength of the test spline (4mm thick) far away from the ablation surface can reach 92%, which indicates that the sampling position has a great influence on the mechanical properties of the residual strength.

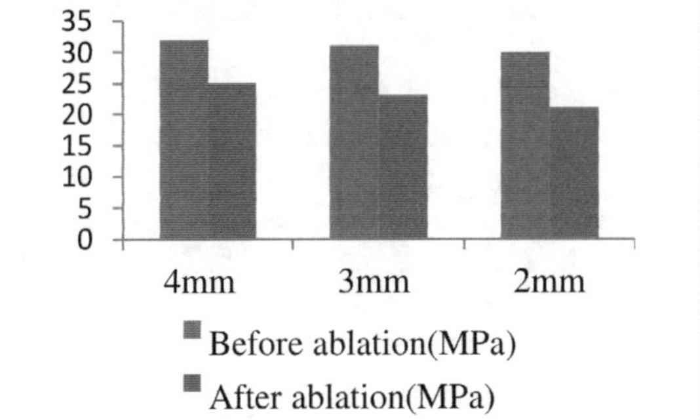

Figure 4. Residual shear strength of composites

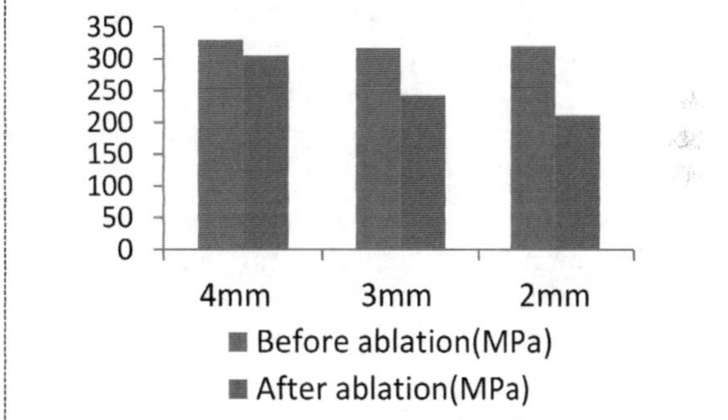

Figure 5. Residual flexural strength of composites

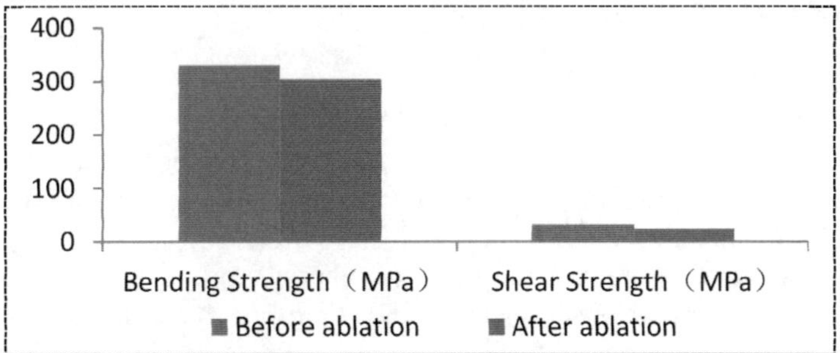

Figure 6. Residual mechanical properties data of composites

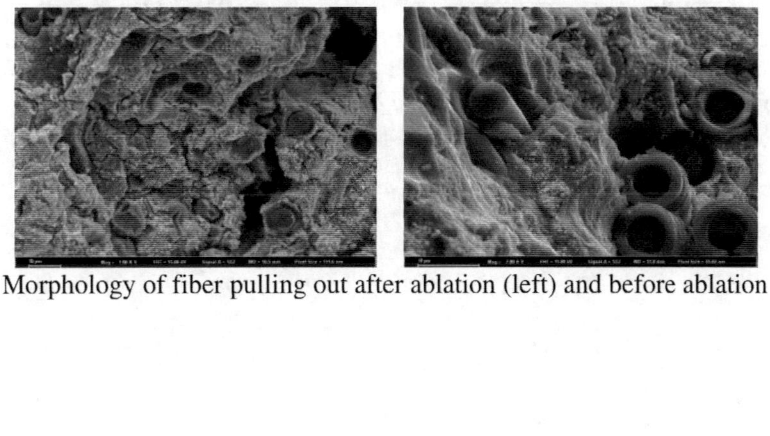

Figure.7 Morphology of fiber pulling out after ablation (left) and before ablation (right) (10μm)

In order to further eliminate the influence of residual ablation thickness on the test results of mechanical properties, the composites (ablation less than 0.1 mm) were sampled (all 4 mm thick), and the mechanical properties of the composites were compared with those of the bulk materials. The mechanical strength data of the body and the debris are shown in the figure 6. From the figure, it can be seen that the mechanical residual strength of the debris is maintained well.

To further explore the influence factors of mechanical properties of materials before and after ablation, the fracture morphology after fracture is analyzed by SEM. See Fig. 7. From the figure, it can be seen that the macro-morphology difference between the two is small, the fracture fiber pull-out morphology is slightly different, the fiber pull-out phenomenon at the fracture surface of the body is obvious, which shows that the matrix protects the fiber well and the fiber is not damaged. The brittle fracture of the fibers is the criterion, which indicates that the mechanical properties of the fibers may be affected by the destruction of the matrix at high temperature, resulting in the decrease of the residual mechanical strength of the materials.

4. Conclusions

(1) The C/ZrC-SiC composites prepared shows good ablation resistance and there were no pits and bare fibers on the surface of the composites.

(2) A denser oxide film mainly composed of SiO_2 and ZrO_2 is formed on the surface of the ablation products under high temperature oxidation environment has a good protective effect on the structure of the material.

(3) The data of residual mechanical properties of ablated materials indicate that the synergistic effect of high temperature and thermal stress under ablation environment may lead to the decrease of mechanical strength.

5. Reference

[1] Han Jiecai, Hu Ping, Zhang Xinghong, Meng Songhe. Advances in Ultra-high Temperature Materials [J]. *Solid Rocket Technology*, 2005, **28**:289-294.

[2] Zhang Qiuping. Research status of high temperature resistant non-metallic structural materials for hypersonic vehicles abroad [J]. *Aerial missiles*, 2011, 11:92-95.

[3] Yan Chunlei, Liu Rongjun, Cao Yingbin, Zhang Changrui and Zhang Deru. Progress in preparation technology of ultra-high temperature ceramic matrix composites [J]. *Aerospace materials technology*, 2012, **42**:7-11.

[4] Wang H., Gao B., Chen X., Wang J., Chen S., Gou Y. Synthesis and Pyrolysis of a Novel Preceramic Polymer PZMS from PMS to Fabricate High-Temperature-Resistant ZrC/SiC Ceramic Composite [J]. *Applied Organometric Chemistry*, 2013, **27**: 166-173.

[5] Grallert H., Vollmer K. Conceptual design of the thermal protection system for the reference concept SAENGER by means of advanced methods [D]. *Deutsche Aerospace AG, Munchen, Germany*, 1993.

[6] Zhang Litong, Chenglaifei. Sustainable Development Strategy of Continuous Fiber Toughened Ceramic Matrix Composites [J]. *Journal of Composite Materials*, 2007, **24**:1-6.

[7] Li Z., Li H., Zhang S., Li K. Microstructure and ablation behaviors of integer felt reinforced C/C-SiC-ZrC composites prepared by a two-step method [J]. *Ceramic International*, 2012, **38**:3419-3425.

[8] Wang Z., Dong S., Zhang X., Zhou H., Wu D., Zhou Q. Fabrication and properties of Cf/SiC-ZrC composites [J]. *Journal of the American Ceramic Society*, 2008, **91**: 3434-3436

Layers obtained on TiCN aluminum nanocomposites by electron-beam treatment

R Lazarova[1], S Valkov[2], V Dyakova[3], and P Petrov[4]

1 Prof. Dr., IMSETHAC – BAS, Sofia, Bulgaria
2 Chief Assistant Prof. Dr., IE – BAS, Sofia, Bulgaria
3 Chief Assistant Prof. Dr., V Dyakova, IMSETHAC – BAS, Sofia, Bulgaria
4 Prof. D.Sc., IE – BAS, Sofia, Bulgaria
E-mail: R.Lazarova@ims.bas.bg

Abstract. The impact of electron beam treatment on aluminum composites containing 0, 2, 5 and 10 wt. % TiCN nanoparticles was studied. Aluminum nanocomposites were produced as rods with 12 mm diameter by means of preliminary cold volume compression and succeeding hot pressing. Cylinders with a height of 10mm were cut from the rods and their flat surfaces treated by scanning electron beam techniques. The electron-beam treatment (EBT) process was carried out using Leybold Heraeus (EWS 300/15–60) electron beam equipment. Two technological regimes were used: with low energy power input Q_1 and with high energy power input Q_2. Composite layers with a strong bond to the substrate were obtained. Light microscopy (LM), scanning electron microscopy (SEM) and energy dispersive X-ray spectroscopy (EDX) were used for characterizing the microstructure. The phase composition of the obtained specimens was studied by means of X-ray diffraction (XRD) using CuKα characteristic radiation (λ=1.54 Å). The microhardness was discussed with respect to the applied technological conditions of the EBT and corresponding microstructure and crystallographic structure of the formed layers. Nanocomposite layers with improved hardness were obtained which is necessary as exploitation surface properties of automobile and aircraft parts. It was found out that the microhardness of samples treated with high energy was much lower than the one of the samples treated with low energy. The strengthening mechanism in the samples was discussed.

1. Introduction

Aluminum matrix nanocomposites are widely used in the modern automotive and aircraft construction because they have higher strength properties than aluminum alloys, maintaining good plasticity and toughness and lightweight [1-8]. Coatings and surface layers presenting aluminum nanocomposites are used as protective materials for different machine parts because of their high hardness, wear resistance and corrosion resistance [9-15]. Researchers are constantly striving to improve the exploitation properties of materials, so they are looking for new technologies to process them. In our previous works [16-19] we have found out that the aluminum nanocomposites improved their mechanical and tribological properties after electron beam treatment (EBT) with defined technological parameters with low energy and the microhardness of surface layers increases up to 21 times. These significant changes of the properties could be explained by the difficulty in the movement of the dislocations provoked by the hard TiCN nanoparticles. The dislocations form loops around the nanoparticles, the number of which increases as the applied strain increases [20]. This mechanism of strengthening known as Orowan mechanism is the main mechanism acting in the aluminum nanocomposites with TiCN nanoparticles up to 10 wt.%. When the nanoparticles weight part exceeds this limit, then begins a process of decreasing strength and softening the material.

The aim of the present study is to investigate the microstructure and microhardness of aluminum nanocomposites layers containing 0%, 2%, 5%, and 10 wt. % TiCN nanoparticles formed after a surface treatment by electron beam irradiation with low and high energy power input.

2. Material and methods

Aluminum nanocomposites containing 0, 2, 5 and 10 wt. % TiCN nanoparticles were produced as rods with 12 mm diameter by means of preliminary cold volume compression and succeeding hot pressing. Cylinders with a height of 10mm were cut from the rods. Their flat surfaces were treated by electron beam irradiation at following parameters: voltage (U) = 50 kV, speed of the specimen motion (V) = 50 mm/sec and scanning frequency (f) = 10 kHz. When the beam current I_b = 26 mA, the regime is low energy, designed as Q_1= 1300W. When the beam current I_b = 30 mA, the regime is high energy and designed as Q_2 =1500W. The experiments were carried out with Leybold Heraeus (EWS 300/15–60) electron beam equipment.

Metallographic samples were prepared by standard procedure: mounted in acrylic resin and wet ground up to 4000 grinding paper. Then the samples were electrolytic polished with the followed electrolyte: 575 ml methyl alcohol, 25 ml $HClO_4$ and 10 ml HNO_3. The metallographic observations were carried out by microscope PolyvarMet at magnifications up to 1000x.

The microhardness was determined by means of device MicroDuromat 4000 with a load of 20g, time for reaching the load - 10s and time for holding the load - 10s.

The investigations by SEM and EDX were carried out by means of SEM/FIB LYRAI XMU.

Plates, with a thickness of ~700 μm, were mechanically thinned by grinding (SiC paper) to a thickness of 20÷30 μm. From the as prepared foils were cut small discs of 3 mm in diameter, which were subsequently electrolytically polished at room temperature (Struers Tenupol Electropolisher, operating at 15V) in a solution of 20% $HClO_4$ and 80% CH_3COOH. These specimens were used for TEM analysis performed by TEM JEOL-JEM 1011 with 100 kV accelerating voltage.

The phase composition of the obtained samples was studied by X-ray diffraction (XRD). The experiments were carried out using CuKα characteristic radiation (1.54 Å). The measurements were conducted in the range from 5° to 90° at the 2θ scale with a step of 0.02° and counting time of 5 sec. per step.

3. Results and discussion

3.1. Temperature distribution during electron beam treatment

Figure 1 presents the results of the calculated temperature field distribution along with the depth of the electron beam treated area. The calculations are made by developing a heat model that is detailed in our previous work [17]. The numerical model is based on an analytical solution of the heat transfer equation using Green's functions.

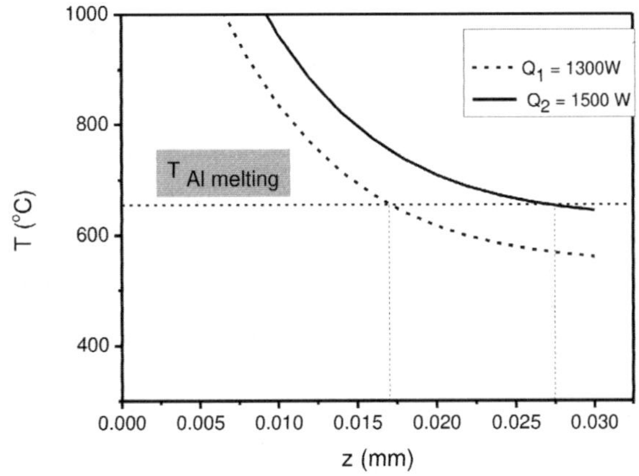

Figure 1. Temperature distributions along the z axis (depth of the treated zone) for different energy power input (V=50 mm/s, f=10 kHz)

The results presented in figure 1 show that increasing the electron beam power leads to an increase in the alloyed layer thickness, which is expected. The molten zone depth decreases from 27 to 17 μm for energy power input variation from 1500 to 1300 W.

3.2. Samples treated with Q_1 (low energy power input)

The microstructures of samples with 0 and 10 wt. % TiCN treated with low energy power input Q_1 are shown in figure 2. The samples with (0 ÷ 10) wt. % TiCN have a clearly formed layer. The layers are too convex, so good focusing of the parent metal is impossible with a metallographic microscope. Such protrusive layers are formed during grinding and polishing samples because the hardness of layers is much higher than the hardness of the parent metal. It could be seen that the layer's microhardness of the EBT aluminum sample is nearly two times higher than the microhardness of the parent metal. The tendency of microhardness alteration with TiCN content rising is shown in figure 3. The microhardness progressively increases with TiCN percentage and in the sample with 10 wt. % TiCN reaches a maximum of 1873HV after which it decreases. Analogic dependence of the strengthening from the weight fraction of TiCN nanoparticles contained in the parent metal is found out and explicated in [20].

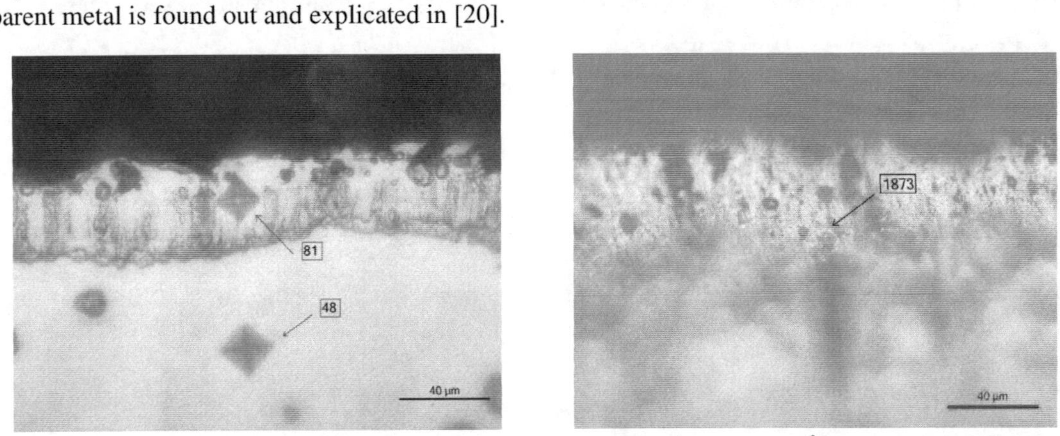

a b

Figure 2. Microstructures of samples treated with Q_1: (a) Al with 0 wt. % TiCN and (b) Al with 10 wt. % TiCN

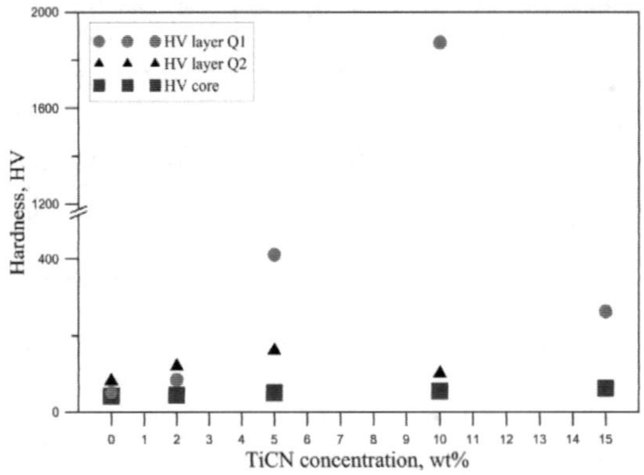

Figure 3. Microhardness alteration with TiCN wt% in investigated aluminum nanocomposites

The EDX analyses show that there is 8.23 wt.% Ti in the sample with 10 wt. % TiCN - figure 4, which is an acceptable result and the observed particles with SEM are precisely nanoparticles of TiCN - figure 5 (a), (b) and (c). The chemical spectrum unambiguously shows the sharp increase in titanium, carbon and nitrogen content in the particles – figure 5 (c). We have chosen to demonstrate larger particles in order to obtain clear results. TEM image – figure 5 (d) shows increased dislocation density around the nanoparticles, the formation of dislocation walls and substructure in the aluminum matrix. The TiCN nanoparticles prevent and stop the movement of dislocations resulting in the reinforcement of composites. TiCN particles can impede the dislocation motion through a variety of interaction mechanisms, some of which can operate simultaneously. The model of internal strain hardening distinguishes two main cases depending on the distribution of the obstacles: (a) the stress fields are closely spaced and the dislocation line has to curl between them, or (b) the fields are widely spaced and the dislocation line has to pass around them, subsequently leaving a dislocation loop, known as Orowan loop [21] around each of them. The strengthening effect of the reinforcement additive on the composite is a result of the formation of multiple dislocation Orowan loops around the nanoparticles, as argued by other researchers [22-24]. When the accumulated stresses around the particles exceed the strength limit of the matrix, the strength reduction begins. The accumulated stress is then relieved by the formation of nanosized cracks usually between the nanoparticles [20]. The strengthening effect depends on the volume fraction of the additive f and is limited by a critical value, f_{max}, beyond which the strengthening decreases. This standpoint explains very well the gradual increase of the microhardness values to the maximum at 10 wt. % TiCN, and the next decrease.

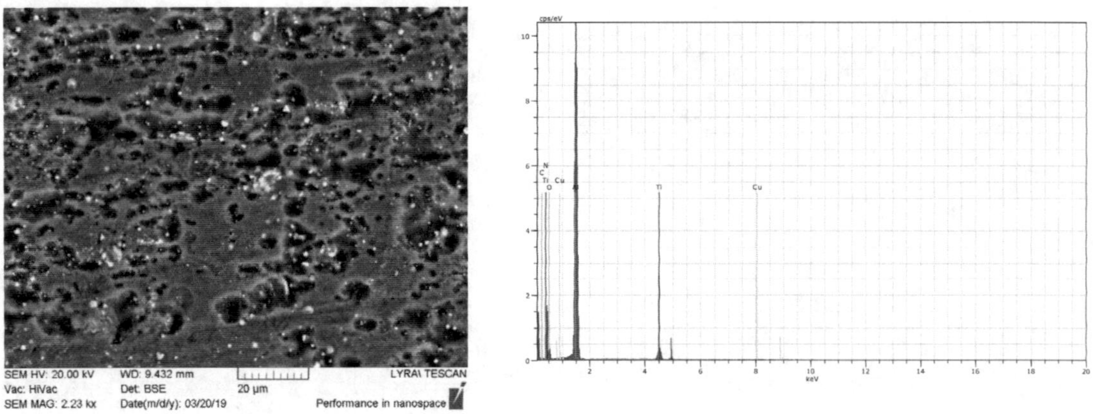

a

IOP Conf. Series: Materials Science and Engineering **733** (2020) 012017 doi:10.1088/1757-899X/733/1/012017

Spectrum: Acquisition 9244

El	AN	Series	unn. C [wt.%]	norm. C [wt.%]	Atom. C [at.%]	Error [%]
Al	13	K-series	90.49	68.59	58.98	4.4
O	8	K-series	25.09	19.02	27.58	40.8
C	6	K-series	4.06	3.08	5.94	1.0
N	7	K-series	3.42	2.60	4.30	6.0
Ti	22	K-series	8.23	6.24	3.02	0.3
Cu	29	K-series	0.62	0.47	0.17	0.1
		Total:	131.92	100.00	100.00	

b

Figure 4. An area in the sample with 10 wt.% TiCN nanoparticles – (a) and (b) EDX spectrum and content of chemical elements

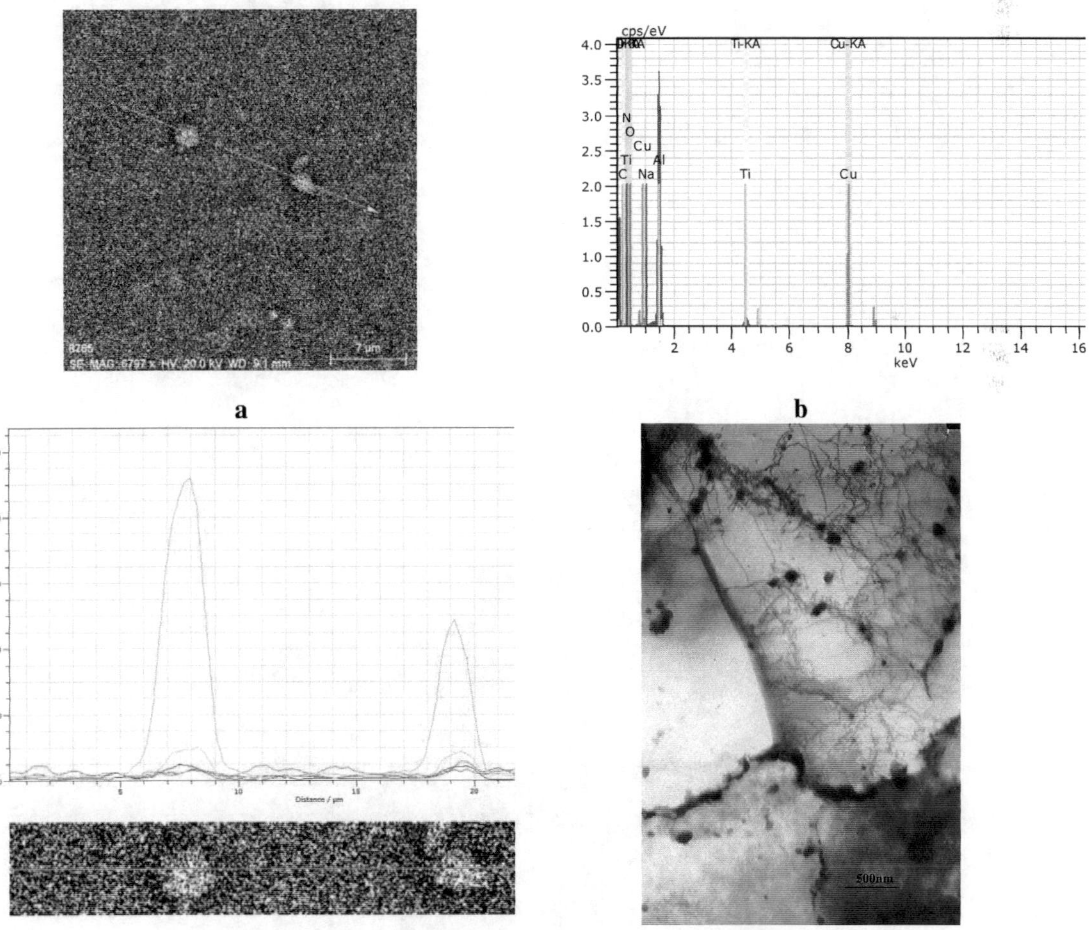

a b

c d

Figure 5. The area around two TiCN particles in the sample with 10wt. %: (a); EDX spectrum of the chemical elements: (b); linear elements distribution: (c) a TEM image of nanoparticles and dislocations around them: (d)

IOP Conf. Series: Materials Science and Engineering **733** (2020) 012017 doi:10.1088/1757-899X/733/1/012017

3.3. Samples treated with Q_2 (high energy power input).

The results from light microscopy investigation of samples treated with Q_2 are shown in figure 6. A layer with a thickness of 25 µm nearly is formed on the aluminum sample with 0 wt. % - figure 6 (a). The layer's microhardness is 78HV which is 2, 6 times higher than the microhardness of the parent metal. This means the EBT itself causes a hardening. The sample with 10 wt. % TiCN nanoparticles layer is with about 80 µm thickness and 102HV microhardness. The maximum micro-hardness after this type of treatment is achieved in the 5 wt. % TiCN sample. This means that the high energy displaces the maximum of microhardness to the lower values of the nanoparticles weight part.
Great particles are optically registered in the sample with 10 wt. % TiCN – figure 6 (c) and (d). The TiCN nanoparticles are in principles refractory but in this case, favorable conditions are probably created for TiCN interfacial reaction with some elements from the molten matrix. Figure 6 (c) and 6 (d) register spherical nanoparticle in the layer and nanoparticles having a core and a cover. We assume that the high temperature of the liquid bath is kept in some places in the sample for a relatively long time under the electron beam impact, during which TiCN reacts with some elements of the melting and nanoparticles or a part of them do not exist yet as reinforcement. This could explain the lower microhardness of the samples treated with high energy - Q_2.

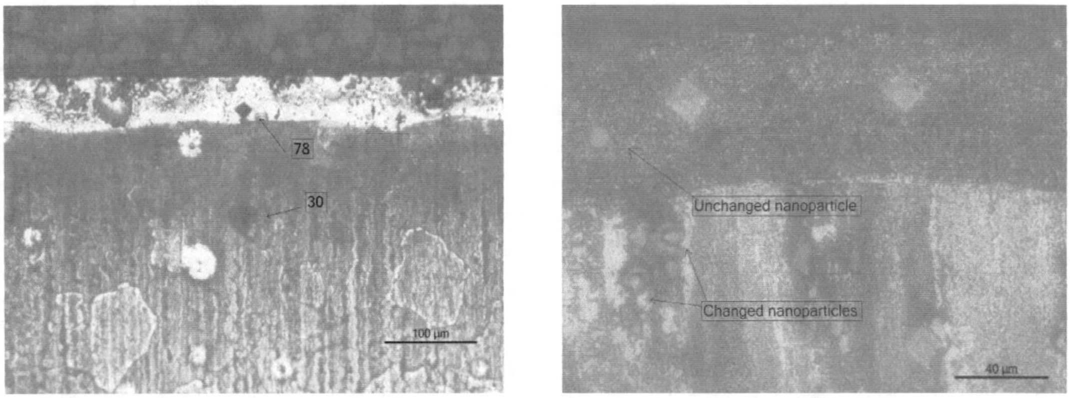

a b

Figure 6. Microstructures of samples treated with Q_2: (a) Al with 0 wt. % TiCN and (b) Al with 10 wt. % TiCN

Figure 6 represents a SEM image of the layer of the sample with 10wt. % TiCN. The layer shows dendritic-like microstructure at higher magnification.

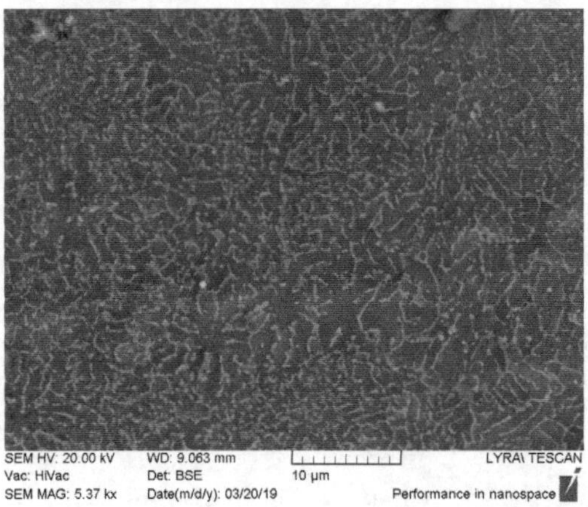

Figure 7. SEM image of the layer of the sample with 10 wt. % TiCN

The XRD patterns of the samples with 10 wt. % TiCN, electron beam treated with Q_1 and Q_2 are shown in figure 8 (a) and figure 8 (b), respectively. Both diffraction patterns exhibit peaks of pure Al and amorphous-like halos. In addition, at the sample treated with Q_1, peaks of TiCN phase, which means that some amount of the nanoparticles exist in the layer formed by the lower energy, contrary to the case of electron beam treatment with Q_2. This result is another proof that the high energy promotes the interaction between particles and some element s from the matrix and converts them almost entirely into an amorphous phase. The TiCN phase has body-centered cubic crystal structure where the carbon and nitrogen atoms occupied the octahedral sites of the closed-packed hexagonal Ti lattice forming TiC_xN_{1-x}. Therefore, the identification of the latter phase was done according to both TiN (ICDD PDF#38-1420) and TiC (ICDD PFD# 32-1383).

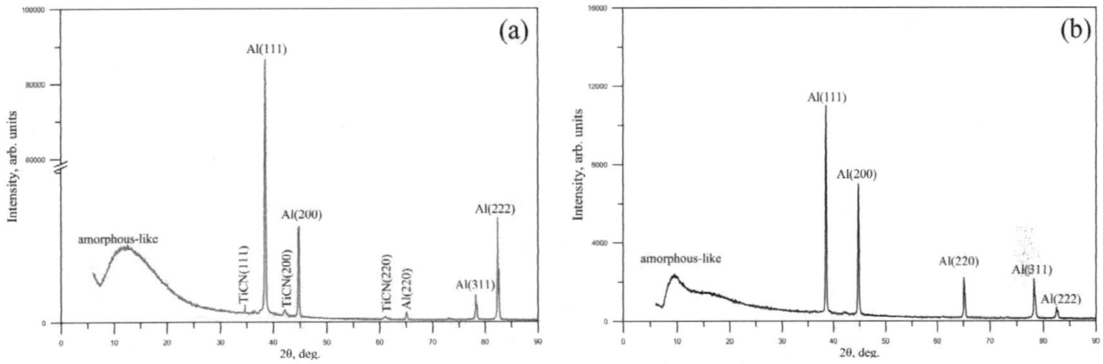

Fig. 8. XRD patterns of the samples with 10 wt. % TiCN and treated with: (a) Q_1 and (b) Q_2

4. Conclusions

- Layers obtained on aluminum nanocomposites containing 0%, 2%, 5%, and 10 wt. % TiCN nanoparticles were produced by electron beam treatment with low and high energy.
- The nanoparticles are retained in the layers obtained by low energy irradiation, whereas in the layers obtained with high energy the interaction between particles and some elements from the matrix occurs and they almost entirely convert into an amorphous phase.
- The microhardness alteration in the samples treated with low energy is expressed in progressively increasing with TiCN content and in the sample with 10 wt. % of TiCN reaches a maximum of 1873HV after which it decreases. The main strengthening mechanism in the aluminum matrix nanocomposites is the Orowan mechanism. This is the cause of higher microhardness in the samples treated with low energy where the nanoparticles remain unchanged in comparison with the samples treated with high energy where the nanoparticles are transformed into an amorphous phase.
- The microhardness alteration in the samples treated with high energy is expressed in progressively slight increasing with TiCN content and in the sample with 5% wt. TiCN reaches a maximum of 161HV after which it decreases. The slight increase in microhardness is explained in this case by the transformation of the hard TiCN nanoparticles into an amorphous phase.
- This investigation confirms the fact that the strengthening effect depends on the volume fraction of the additive f and is limited by a critical value, f_{max}, beyond which the strengthening decreases.

Acknowledgments
This work is financially supported by the Fund "Scientific Researches" at the Ministry of Education and Science, Contract DH07/16

References

[1] Ibrahim I A, Mohamed F A, Lavernia E J, 1991 Particulate reinforced metal matrix composites - a review, *Journal of Materials Science*, 26, pp 1137-56.

[2] Konstantin Borodianskiy, Alexey Kossenko, and Michael Zinigrad 2013 Improvement of the Mechanical Properties of Al-Si Alloys by TiC Nanoparticles *Metall and Mat Trans* A 44 pp 4948-53.

[3] Wang K, Jiang H Y, Wang Y X, Wang Q D, Ye B and Ding W J 2016 Microstructure and mechanical properties of hypoeutectic Al–Si composite reinforced with TiCN nanoparticles, *Materials and Design* 95 pp 545–554.

[4] Dasgupta R, 2012 Aluminium Alloy-Based Metal Matrix Composites: A potential Material for Wear Resistant Applications *ISRN Metallurgy* pp 1-14.

[5] Kale V, 2015 Aluminium Based Metal Matrix Composites for Aerospace Application: A Literature Review *IOSR Journal of Mechanical and Civil Engineering*, vol. **12**, Issue 6 Ver. V pp 31-36.

[6] Zheng Wu et al. 2003 “In-Situ Processing of Al Alloy Composites” *Advanced Engineering Materials* **5**, No. 3, , pp. 167-173.

[7] Das D et al. 2014 Fabrication and heat treatment of ceramic-reinforced aluminium matrix composites – a review *Int. J. Mech. Mater. Eng.* **9** (1) pp 1–15.

[8] Pan Ma et al. 2017 Effect of Al2O3 Nanoparticles as Reinforcement on the Tensile Behavior of Al-12Si *Composites Metals* **7** p 359.

[9] Murzakov M A, Chirikov S N and Markushov Y V 2016 Research on microstructure and wear resistance of coatings obtained by adding nanoparticles of refractory compounds in laser cladding *Journal of Physics: Conference Series* **747** p 12062.

[10] Saoula N Madaoui N Tadjine R Erasmus R M Shrivastava S and Comins J D 2016 Influence of substrate bias on the structure and properties of TiCN films deposited by radio frequency magnetron sputtering *Thin Solid Films* **616** pp 521–529.

[11] Caicedo JC Amaya C Yate L Gomez M E Zambrano G Alvarado-Rivera J Mu˜noz-Salda˜na J and Prieto P 2010 TiCN/TiNbCN multilayer coatings with enhanced mechanical properties *Applied Surface Science* **256** 5898–04.

[12] Beresnev V M Pogrebnyak A D Azarenkov N A Farenik G V and Kirik V I 2007 nanocrystal and Nanocomposit coatingsmicrostructure and properties *PSE* vol **5**, No1-2

[13] Ping Chuan Siow et al. 2013 Characterization of TiCN and TiCN/ZrN coatings for cutting tool application *Ceramics International* **39** 1293–98.

[14] Sehrt J and Kleszczynski S 2017 Nanoparticle Improved Metal Materials for Additive Manufacturing *Progress Add. Manuf* **2** 179–191.

[15] Ivanov Yu F et al. 2015 Electron-beam modification of coating - aluminum substrate systems *IOP Conf. Series: Materials Science and Engineering* **71** pp 1-6

[16] Lazarova R Dimitrova R Murdjeva Y Valkov St and Petrov P 2018 Layers obtained on aluminum by nanopowder deposition and subsequent electron beam scanning, Materials and *Manufacturing Processes* (Taylor&Francis) pp 1-5.

[17] Angelov V Ormanova M Kaisheva D Lazarova R Dimitrova R Petrov P 2019 Selective electron beam surface alloying of aluminum with TiCN nanoparticles *Nuclear Instruments and Methods in Physics Research B* **440** pp 88-94.

[18] Kolev M Dimitrova R Parshorov St Valkov St Lazarova R and Petrov P 2018 Surface modification of an aluminum alloy by electron beam introducing TiCN nanoparticles *Journal of Physics: Conf. Series* **992** (IOP Publishing) pp 1-5.

[19] Valkov S Bezdushnyi R Lazarova R Dimitrova R and Petrov P 2019 Surface modification of Al substrate with TiCN nanopowder by electron-beam treatment *AIP Conference Proceedings* 2075 160017.

[20] Anestiev L Lazarova R Petrov P Dyakova V and Stanev L On the strengthening and the strength reducing mechanisms at Aluminium Matrix Composites reinforced with TiCN nano-sized particulates *Metallurgical and Materials Transactions A* under review.
[21] Orowan E 1934 *Z. für Kristall.- Crystalline Materials* vol **89** No1 pp 327–343.
[22] Fisher J C Hart E W and Pry R H 1953 *Acta Metallurgica* vol **1** p 336.
[23] Hart E W 1972 *Acta Metallurgica* vol **20** p 275.
[24] Ashby M F 1970 *Philosophical Magazine* vol **21** No 170 pp 399–424.

IOP Conf. Series: Materials Science and Engineering **733** (2020) 012018 doi:10.1088/1757-899X/733/1/012018

A monolithic integration scheme for GaN-based power converter integrated circuit using fully-Schottky versatile HEMTs

Zeheng Wang[1,2], Yi Luo[3], Chao Chen[1], Zhanzhuang Chen[1], Xuewei Feng[4] and Yuanzhe Yao[1]

[1] School of Information and Software Engineering, University of Electronic Science and Technology of China, Chengdu 610054, China
[2] School of Electrical Engineering and Telecommunications, The University of New South Wales, Sydney, New South Wales 2052, Australia
[3] Department of Electrical Engineering, Columbia University, West 116th Street, New York, USA
[4] Department of Electrical and Computer Engineering, National University of Singapore, Kent Ridge119077, Singapore

E-mail: zenwang@outlook.com and yzyao@tsinghua.edu.cn

Abstract. A monolithic integration scheme for GaN-based power converter integrated circuit is proposed in this paper, wherein the scheme is based on a novel versatile HEMT architecture featuring full-Schottky contacts. By simply introducing only one type of device for building both field effect transistors and diodes, the power converter ICs could be realized in a compact way. In addition, this scheme avoids harmful high temperature process which is for forming ohmic contacts in some nitride semiconductors. In this sense, this scheme could significantly reduce the complexity of the integration, and therefore, promisingly enable higher device performance.

1. Introduction

Gallium Nitride (GaN) devices have been widely adopted in power applications such as wireless power transmission and charging station due to its superior material performances in transporting current [1–3]. The high electron mobility transistors (HEMTs) and heterojunction lateral diodes, which are fabricated on the stressed AlGaN/GaN heterojunction and hereby intrinsically owns high density two dimensional electron gas (2DEG), have the most promising potential among all kinds of GaN devices in 5G and post 5G applications that require higher capabilities of high-frequency response, power transmission and thermal dispersion [4,5].

To fulfill such requirement, many discrete device architectures such as composite buffer diode [6], Gama-shaped anode diode [7], field-coupled HEMT [8], HEMT with quantum well field plate [9–11], HEMT with patterned substrate [12], and HEMT with local charge modulation trench [13], are all aiming to enhance the breakdown voltage and cut-off frequency of the device without dramatically shrinking the power transmission capability. However, these discrete devices fail to realize the monolithic integration for further reducing the chip size, and therefore, the total cost.

Fortunately, besides few strategies that are directly integrating GaN devices on-chip, some novel concepts which are the functional integration of GaN devices has been merged recently [14]. The light emission diode integration-compatible regulator, for instance, is an attempt to integrate the power devices and other functional devices on one chip [14]. Moreover, some smart techniques such as field assembled HEMT has been proposed to realize the enhancement functionality with new mechanism and high merits, which paved the way to establish the novel integration platform containing all kinds of devices [15].

In this paper, a monolithic integration scheme for GaN-based power converter integrated circuit is proposed. The scheme is based on a novel field assembled HEMT (FA-HEMT) architecture featuring full-Schottky contacts, which is utilized to realize the power converter ICs by simply introducing only one type of device for building both field effect transistors and diodes. Additionally, this scheme avoids high temperature process for forming ohmic contacts in nitride semiconductors and will be harmful for nitride alloy containing Indium fraction. Therefore, this scheme could significantly reduce the complexity of the integration and promisingly enable higher device performance by choosing InAlN as the barrier material.

2. Method and mechanism of GaN-based versatile HEMT

The proposed GaN-based versatile FA-HEMT is generated on the AlGa(In)N/GaN heterojunction wafer, shown in Figure 1. The fabrication process flow could be as follows: (i) passivation; (ii) window opening; (iii) Schottky drain formation; (iv) Schottky source formation; (v) dielectric deposition; (vi) gate deposition. The source metal could be Titanium or Aluminum which have relatively intermediate high work-function, and the drain contact must be the metal or the alloy that owns relatively low work-function. The reason is that Ti/Al is used for the modulation of the reverse current to realize an enhancement-type function [15]. And, the drain contact, which is not required to have the performance of rectifier, should have the electron barrier as low as possible in term of energy band.

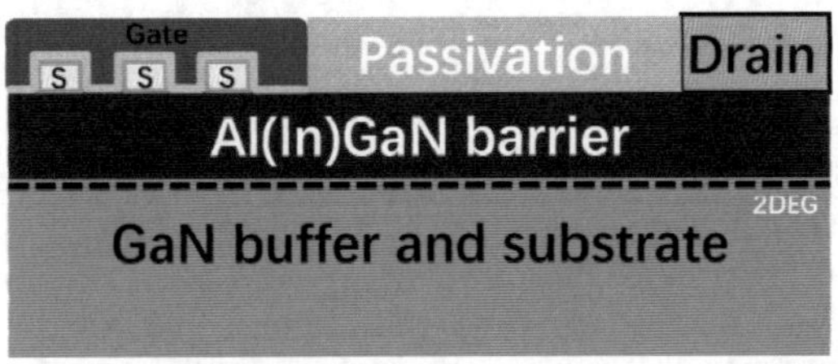

Figure 1. The schematic structure of the proposed GaN-based versatile FA-HEMT.

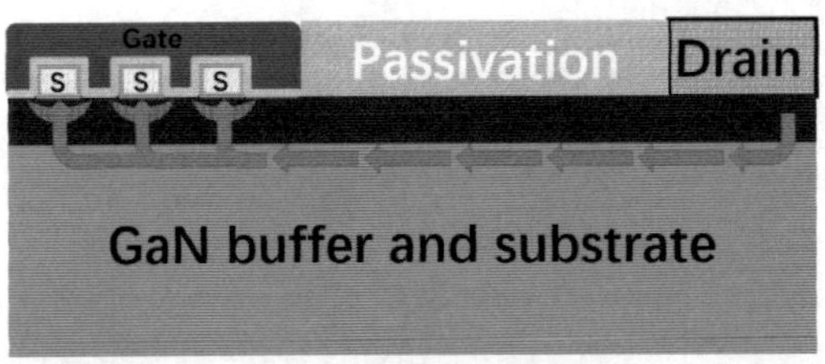

Figure 2. The field-assembled effect modulates the Schottky barrier of the source in the proposed HEMT. (The red arrow indicates the current flow and the purple shadow is the gate-modulated region)

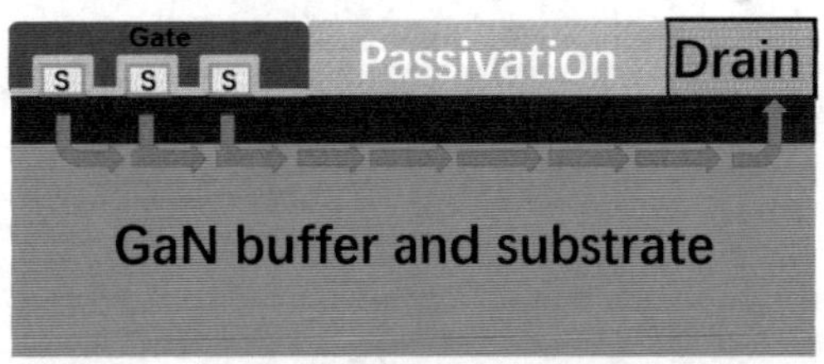

Figure 3. The reverse conduction of the proposed HEMT by using the device as a diode, the source being the anode. (The red arrow indicates the reverse current flow.)

In detail, the procedure of switching on and off could be illustrated in Figure 2 and 3, where the Figure 2 depicts the field-assembled effect modulating the Schottky barrier of the source, and the Figure 3 shows the reverse conduction of the proposed HEMT by using the device as a diode, the source being the anode. In this sense, when the HEMT is used as a transistor, the drain and the source should be put a high and low voltage respectively. The gate could modulate the barrier of the Schottky source by pulling down or building up the conduction band energy. When the HEMT is used as a diode, the voltage polarity between source and drain should be swap around, by which the Schottky source would act as a rectifier. However, due to the relatively low Schottky barrier of the source, the intrinsic reverse blocking is relatively low as well. While, the gate exhibits the feature of field plate, which would further enhance the breakdown voltage of the device used in diode mode.

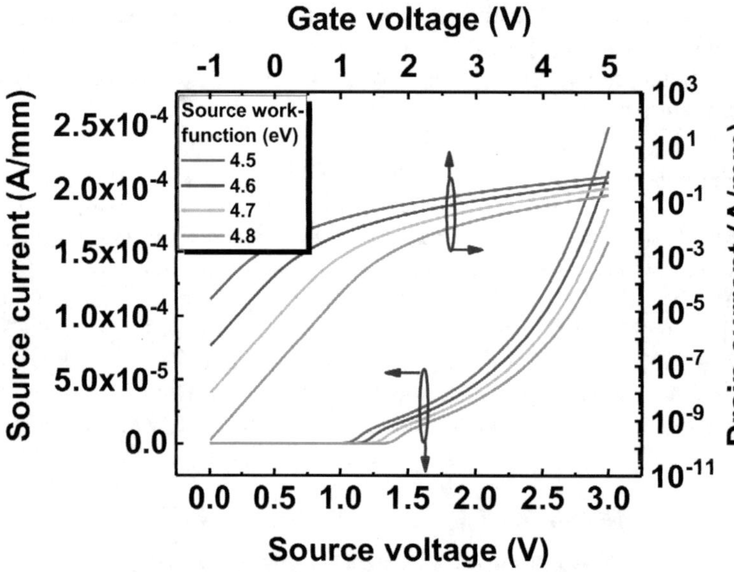

Figure 4. In this case simply justify the caption so that it is as the same width as the graphic.

The proposed device is investigated via TCAD simulation. The details of the configurations could be found in our previous work [15]. It should be noted that in this work the drain contact in this design is a Schottky contact with very low work-function. As shown in Figure 4, the proposed HEMT could exhibit a high-performance in switching mode. Besides, the device successfully features the reverse conductivity, making the device a promising versatile component.

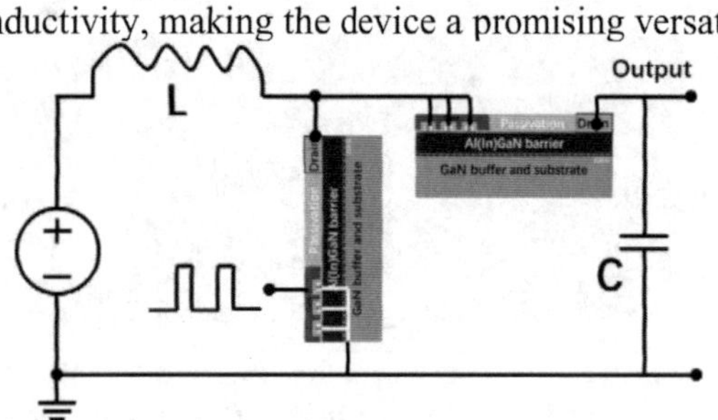

Figure 5. The monolithic integration scheme of boost converter.

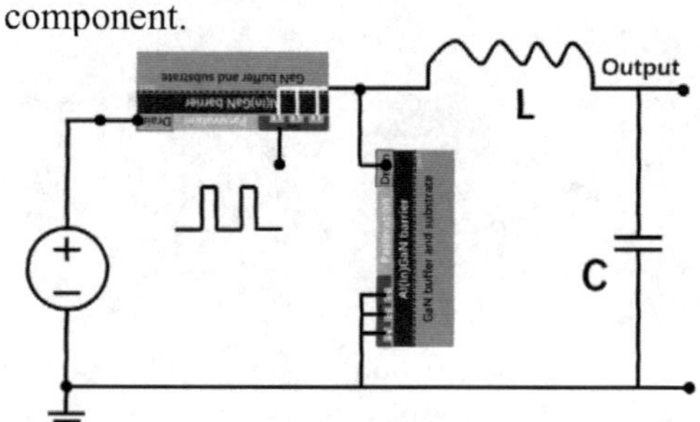

Figure 6. The monolithic integration scheme of buck converter.

3. Monolithic integration scheme

By the virtue of the versatile conductivity, the device could be monolithic integrated on a single chip, the scheme shown in Figure 5 and Figure 6. Figure 5 and Figure 6 show the scheme of monolithic integration of boost converter and buck converter, respectively. It is simple to connect the proposed HEMTs by one layer of metal interconnect, suggesting the promising potential of the HEMT and its integration scheme.

Conclusion

In this paper, we demonstrated a novel scheme for monolithic integration of GaN-based power converter ICs. The core of this design is the proposed new HEMT. This kind of HEMT features full-Schottky contacts and versatile functionality, enabling itself a potential in next generation GaN applications such as 5G communication and wireless power transmission.

Acknowledgement

This work was supported in part by Sichuan Science and Technology Program under Grant 2019YFH0006, and in part by Sichuan Incubation Project of Scientific and Technological Innovation under grant 2019032. All the authors except Yi Luo are associated under the Undergraduate Training Program for Innovation and Entrepreneurship (No. 201810614067) within University of Electronic Science and Technology of China.

References

[1] Sheppard S T, Doverspike K, Pribble W L, Allen S T, Palmour J W, Kehias L T and Jenkins T J 1999 High-Power Microwave GaN / AlGaN HEMT ' s on Semi-Insulating Silicon Carbide Substrates *IEEE Electron Device Lett.* **20** 161–3

[2] Mishra U K, Parikh P and Wu Y F 2002 AlGaN/GaN HEMTs - An overview of device operation and applications *Proc. IEEE* **90** 1022–31

[3] Oka T 2019 Recent development of vertical GaN power devices *Jpn. J. Appl. Phys.* **58** SB0805

[4] Jones E A, Wang F F and Costinett D 2016 Review of Commercial GaN Power Devices and GaN-Based Converter Design Challenges *IEEE J. Emerg. Sel. Top. Power Electron.* **4** 707–19

[5] Smorchkova I P, Elsass C R, Ibbetson J P, Vetury R, Heying B, Fini P, Haus E, DenBaars S P, Speck J S and Mishra U K 1999 Polarization-induced charge and electron mobility in AlGaN/GaN heterostructures grown by plasma-assisted molecular-beam epitaxy *J. Appl. Phys.* **86** 4520–6

[6] Wang Z, Wang F, Guo S and Wang Z 2017 Simulation study of high-reverse blocking AlGaN/GaN power rectifier with an integrated lateral composite buffer diode *Micro Nano Lett.* **12** 660–3

[7] Wang Z, Chen W, Wang F, Cao J, Sun R, Ren K, Luo Y, Guo S, Wang Z, Jin X, Yang L and Zhang B 2018 Simulation study on AlGaN/GaN diode with Γ-shaped anode for ultra-low turn-on voltage *Superlattices Microstruct.* **117** 330–5

[8] Wang Z, Cao J, Wang F, Chen W, Zhang B, Guo S and Yao Y 2018 Proposal of a novel enhancement type AlGaN/GaN HEMT using recess-free field coupled gate *Superlattices Microstruct.* **122** 343–8

[9] Wang Z, Feng X, Yang D, Chen C, Wang Z, Chen X, Wang C, Wang S, Cao J and Yao Y 2019 A Novel High Performance Lateral AlGaN/GaN Schottky Barrier Diode Using Highly Effective Field Plate with Polarization Enhanced Channel *2019 Electron Devices Technology and Manufacturing Conference (EDTM)* (IEEE) pp 377–9

[10] Wang Z, Yang D, Cao J, Wang F and Yao Y 2019 A novel technology for turn-on voltage reduction of high-performance lateral heterojunction diode with source-gate shorted anode *Superlattices Microstruct.* **125** 144–50

[11] Wang Z, Cao J, Sun R, Wang F and Yao Y 2018 Numerical investigation on AlGaN/GaN short channel HEMT with AlGaN/InGaN/AlGaN quantum well plate *Superlattices Microstruct.* **120** 753–8
[12] Wang Z, Wang Z, Zhang Z, Yang D and Yao Y 2019 On the Baliga's Figure-Of-Merits (BFOM) Enhancement of a Novel GaN Nano-Pillar Vertical Field Effect Transistor (FET) with 2DEG Channel and Patterned Substrate *Nanoscale Res. Lett.* **14** 128
[13] Wang Z, Zhang Z, Wang S, Chen C, Wang Z and Yao Y 2019 Design and Optimization on a Novel High-Performance Ultra-Thin Barrier AlGaN/GaN Power HEMT With Local Charge Compensation Trench *Appl. Sci.* **9** 3054
[14] Wang Z, Wang S, Zhang Z, Wang C, Yang D, Chen X, Wang Z, Cao J and Yao Y 2019 A High-Performance Tunable LED-Compatible Current Regulator Using an Integrated Voltage Nanosensor *IEEE Trans. Electron Devices* **66** 1917–23
[15] Wang Z 2019 Proposal of a novel recess-free enhancement-mode AlGaN/GaN HEMT with field-assembled structure: a simulation study *J. Comput. Electron. https://doi.org/10.1007/s10825-019-01383-7*

Construction of high sensitivity non-enzymatic glucose sensor based on three-dimensiona nickel foam supported $Ni_2P/NiO/CeO_2$ nanoflake arrays

Yongqiang Qin*, Yan Wang, Cuiping Yu, Jiewu Cui, Dongbo Yu and Yucheng Wu
School of Materials Science and Engineering, Hefei University of Technology, Hefei 230009, China
E-mail: albon@hfut.edu.cn

Abstract. Three-dimensional (3D) nickel foam supported $Ni_2P/NiO/CeO_2$ nanoflake arrays (NFAs) were synthesized by hydrothermal, anneal and low-temperature phosphatizing processes. As-prepared samples were systematically researched by SEM, XRD, XPS, TEM, EDS and HRTEM technology. Cyclic Voltammgrams and Chronoamperometry technology were used to measure the sensors performance of $Ni_2P/NiO/CeO_2$ NFAs. The results indicated that non-enzymatic glucose sensors based on $Ni_2P/NiO/CeO_2$ NFAs exhibited a high sensitivity of 28.23 mA cm^{-2} mM^{-1} and good anti-interference ability, suggesting that the 3D nickel foam supported $Ni_2P/NiO/CeO_2$ NFAs as a non-enzymatic glucose sensor is promising for practical applications.

1.Introduction

Glucose, as the most direct source of energy existed in blood for normal activities, is important for human[1]. However, too high or too low blood sugar will cause direct or indirect harm to the health of the human body. Especially high blood sugar might cause obesity, diabetes and so on. Hence, fast and accurate detection of glucose are of great importance for humans health[2]. Electrochemical glucose sensors have received much intensive attention with advantages simplicity, fast response, high sensitivity and suitability for real-time detection[3]. Conventional electrochemical glucose sensors can be divided into two types: enzyme glucose sensors and non-enzymatic glucose sensors[4]. Electrochemical enzyme glucose sensors use various bio-enzyme, such as glucose oxidase and glucose dehydrogenase, with advantages of high selectivity and a lower detection limit[5]. However, instability under extreme condition (higher temperature and very acidic/basic pH) and difficultly in preparation, purification in large quantities, complex enzyme immobilization procedures and high cost have limited their wide range application[6]. To overcome existed disadvantages, direct oxidation of glucose using non-enzymatic sensors has been explored in recent decades.

Nickel based compounds, such as oxides, hydroxides and phosphides, have received intensive research attention in the past decades based on their good electrochemical properties[7]. Especially for phosphides, with their good electrical conductivity and high activity, have widely been researched in recent years[8]. At the same time, CeO_2 is one of the promising candidates to build the glucose electrochemical sensors because of its environment friendly, biocompatibility and oxygen defect.

In this work, we have designed $Ni_2P/NiO/CeO_2$ NFAs grown on nickel foam by hydrothermal, anneal and phosphorization processes for nonenzymatic glucose detection. 3D nickel foam was chose as the substrate with their large surface area for growth of $Ni_2P/NiO/CeO_2$ NFAs. The results indicated

IOP Conf. Series: Materials Science and Engineering **733** (2020) 012019 doi:10.1088/1757-899X/733/1/012019

that the $Ni_2P/NiO/CeO_2$ NFAs presented a high sensitivity to glucose detection, suggesting the $Ni_2P/NiO/CeO_2$ NFAs were excellent substrate for nonenzymatic glucose detection.

2.Experiment

2.1 Materials

Synthesis of NiO/CeO_2 NFAs was carried out by the hydrothermal method[9]. 173 mg $Ce(NO_3)_3 \cdot 6H_2O$, 30 mg NH_4F and 121 mg $CO(NH)_2$ were dissolved in 40 ml Milli-Q water (18.25 $M\Omega \cdot cm^{-1}$) and transferred to 50 mL Telfon stainless steel autoclave. Nickel foam (about 2 cm×3 cm) was etched with 3 M HCl with ultra-sonication for 20 minutes to remove the oxide layer on the surface of nickel foam, and then rinsed with ethanol and Milli-Q water alternatively. The pre-treated nickel foam was immersed into the aforementioned mixture solution. Subsequently, the autoclave was transferred to a conventional oven at 100 °C for 12 h and the autoclave was cooled down to room temperature naturally. The samples were washed by Milli-Q water and ethanol alternatively and dried under 70 °C for 24 h. Finally, the precursors were heated in a tube furnace with a heating rate of 2 °C min^{-1}, maintained at 500 °C for 2 h in air and then cooled down naturally. Then the as-prepared NiO/CeO_2 NFAs in the combustion boat was placed at the downsteam and 400 mg NaH_2PO_2 in the combustion boat was placed at the upsteam of the tube furnace. The samples were heated at 300 °C for 2 h with a heating rate of 5 °C min^{-1} in Ar atmosphere, and then cooled down to ambient temperature naturally.

2.2 Characterization

The as-obtained samples were characterized by X-ray diffraction (XRD, D/MAX2500V, Rigaku, Japan), field emission scanning electron microscopy (FESEM, SU8020, Hitachi, Japan) and transmission electron microscopy (TEM, JEM-2100F, JEOL, Japan). The electrochemical measurements of as-obtained materials were tested by CHI760E electrochemical workstation (Chenhua, China) in a conventional three-electrode system in 0.1 M NaOH aqueous solution. A Ni foam supported active material (around 0.2 cm^2 in geometry area) was used as working electrode. Ag/AgCl (3 M KCl) and Pt/Ti electrode were employed as reference electrode and counter electrode, respectively. Cyclic Voltammetry (CV) was carried out in the quiescent solution and Chronoamperometry measurement was performed under continuous stirring.

3.Results and discussion

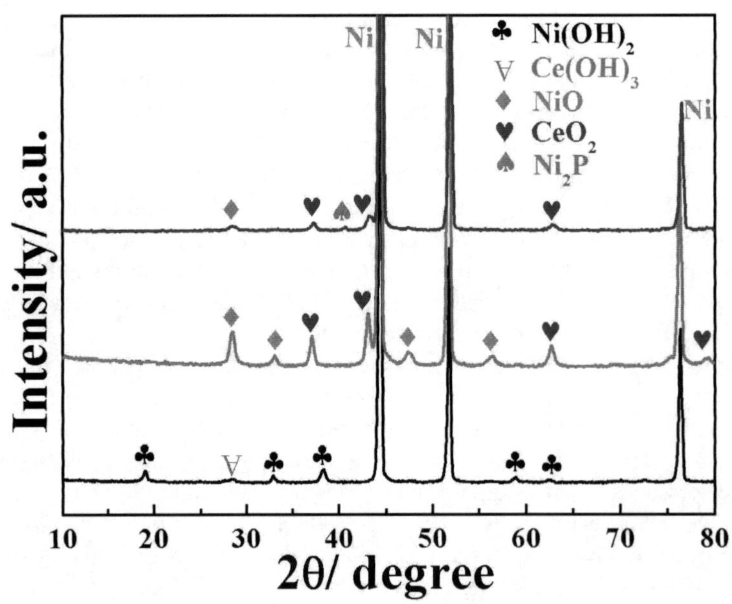

Figure 1. XRD patterns of the as-prepared samples on Ni foam.

The phase structure of as-prepared sample is characterize by XRD technique, as displayed in Figure 1. The black, red and blue curves correspond to as-prepared precursors, NiO/CeO_2 NFAs and $Ni_2P/NiO/CeO_2$ NFAs, respectively. The XRD pattern of precursors exhibit the diffication peaks of $Ni(OH)_2$ (JCPDS 14-0117)[10] and $Ce(OH)_3$ (JCPDS 19-0284)[9], and the diffication peaks of NiO/CeO_2 NFAs can be assigned to NiO (JCPDS 47-1049)[11] and CeO_2 (JCPDS 34-0394), which are corresponding to our previous study[9]. After phosphorization, the peaks of NiO weaken obviously compared to that of NiO in NiO/CeO_2 NFAs, at the same time, a new diffraction peak is presented around 40.8°, which is assigned to the (111) plane of Ni_2P (JCPDS 65-3544)[12], indicating the co-existence of NiP, NiO and CeO_2 in the sample.

The FESEM technology is used to observe the surface morphology of the samples. Figure 2 (a) and (b) display the FESEM images of as-obtained samples after phosphorization process with different magnifications. Apparently, the Ni foam is covered by $Ni_2P/NiO/CeO_2$ NFAs uniformly. It is found that the amount of pores are formed by an anneal process in air. To obtain more detailed information, the TEM and high resolution transmission electron microscopy (HRTEM) technology are adopt to measure the as-obtained samples, and the results are presented in Figure 2 (c) and (d). It can be observed that the diameter of individual nanoflake is around 300 nm. Figure 2. (d) presents the HRTEM image of nanoflake selected from the marked area in Figure 2(c), the calculated lattice spacing about 0.19 nm, 0.24 nm and 0.31 nm can be attributed the (210) plane of Ni_2P, (111) plane of NiO and (111) plane of CeO_2, respectively. The results of HRTEM is consistent with XRD pattern showed in Figure 1. And the composition distribution of the hybrid nanoflake is analysized by energy dispersive spectroscopy (EDS), as depicted in Figure 3. It is obvious that Ni, Ce, O and P elements are distributed in individual nanoflake uniformly.

To further investigate the sensor performance of as-obtained NFAs, the nickel foam supported $Ni_2P/NiO/CeO_2$ NFAs are utilized as working electrode directly. Figure 4 (a) depicts the CV curves of non-enzymatic glucose sensors based on $Ni_2P/NiO/CeO_2$ NFAs in 0.1 M NaOH with different glucose concentrations at the scan rate of 10 mV s^{-1} within the potential range of 0 to 0.8 V. As showed in Figure 4 (a), $Ni_2P/NiO/CeO_2$ NFAs shows a pair of redox peaks at about 0.55 and 0.30 V, which could be assigned to the interconversion of Ni^{2+} and Ni^{3+} [1]. The results of CV curves suggest that the current increases with the increasing of glucose concentrations. The working potential is further investigated

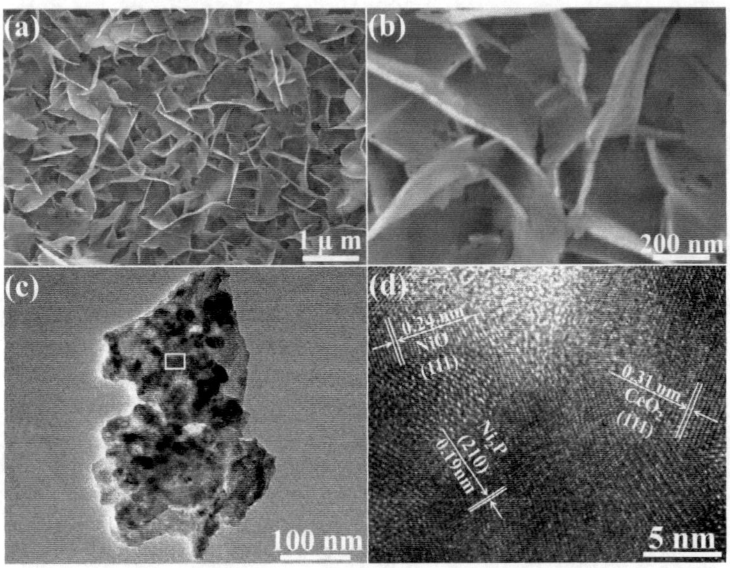

Fig. 2 (a) and (b) FESEM images of $Ni_2P/NiO/CeO_2$ NFAs on Ni foam at different magnifications. (c) TEM images of an individual $Ni_2P/NiO/CeO_2$ nanoflake. (d) HRTEM image of $Ni_2P/NiO/CeO_2$ from (c).

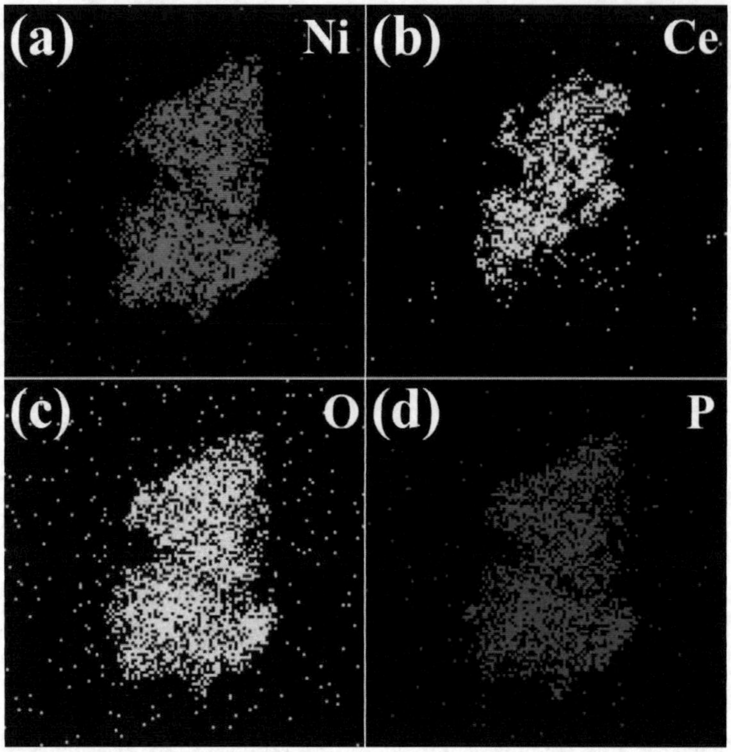

Figure 3. EDX mapping of $Ni_2P/NiO/CeO_2$ NFAs depicted in Fig. 2(c).

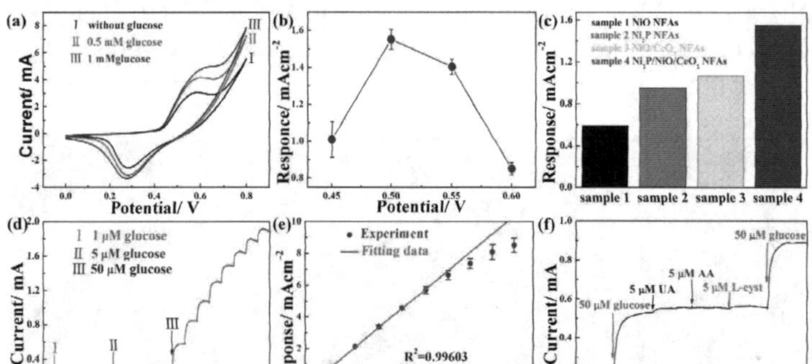

Figure 4. (a) Cyclic voltammograms of $Ni_2P/NiO/CeO_2$ NFA-based glucose sensors in 0.1 M NaOH with different glucose concentrations. (b) Working potential optimization. (c) Comparison of different materials. (d) I-t curve of $Ni_2P/NiO/CeO_2$ NFA-based glucose sensor with successive injection of different concentrations of glucose. (e) Calibration curve of the as-prepared glucose sensor. (f) I-t curve of as-prepared glucose sensor with successive injection of 50 μM glucose,5 μM UA,5 μM AA,5 μM L-cyst, and 50μM glucose.

by chronoamperometry technology, and the response of as-prepared sample at different working potentials are depicted in Figure 4 (b). Eventually, 0.5 V is chose as working potential and utilized in the following experiments. Furthermore, the catalytic performance of NiO NFAs, Ni_2P NFAs, NiO/CeO_2 NFAs and $Ni_2P/NiO/CeO_2$ NFAs are measured under the same conditions for comparison as shown in Figure 4 (c). The results indicate that $Ni_2P/NiO/CeO_2$ NFAs is the optimum material for detect of glucose. Figure 4 (d) presents the typical amperometric response of the glucose sensor based on $Ni_2P/NiO/CeO_2$ NFAs with successive injection of glucose with various concentrations. The calibration curve for our glucose sensor is presented in Fig. 4 (e), which gives a linear range from 1 to 250 μM (R^2=0.996). The detection limit and sensitivity are calculated to be 0.5 μM and 28.23 mA cm^{-2} mM^{-1}, respectively.

Selectivity is one of the major factors to evaluated the nonenzymatic glucose detection perfromance. Uric acid (UA), ascorbic acid (AA) and L-cysteine (L-cyst) are usually co-existence with glucose in human blood serum. Hence, UA, AA and L-cyst are used to check the anti-interference capability. However, the normal physiological level of glucose concentration (4.4-6.6 mM) is much higher than those of UA (~0.02 mM), AA (~0.1 mM) and L-cyst (~0.1 mM). Figure 4 (f) shows the amperometric responses of the as-prepared samples at 0.5 V in 0.1 M NaOH with successive inject of 50 μM glucose, 5 μM UA, 5 μM AA, 5 μM L-cyst and 50 μM glucose. Compared to the response of glucose, the response of UA, AA and L-cyst are negligible, suggesting good anti-interference capability of glucose sensors constructed by $Ni_2P/NiO/CeO_2$ NFAs.

Conclusion

In summary, 3D nickel foam supported $Ni_2P/NiO/CeO_2$ NFAs are successfully synthesized and used to fabricate the non-enzymatic electrochemical sensor for glucose detection. The results indicate that as-obtained electrodes exhibit high selectivity of 28.23 mA cm^{-2} mM^{-1} and good anti-interference ability. The $Ni_2P/NiO/CeO_2$ NFAs also shows excellent selectivity, reproducibility and long-tern stability. Hence, 3D nickel foam supported $Ni_2P/NiO/CeO_2$ NFAs as a non-enzymatic glucose sensor is promising for practical applications.

References

[1] Chen T, Liu D, Lu W, Wang K, Du G, Asiri A M and Sun X 2016 Analytical chemistry, 88 7885.
[2] Fang B, Gu A, Wang G, Wang W, Feng Y, Zhang C and Zhang X 2009 ACS applied materials & interfaces, 1 2829.
[3] Xie L, Asiri A M and Sun X 2017 Sensors and Actuators B: Chemical, 244 11.
[4] Wang G, Lu X, Zhai T, Ling Y, Wang H, Tong Y and Li Y 2012 Nanoscale, 4 3123.
[5] Kang X, Mai Z, Zou X, Cai P and Mo J 2007 Analytical biochemistry, 369 71.
[6] Wei H and Wang E 2013 Chemical Society reviews, 42 6060.
[7] Yang J, Cho M and Lee Y 2016 Sensors and Actuators B: Chemical, 222 674.
[8] Bai Y, Zhang H, Fang L, Liu L, Qiu H and Wang Y 2015 Journal of Materials Chemistry A, 3 5434.
[9] Cui J, Luo J, Peng B, Zhang X, Zhang Y, Wang Y, Qin Y, Zheng H, Shu X and Wu Y 2016 Nanoscale, 8 770.
[10] Hu X, Liu S, Li C, Huang J, Luv J, Xu P, Liu J and You X Z 2016, Nanoscale, 8 11797.
[11] Bao M, Chen Y, Li F, Ma J, Lv T, Tang Y, Chen L, Xu Z and Wang T 2014 Nanoscale, 6 4063.
[12] Wang A, Lin J, Xu H, Tong Y and Li G 2016 Journal of Materials Chemistry A, 4 16992.

Acknowledgments

This work is financially supported by National Natural Science Foundation of China (Nos. 51402081, 51502071 and 51372063).

Preparation of doped microstructure diffusing film by photocuring micro-imprint

Wang Jiajia[1], He Jianyun[1], Zhao Changsong[2], XiePengcheng[1], Tong Jinge[1]

[1]Beijing University of Chemical Technology,Beijing 100029, China

[2]The 11th Research Institute of China Electronics Technology Group Corporation, Beijing 100015, China

Email: 2805156949@qq.com

Abstract. In this paper, the process of preparing of doped diffusion film by ultraviolet curing micro-embossing was studied. The preparation method of microstructure diffusion film by photo-curing micro-embossing is different from the traditional methods. The effect of irradiation time on the apparent quality and the effect of the percentage of light diffusing agent on the optical properties and tensile properties of the diffusion film were investigated. The results shown that under the condition of 365 nm ultraviolet light wavelength and a light intensity of 800 mW/cm^2, the apparent quality of the diffusion film was good. When the irradiation time was 6 s; and the content of the diffusing agent was 30%, the optical performance was relatively good; the larger the content of the light diffusing agent, the smaller the tensile strength of the diffusing film.

1. Introduction

The light diffusing film is an optical film which can uniformly distribute the emitted light. The light-diffusing film is based on the principle of reflection, refraction and scattering between different dielectric materials when light passes through the light-diffusing film, thereby diffusing the light [1], eliminating glare, and make the light softer. With the advancement of science and technology, light diffusion films have been widely used in liquid crystal display technology, beam shaping [2-3], inorganic light emitting diodes [4-8] and photovoltaic devices [9-11]. Therefore, it has become the focus of research in the related fields.

At present, there are various methods for preparing microstructures on the surface of a diffusion membrane, mainly including chemical etching, laser engraving, micro-injection [12], and hot stamping[13] and so on. However, there are still some problems with these methods. Such as using the chemical etching method, the surface of the diffusion film is not smooth enough, and the uniformity is difficult to control for large-area preparation. Laser engraving has the disadvantages of low production efficiency and difficulty in mass production. The disadvantages of the micro-injection method is that the mold is complex and costly, and is difficult in manufacturing a large-area diffusion film. And the hot stamping process is complicated and requires heating and cooling during processing, which easily causes thermal deformation of the diffusion film, resulting in low molding precision of the product. Therefore, in this paper a UV-curing micro-imprinting method to prepare a doped diffusing film was proposed. The UV-curing micro-imprinting process uses the photosensitive material which is a liquid at room temperature and there is no need to heat and cool during the molding process, thus high molding efficiency, energy saving and environmental protection can be achieved. Therefore, the UV- curing micro-imprinting process provides a way to prepare a diffusing film with good light diffusibility, good tensile properties and good apparent quality.

In the experiment, the doped micro-structured light-diffusion film was studied. The light-diffusing agent exists on the surface or inside of the diffusion film. So the light can shift between the materials with different refractive indexes to achieve the diffusion effect and optical performance. The method of making diffusing film by UV-curing micro-imprinting has the advantages of fast forming, high efficiency, low cost, green environmental protection and has a broad market application prospect.

2. Experimental Preparation

2.1. Main Raw Materials

Ethyl ethoxyacrylate: EOEOEA, Changshu Hengrong Trading Co., Ltd.; Trimethylolpropane triacrylate: TMPTA, Changshu Hengrong Trading Co., Ltd.;Bifunctional urethane acrylate, LE-6702, Inspirational Chemical Co., Ltd.; Hexafunctional urethane acrylate, LE-6706, Inspirational Chemical Co., Ltd.; 2-hydroxy-2-methyl-1-phenyl-1-propanone: 1173, Changshu Hengrong Trading Co., Ltd.;3-leveling agent, BYK-333, Guangzhou Zhongwan New Material Co., Ltd.; 4-organosilicon light diffusing agent: Dongguan Kemai New Material Co., Ltd.

2.2. Main Equipment

Ultraviolet light source radiation system, homemade; Light curing micro-imprinting machine, homemade; UV light source, Shanghai Runzhu Electronic Technology Co., Ltd.; 10,000-point electronic scale, ME104E, Olympus Industrial Co., Ltd.; High-speed spin coating machine, KW-4A, Institute of Microelectronics, Chinese Academy of ciences; Laser confocal microscope, OLS5000, Olympus Industrial Co., Ltd.; Scanning electron microscope, HITACHI-S4700, Hitachi, Japan.

2.3. Preparation of diffusion membrane

2.3.1. Preparation of Photocurable Materials

First, urethane acrylate, ethoxyethyl acrylate (EOEOEA), tripropylene glycol diacrylate (TPGDA), photoinitiator (1173) and leveling agent (BYK-333) were thoroughly mixed under a certain ratio. Second, a certain amount of a silicone light diffusing agent was added in.Then uniformly mixed by a magnetic stirrer. At last the photocurable materials containing the light diffusing agent was obtained.

2.3.2. Experimental Steps

(1) The prepared photocurable materials is vacuumed and defoamed, and the bubbles are removed and taken out for use.

(2) uniformly coating the prepared photocurable resin on a PC substrate and imprinting it with a smooth PDMS mold.

(3) Set the illumination intensity and irradiation time of the ultraviolet light source, turn on the ultraviolet light source, irradiate the mold, and the coating under the mold will be solidified.
(4) After the coating is completely cured, the PDMS template is released from the surface of the substrate to obtain a photocured microembossed product.
(5) Microscopic analysis and characterization of the diffusion membranes obtained under different process conditions.

3.Results and Discussion

Fig. 1is the result of microscopic investigation of the cross section of the diffusion film by a scanning electron microscope. As it can be seen from Fig.1, the diffusion particles are uniformly distributed in the photocurable material.

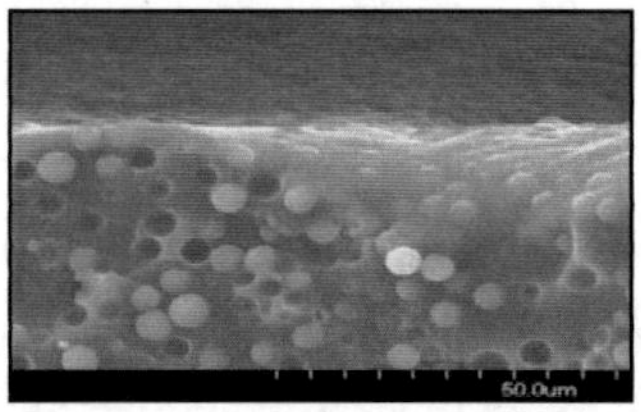

Figure 1. Dispersion of light diffusing agent

3.1. Influence of irradiation time on the apparent quality of the diffusion film

In this study, an ultraviolet light source with a wavelength of 365 nm and a light intensity of 800 mW/cm^2 was used to investigate the effect of irradiation time on the apparent quality of the diffusion film. The irradiation time was 2s, 3s, 4s, 5s, and 6s, respectively, a microscopic morphology of the surface of the diffusing film is shown in Fig.2.

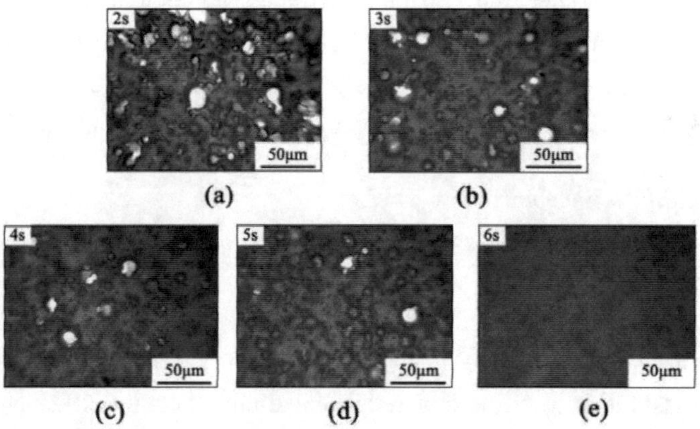

Figure 2. Micrograph of the light diffusing film at different irradiation times
(a)2s (b)3s (c)4s (d)5s (e)6s

It can be seen from Fig.2 that when the irradiation time was short, the rugged shape appeared, the curing of the photocurable material was incomplete. And with the increasing of irradiation time, the curing degree of the photocurable material was improved. When the irradiation time reached to 6s, the photocurable material was completely cured, the apparent quality of the diffusion film was good, and at the same time the molding efficiency was ensured.

3.2. The effect of the percentage of light diffusing agent on the optical properties

The percentage of light diffusing agent has a great influence on the optical properties of the diffusing film. In the experiment 5 μm particle size of the light diffusing agent was used, the adding amount of the diffusing agent was 10%, 20%, 30%, 40%, 50% , respectively. The optical properties of the diffusion film were evaluated by the light transmittance and haze, and the results were shown in Fig.3.

As shown in Fig.3, when the percentage of the diffusing agent added was 10%, the transmittance of the diffusing film was 82.53%, and the haze was 69.81%. With the increasing of the percentage of the diffusing agent added , the haze gradually increased and the light rate decreased. However, when the content of diffusing agent was too high, the light diffusing agent scattered and reflected the light multiple times, and the light energy loss was severe. When the percentage of the light diffusing agent increased to 40%, the haze of diffusion film reached to 90.49%, increased by 20.64%, and the optical performance was better.

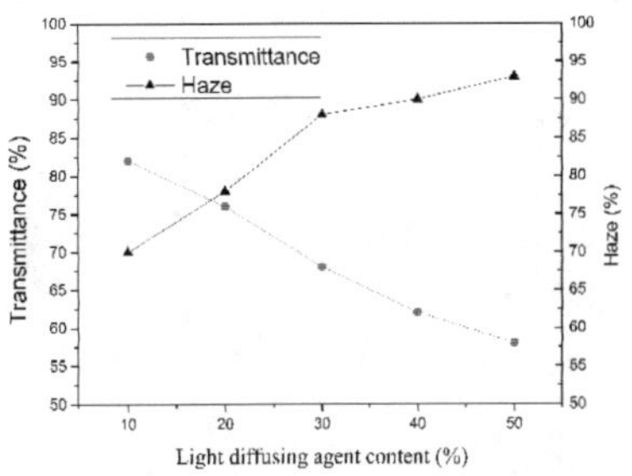

Figure 3 . Effect of the amount of light Diffusing agent on the optical properties of the diffusion film

3.3. The effect of percentage of light diffusing agent on mechanical properties

In practical applications, not only the optical properties of the diffusion film are required, the mechanical properties are also required. This section explored the effect of the percentage of light diffusing agent on mechanical properties. The effect on the tensile strength of the diffusion film was tested under different diffusion agent contents.The result was shown in Figure 4.

As shown in Fig.4, with the increase of diffuser content, the mechanical properties of the diffusing film gradually decreased. When the content of the light diffusing agent was 20%, the tensile strength was 18.8 MPa, and when the percentage of the light diffusing agent increased to 40%, the tensile strength was 14.8 MPa, which reduced by 4 MPa. Because with the increase of diffuser content, some of the particles may be mixed unevenly and resulted in agglomeration.

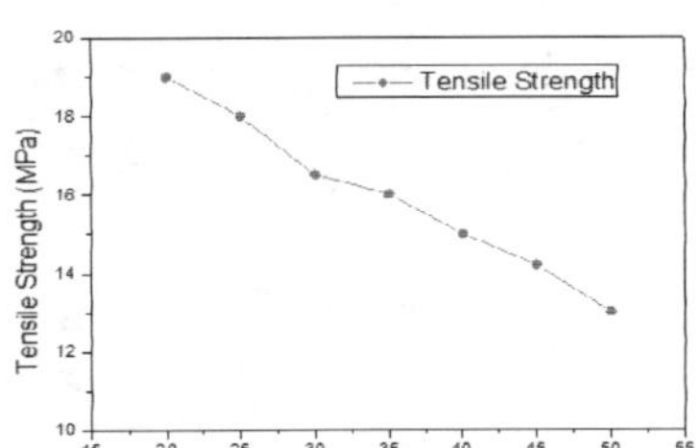

Figure 4 . Effect of diffusing agent content on tensile properties of diffusion film

4.Conclusions

In this paper, the effects of illumination time and diffuser content on the apparent quality, optical properties and mechanical properties of the diffusion film were studied.

(1) When the illumination time was short, the curing of the material is incomplete. With the increasing of illumination time, the curing degree of the photocurable material increased. When the illumination time was 6s, the photocurable material was completely cured and the apparent quality was good.

(2) As the percentage of diffusing agent increases, the haze gradually increases, and the light transmittance decreases. However, when the diffusing agent content is too high, the optical energy loss is severe. When the percentage of the light diffusing agent is 30%, Good optical and mechanical properties are obtained.

(3) With the increasing of diffuser content, the mechanical properties of the diffusing film gradually decreased.

References

[1] E ndo T. Method of designing a diffusion film, Process for Producing The Same and a diffusion film obtained thereby:, EP 1852719 B1[P]. 2012.

[2] H Hsiao Y N , Wu H P , Chen C H , et al. One-dimensional reflective diffuser for line beam shaper withmicrolens arra homogenizer[J]. Optical Review, 2014, 21(5):715-718.

[3] Voelkel R, Weible K J. Laser beam homogenizing: limitations and constraints[J]. Proc Spie, 2008, 7102(4):71020J-71020J-12.

[4] Suh M C , Pyo B , Kim H S . Suppression of the viewing angle dependence by introduction of nanoporous diffuser film on blue OLEDs with strong microcavity effect[J]. Organic lectronics, 2016, 28:31-38.

[5] Pyo B , Joo C W , Kim H S , et al. A nanoporous polymer film as a diffuser as well as a light extraction component for top emitting organic light emitting diode with strong microcavity structure[J]. Nanoscale, 2016, 8(16):8575-8582.

[6] Cho D H , Shin J W , Joo C W , et al. Light diffusing effects of nano and micro-structures on OLED with microcavity[J]. Optics Express, 2014, 22(S6):A1507.

[7] Chen P L , Peng W J , Chen H T , et al. The illumination uniformity study of diffuser plates for reflective LED desk lighting[C]// 2014 9th IEEE International Conference on Nano/Micro Engineered and Molecular Systems (NEMS). IEEE, 2014.

[8] Lee X H , Wu C S , Lee K H , et al. An optical-adjustable illumination pattern with a surface-structured diffuser compensated by index-matching liquid[J]. Optics & Laser Technology, 2013, 49:153-155.

[9] Fang Z , Zhu H , Yuan Y , et al. Novel nanostructured paper with ultrahigh transparency and ultrahigh haze for solar cells.[J]. Nano Letters, 2014, 14(2):765.

[10] Mahpeykar S M , Xiong Q , Wang X . Resonance-induced absorption enhancement in colloidal quantum dot solar cells using nanostructured electrodes[J]. Optics Express, 2014, 22(S6):A1576

[11] Brongersma M L , Cui Y , Fan S . Light management for photovoltaics using high-index nanostructures[J]. Nature Materials, 2014, 13(5):451-460.

[12] Mcormick R. M., Nelson R. J., Alonso M. Cz, et al, Micro-channel electro-phoretic separations of DNA in injection molded plastic substrates [J], Analytical Chemistry, 2007, 69: 2626-2630.

[13] Chien R D . Hot embossing of microfluidic platform[J]. International Communications in Heat & Mass Transfer, 2006, 33(5):645-653.

IOP Conf. Series: Materials Science and Engineering **733** (2020) 012021 doi:10.1088/1757-899X/733/1/012021

Study on Preparation of Microlens Array Hydrophobic Flim by Ultraviolet Curing Micro-imprint

Wang Jiajia[1], He Jianyun[1]*, Shen Zengqiang[1], Ma Xu[1], Yang Weimin[1]*

[1]Beijing University of Chemical Technology, Beijing 100029, China
Email: 2805156949@qq.com

Abstract. In this paper, the formation process of microstructure array hydrophobic film prepared by ultraviolet curing imprinting was studied. The UV-curing micro-imprinting method was used to replace the traditional hot stamping method to prepare the micro-structured hydrophobic film. The effects of illumination time and intensity on the surface hydrophobicity were investigated.Results shown that the micro-structure hydrophobic film prepared by UV-curing micro-imprint has high replication precision and good hydrophobic effect. With the increase of illumination time, the water contact angle of the product increased and reached to 125° at 30s illumination time. With the increase of the illumination intensity, the water contact angle of the micro-imprinted product was also gradually increased.

1. Introduction

With the development of science and technology, the living standards of Chinese people are improved continuously. Their demand for functional materials continues to increase. Hydrophobic film has excellent properties such as hydrophobicity[1], self-cleaning property[2], ice-suppressing property[3], drag reducing property[4] and corrosion resistance[5] and so on. It makes it widely used in People's daily life. For example, it can be applied to the surface of building glass to achieve self-cleaning effect, save water resources and avoid the danger of workers working at heights. It can be used in aircraft and outdoor buildings to provide anti-icing function. It can be applied to metal surfaces to make it corrosion resistant. It can also be used in medical protective fabrics to reduce bacterial transmission and so on. Therefore, hydrophobic materials have a very broad application prospect in the fields of construction, daily life, and aerospace and so on[6].

Surface microstructure has an important influence on the hydrophobicity of the product. There are various ways to prepare microstructure on the surface of the product, mainly includes template[7], electrospinning[8], spraying[9], etching[10], self-assembly[11] and so on. However, there are some short-comings with the above preparation methods. For example, equipment and materials are expensive and the production cycle is long, and so on.

IOP Conf. Series: Materials Science and Engineering **733** (2020) 012021 doi:10.1088/1757-899X/733/1/012021

In this study, a hydrophobic film was prepared by ultraviolet curing micro-imprinting. The UV- curing micro-imprinting process is a novel method with simple equipment and easy operation. The micro-structure is obtained by rapid imprinting and UV-cured under normal temperature and pressure. It has the advantages of rapid molding, low cost, environmental protection, high molding precision, and has a good application prospect.

2. Experimental Preparation

2.1. Main Raw Materials

Silicone resin, YL-1, Laiyang Yulin Chemical Co., Ltd.; Polyurethane acrylate, prepolymer, Jiangmen City Inspiration Co., Ltd.; Ethoxyethyl acrylate, monomer, Taiwan Changxing Chemical Industry Co., Ltd.; 2-hydroxy-2-methyl-1-phenyl-1-propanone, 1173, Taiwan Changxing Chemical Industry Co., Ltd.; BYK-333, leveling agent, Guangzhou Zhongwan New Material Co., Ltd.; Polydimethylsiloxane, PDMS, Dow Corning, USA.

2.2. Main Equipment

Light curing micro-imprinting machine, homemade; Ultraviolet light source radiation system, homemade; UV light source, Shanghai Runzhu Electronic Technology Co., Ltd.; Electronic balance, FR124CN, Ohaus, USA; Vacuum drying oven, ZK-35ASB, Beijing Kewei Yongxing Instrument Co., Ltd.; Contact angle measuring instrument, JC2000D, Beijing Zhongyi Kexin Technology Co., Ltd.; High-speed spin coating machine, KW-4A, Institute of Microelectronics, Chinese Academy of Sciences; Laser confocal microscope, OLS5000, Olympus Industrial Co., Ltd.; Scanning electron microscope, HITACHI-S4700, Japan Electronics Co., Ltd..

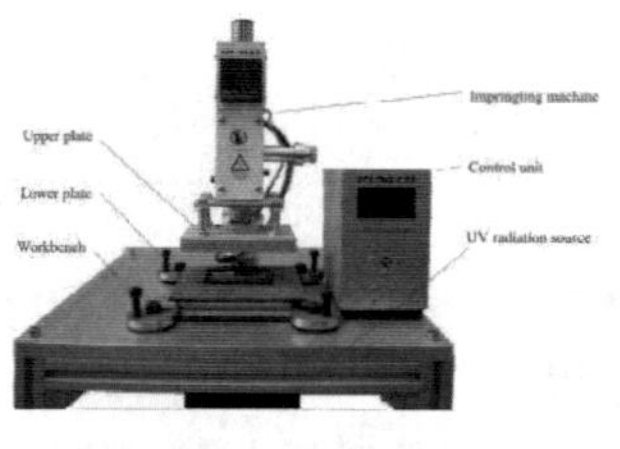

Figure1. Homemade light curing micro-imprinting equipment

2.3. Experimental Process

2.3.1. Preparation of Photocurable Materials

The urethane acrylate, ethoxyethyl acrylate, 2-hydroxy-2-methyl-1-phenyl-1-propanone, and a leveling agent were thoroughly mixed under a certain ratio, then defoamed under vacuum for one hour.

2.3.2. Experimental Steps

(1) UV-curable material was evenly coated on the PC chip, then the PDMS mold was pressed onto the UV curable material.

(2) The illumination intensity and time of the ultraviolet light source were set, the above PC chip with UV-curable material was irradiated.

(3) After the UV-curable material was completely cured, the PDMS soft template was released from the surface of the substrate to obtain a micro-structured film product.

(4) The properties of the films prepared under different conditions were analyzed separately, and the water contact angle of the micro-imprinted products was measured by a contact angle measuring instrument.

3.Results and Discussion

The Surface morphology of PDMS mold observed by laser confocal microscope and SEM was shown in Fig.2.

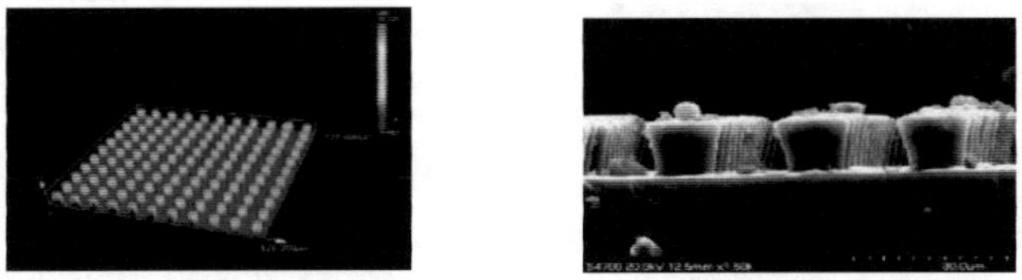

(a) three-dimensional topography (b)SEM

Figure2. Surface morphology of PDMS mold

3.1. Effect of illumination time on surface hydrophobic properties

At 0.2Mpa imprinting pressure and 1000mW/cm^2 UV light intensity, and the UV irradiation time 15s, 20s, 25s, 30s and 35s, respctively, the corresponding water contact angle on the products with the micro array structure prepared in this research was shown in Fig.3.

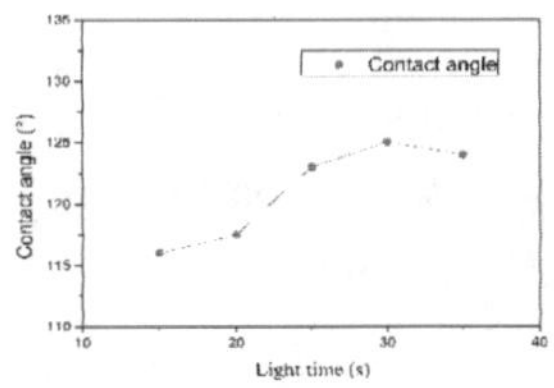

Figure3. The effect of different illumination time on the water contact angle of micro-imprinted film

From Fig.3the change of the water contact angle of the surface with array micro-structure under different illumination times can be obtained. It shown that with the illumination time increasing, the water contact angle increased, at 30 s illumination time, the water contact angle reached to a maximum of 125°. The water contact angle images of the product measured by the water contact angle measuring instrument were shown in Fig.4.

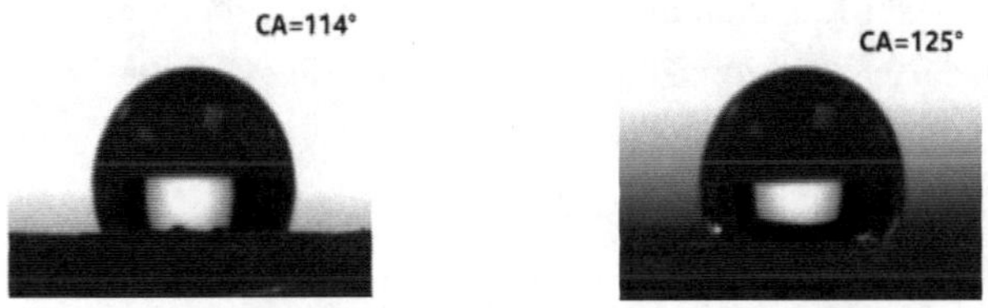

(a)15s (b)30s

Figure4. The contact angle of film products with different illumination time

IOP Conf. Series: Materials Science and Engineering **733** (2020) 012021 doi:10.1088/1757-899X/733/1/012021

3.2. Effect of light intensity on surface hydrophobic properties

The imprinting pressure was set to 0.2Mpa, the illumination time was 30s, and the maximum ultraviolet intensity was 1000mW/cm^2. The UV intensity was set to 60%, 70%, 80%, 90%, 100% of the maximum light intensity, respectively, during the imprinting processing. The change of the water contact angle of the films with micro array structure was shown in Fig5.

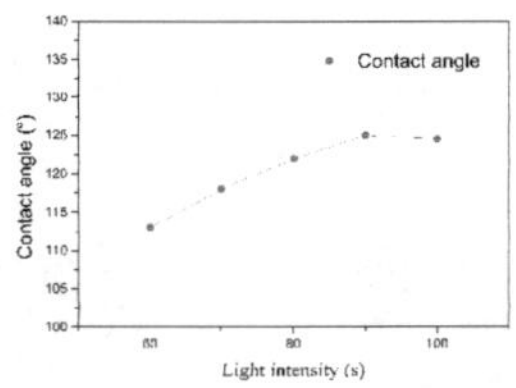

Figure5. Effect of UV light intensity on the water contact angle of micro-imprinted film

It can be seen from Fig.5 that as the UV intensity increases, the water contact angle of the film with microstructure increases. Because the microstructure on the film needs a certain amount of irradiation energy to be fully cured. If the microstructure on the article is not fully cured, the replication accuracy is low, and resulting in the low water contact angle. As the light intensity increases, the microstructure gradually completely cured, and the replication accuracy gradually improved. Therefore, the contact angle of the product increase as UV light intensity increases.

4.Conclusions

In this paper, the microstructure array hydrophobic films were prepared by UV curing micro imprinting. The effects of illumination time and light intensity on the hydrophobic properties of the film surface were investigated. Conclusions were as the follows:

(1) The surface water contact angle of the product increases with the increase of illumination time. The optimal irradiation time under the conditions in this research was 30s;

(2) Irradiation intensity affects the rate of UV-curing reaction. As the light intensity increases, the water contact angle of the film increases. At the condition of 0.2Mpa imprinting pressure, 30s the illumination time, 1000mW/cm^2 the light intensity, the water contact angle of the film surface was 125°,which achieved a hydrophobic effect.

References

[1] Ma M,Hill R M.Superhydrophobic surfaces[J].Current Opinion in Colloid &Interface Science, 2006,11(4):193-202

[2] Motlagh N V,Sargolzaei J,Shahtahmassebi N.Super-liquid-repellent coating on the carbon steel surface[J]. Surface and Coating Technology,2013,235:241-249

[3] Wang F,Li C,Lv Y,etal.Ice accretion on superhydrophobic aluminum surfaces under lowtemperature conditions[J]. Cold Regions Science and Technology, 2010,62(1):29-33

[4] Kraj A G,Bibeau E L.Phases of icing on wind turbine blades characterized by ice accumulation[J].Renewable Energy,2010,35(5):966-972

[5] Wang J,Li D,Gao R,etal.Construction of Superhydrophobic hydromagnesite films on the Mg alloy[J].Materials Chemistry and Physics,2011,129(1):154-160

[6] Wang S,Liu K,Yao X, etal.Bioinspired Superfaces with Superwettability:New Insight on Theory ,Design,and Applications[J].Chemical Review,2015,115(16):8230-8293
[7] Zheng Jianyong, Feng Jie, Zhong Mingqiang. Preparation of polymer super-hydrophilic/superhydrophobic surface by CaCO3 particle template method [J]. Journal of Polymer Science, 2010, 1(10):1186-1192.
[8] Li Fang,Li Qiming. Research Progress in Superhydrophobic Materials Based on Micro / Nano Fibers Produced by Electrospinning [J]. Journal of Liaoning University of Petroleum and Chemical Technology, 2018, v.38;No.140(04):3-11.
[9] Deng T , Varanasi K K , Hsu M , et al. Nonwetting of impinging droplets on textured surface s[J]. Applied Physics Letters, 2009, 94(13):481.
[10] Liu Y , Li X , Jin J , et al. Anti-icing property of bio-inspired micro-structure superhydrophobic surfaces and heat transfer model[J]. Applied Surface Science, 2017, 400:498-505.
[11] Lai Y , Tang Y , Gong J , et al. Transparent superhydrophobic/superhydrophilic TiO2-based coatings for self-cleaning and anti-fogging[J]. Journal of materlals chemistry, 2012, 22 (15):7420-0.

IOP Conf. Series: Materials Science and Engineering **733** (2020) 012022 doi:10.1088/1757-899X/733/1/012022

Synthesis and characterization of a luminescence podand bridged dipyridyl platinum(II) complex

Xiaorui Zheng[1,2], Ming-Zhen Mao[1,2], Xiao-ying Huang[1,2], Lieping Wang*[1,2], Binke Ning*[1,2]
1 State Key Laboratory of Fluorine & Nitrogen Chemicals, Xi'an 710065, China
2 Xi'an Modern Chemistry Research Institute, Xi'an 710065, China,
E-mail: zhxrui@qq.com, bkn204204@126.com

Abstract. We have designed and synthesized an aromatic podand containing two dipyridyl plantinum complexes in both ends of the podand, of which the luminescence in solution is reversibly fabricated by adding equal equivalent of dibenzylammonium hexafluorophosphate and base. The results are obtained by absorption and emission spectroscopy and 1H NMR spectroscopy.

1. Introduction

It is well established that one of the most used means to modulate the luminescence is to control component molecule packing and consequent intermolecular interactions. Much attention is being devoted currently to the host-guest recognition, in which the photochemical and photophysical properties of each component can be modified profoundly upon complexation[1-7]. Such recognition interactions have been used in the design of crown ethers and dialkylammonium ion systems for a variety of applications, on the basis of a cooperative combination of [N^+–H···O] and [C–H···O] hydrogen bonds and π–π stacking interactions. On the other hand, in the past few decades square-planar platinum(II) complexes have been extensively explored due to their capability to exhibit rich photophysical properties[8-10]. Scientist have reported a square-planar platinum(II) complex based on dibenzo[24]crown-8[11] (Pt-Crown), as shown in scheme 1, which exhibited that upon addition of trifluoroacetic acid (TFA) to a solution of Pt-Crown in CH_2Cl_2, significant decrease in the luminescence intensities was observed.

However, it is well-known that podands are open chain crown ethers, which has the various potential applications. Owing to the ion-binding properties, ionophoric properties, podands have been widely studied.[12-14] For example, podands have been studied as delivery vehicles for biological molecules and alkali metal ions, ligands for organometallic catalysts, precursors for anticancer drugs, fluorescent sensors for heavy metals, precursors of nanomaterials with novel electronic, magnetic and optical properties, and reagents for the extraction of radionuclides and rare-earth elements and chiral amino acids. However, host-guest recognition is rarely explored in podand-based system.

In this context, we have combined podand with platinum(II) complexes to design a podand bridged dipyridyl platinum(II) complex (compound **1**), as shown in scheme 1. The luminescence responses can be triggered by adding acid/base. Upon treatment of the solution of compound **1** with acid and base, the luminescence intensity was enhanced and decreased, respectively, together with a reversible change of NMR signals. The change of luminescence intensity is in sharp contrast to the situation of Pt-Crown system mentioned in the previous literature[11].

2. Experiment

The synthetic pathways of compound **1** are shown in scheme 2. The synthesis of 5-(bromomethyl)-2, 2'-bipyridine (bipyBr) and dibenzylammonium hexafluorophosphate (2-H·PF_6) (scheme 1) are according to the previous literatures[15,16]. Compound **4** is reacted with bipyBr to produce compound 3. Compound 3 is stirred with $Pt(DMSO)_2Cl_2$ to produce compound 2, which is followed by CuI catalyzing to produce the compound **1**. The structure of the compound **1** was investigated by NMR and ESI-MS spectra. The photophysical features have been studied by ^1H NMR spectroscopy and absorption and emission spectroscopy to explain the interactions of compound **1** and 2-H·PF_6.

2.1. Materials and Reagents:

Tetrakis(triphenylphosphine)platinum(0) (99%), phenylacetylene, methacrolein, 2-acetylpyridine, 4-toluene sulfonyl chloride (98.5%), iodine (99.8%) were commercially available and used as received. Triethylamine and all solvents were purified by standard procedures. NMR spectra were recorded with a 400 MHz spectrometer. Chemical shifts are reported on the δ scale relative to tetramethylsilane. Emission spectra were recorded in an air-equilibrated solution. ^1H NMR titrations of the samples was measured on a 400 MHz spectrometer, and deuterated solvents were used from freshly opened bottle. Titration experiment was performed in chloroform-d solutions. ESI-MS Experiments were recorded on a solariX Fourier transform ion cyclotron mass spectrometer resonance.

*2.2. The synthesis of compound **1**:*

The suspension compound 2(60 mg, 0.048 mmol), CuI(2.1 mg, 0.011 mmol) in dichloromethane was under N_2 atmosphere, followed by addition of triethylamine (0.7 mL), and the mixture was stirred at rt for 12 h. Solvent was removed under reduced pressure, and the crude product was purified by flash chromatography with CH_2Cl_2/MeOH (15:1, v/v) mixture solution as the eluent to get orange solid. The orange solid was recrystallized by dichloromethane/ether to yield orange powder (47 mg), in 65% yield. ^1H NMR (400 MHz, $CDCl_3$): δ 8.81 (s, 4H), 8.15 (d, J = 8 Hz, 2H), 8.05 (d, J = 8 Hz, 2H), 7.93 (t, J = 8 Hz, 2H), 7.80 (d, J = 8 Hz, 2H), 7.57 (q, J = 8 Hz, 8H), 7.31-7.22 (m, 14H), 6.67 (m, 4H), 4.53 (s, 4H), 4.01 (t, J = 4 Hz, 4H), 3.81 (t, J = 4 Hz, 4H), 3.76 (m, J = 4 Hz, 8H), 3.61 (m, 8H). High resolution mass spectrum m/z: caculate $[C_{72}H_{66}N_4 NaO_8Pt_2]^+$: 1527.4069; found: 1527.4032 (Figure S6). See the Supporting Information for the details on their synthesis and characterization of compound **2-5**, and the experiments.

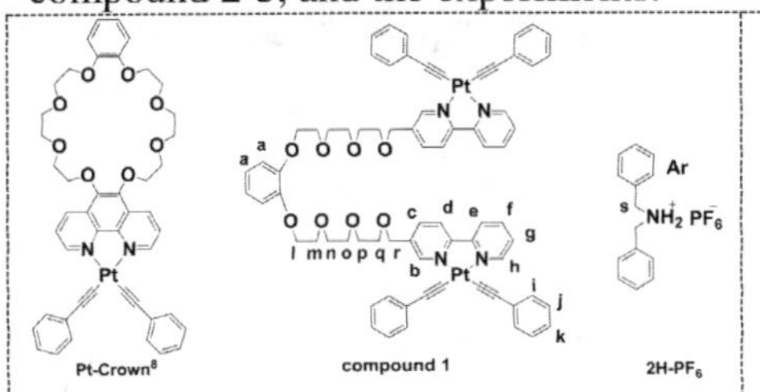

Scheme 1. Structure of compound Pt-crown[8], compound **1** and 2-H·PF_6.

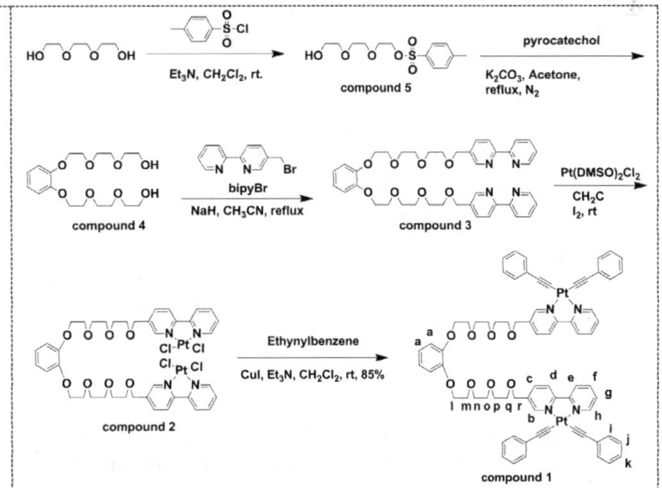

Scheme 2. Synthetic routes for compound **1** and compound **2**.

IOP Conf. Series: Materials Science and Engineering **733** (2020) 012022 doi:10.1088/1757-899X/733/1/012022

3. Results and Discussions

3.1. ^{1}H NMR Spectroscopy:

The spectrum of the compound **1** in $CHCl_3$ solution is shown in Figure 1a. Upon addition of equivalent 2-H·PF_6 in CD_3CN solution to the NMR tube, the spectrum shown in Figure 1b was obtained, in which, the methylene proton H_γ on compound **1** were shifted upfield from $\delta = 4.50$ to 4.40 ppm. The aromatic protons H_i, H_j, H_k in phenylacetylene ligand are shifted upfield, which was pointed out in Figure 1. And, the signals of dipyridyl ligand protons of compound **1** were broadened and aggregated to such an extent. Furthermore, a broad peak at 7.18 ppm was clearly assigned to the protons H_{Arc} of 2-H·PF_6 hosted by the compound **1**, whereas the signal at 7.44 and 7.39 ppm was assigned to the protons H_{Aruc} adjacent to the uncomplexed 2-H·PF_6.17 This phenomenon was assigned to the presence of strong π–π stacking interactions in the system of compound **1** and 2-H·PF_6. Subsequently, upon the addition of one equivalent of N-tert-butyl-N',N',N'',N'',N''',N'''-hexamethylphosphorimidic triamide (P_1-t-Bu) to the same NMR tube, followed by spectroscopic analysis, resulted in the spectrum shown in Figure 1c, which is essentially the same as that illustrated in Figure 1a, except that there are additional peaks arising from the presence of protons in the dibenzylammonium, indicating that the recognition was reversible. These observations suggest that addition of 2-H·PF_6 to the compound **1** results in aggregation through host–guest interactions, in which, π–π stacking interactions play a key role, and that can be reversed by adding base.

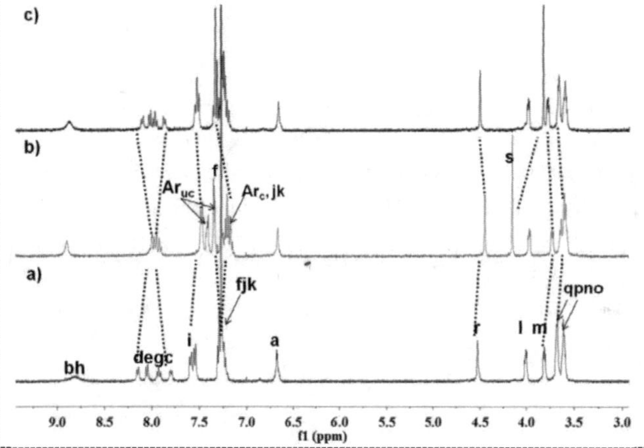

Figure 1. ^{1}H NMR (400 MHz, 25 °C) spectra of (a) compound **1** (in $CDCl_3$, 2.5×10^{-3}mol·L^{-1}), (b) of a 1:1 mixture of 2-H·PF_6 and compound **1** ($CDCl_3$/CH_3CN, 100:4) and (c) of a 1:1:1 mixture of, compound **1** and P_1-t-Bu ($CDCl_3$/CH_3CN, 100:4).

3.2. The photophysical properties of the system:

The UV-vis absorption spectra (Figure 2) of compound **1** in $CDCl_3$ solutions exhibit intense ligand-centered absorption bands at ca. 270 nm along with a broad low-energy absorption at 403 nm, arising likely from MLCT and LLCT states[17]. Upon addition of equivalent 2-H·PF_6, the absorption bands between 390 and 500 nm of compound **1** showed a slight blue shift in comparison with that in compound **1**, which could be basically recovered by further adding equivalent of P_1-t-Bu.

The emission decay trace of the $CHCl_3$ solution recorded at 662 nm was deconvoluted as a bi-exponential fuctions (FigureS8), giving a luminescence lifetime of 8.99 ns(α = 20.82%) and 57.14 ns (α = 79.18%). The bi-exponential decay recorded at 662 nm is due to the superimposttion of the fluorescence band (which is responsible for the measured short time of 8.99 nm) with the phosphorescence one (responsible for the measured long time of 57.14 ns). The internal quantum yields of compound **1** in CHCl3 solution is 4.55%.

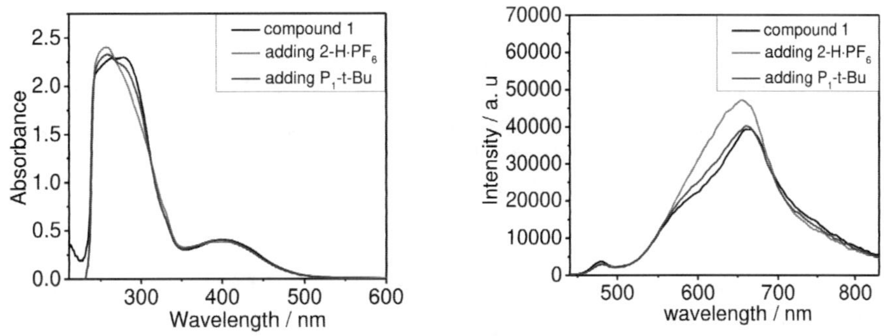

Figure 2. UV-vis absorption and emission spectra of compound **1** (2.5×10^{-5} mol·L^{-1}) in $CHCl_3$ solution (black), 1:1 mixture of 2-H·PF_6 and compound **1** ($CDCl_3/CH_3CN$, 100:1) (red), a 1:1:1 mixture of 2-H·PF_6, compound **1** and P_1-t-Bu ($CDCl_3$/ CH_3CN, 100:1) (blue).

Upon excitation at λex = 420 nm, the $CHCl_3$ solution of compound **1** exhibited a strong luminescence band centered at 662 nm in Figure 2, which was assigned to the mixtures of MLCT and ^{3}MLCT luminescence. Upon addition of equivalent 2-H·PF_6, the band was blue shifted to 655 nm together with an increase in its intensity. The enhancement of emission intensity is in contrast to the situation of Pt-Crown system, of which the emission intensity decreased, mentioned in the previous literature[11]. Besides, and the blue shift of ^{3}MLCT emission has been confirmed to be the aggregation of luminescence compound[11]. Further treatment of this solution with equivalent P_1-t-Bu completely restored the luminescence to the original level. We deduce that the novel increase of emission intensity was attributed to the podand-like structure, and the slight blue shift was attributed to aggregation of compound **1**.

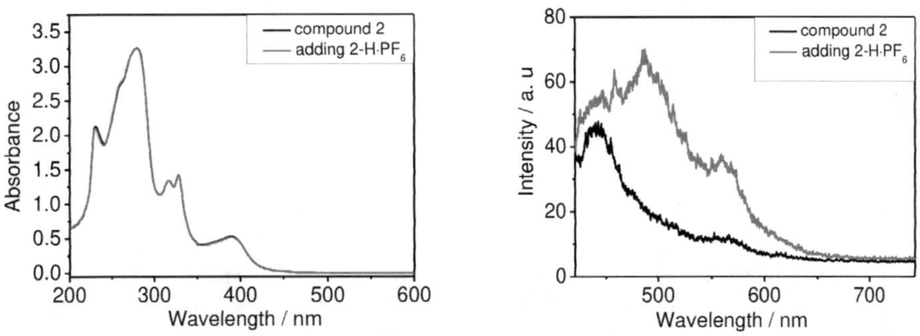

Figure 3. UV-vis absorption and emission spectra of compound **2** (2.5×10^{-5} mol·L^{-1}) in $CHCl_3$ solution (black), 1:1 mixture of 2-H·PF_6 and compound **2** ($CDCl_3/CH_3CN$, 100:1) (red) at rt.

To clarify the mechanism of this unusual luminescence behavior, we further studied interactions of the model compound **2** with 2-H·PF_6. The UV-vis absorption spectra of compound **2** are shown in Figure 3. The spectra consist of high intensity absorption bands below 350 nm, which can be attributed to ligand $^1\pi$-π^* transitions, and a broad, moderately intense absorption band at 390 nm assign to metal-to-ligand charge transfer (MLCT) transition. Upon addition of equivalent 2-H·PF_6, the absorption bands has a very slight change in comparison with that in compound **1**. Upon excitation at λ_{ex} = 390 nm, the solution of compound **2** exhibited a very weak MLCT luminescence band centered at 450 nm and a ^{3}MLCT luminescence band centered at 550 nm. Upon addition of equivalent 2-H·PF_6, the luminescence intensity is increased slightly, which was in contrast to the situation of compound **1** and 2-H·PF_6 mentioned above. Considering the structure difference between compound **1** and compound **2**, the change of emission bands in compound **1** and 2-H·PF_6 were attributed to the presence of π–π stacking interactions mainly at the part of phenylacetylene ligand, as revealed by its ^{1}H NMR spectrum.

4. Conclusion

In summary, we have fabricated a luminescence system based on podand bridged dipyridyl platinum(II) complexes, by adding equivalent 2-H·PF_6 to a solution of compound **1** in $CHCl_3$ on the basis of the π-π interactions between phenylacetylene ligand and 2-H·PF_6. The host–guest recognition leads to a novel enhancement of emission intensity and a slight aggregation of the compound **1** and 2-H·PF_6, as shown in NMR spectrum and emission spectrum. And the changes can be fully reversed upon the addition of equal equivalent of base. We conclude that podand-like structure plays an important role in the unusual luminescence change. Therefore, podands can be promising candidates for constructing smart materials through functionalizing. We believe that the novel synthetic host can attract more attention, and its synthesis and applications will become a new and attractive research topic in the near future.

References

[1] Balzani V, Gómez-López M and Stoddart J F 1998 *Acc. Chem. Res.* **31** 405-14
[2] Xue M, Yang Y, Chi X, Yan X and Huang F 2015 *Chem. Rev.* **115** (15), 7398-501
[3] Li L, Zheng X, Yu B, He L, Zhang J, Liu H Cong Y and Bu W 2016 *Polym. Chem.* **7** 287-91
[4] Ashton P R, Ballardini R, Balzani V, Fyfe M C T, Gandolfi M T, Martínez-Díaz M-V, Morosini M, Schiavo C, Shibata K, Stoddart J F, White A J P and Williams D J 1998 *Chem. Eur. J.* **4** 2332-41
[5] Liu D, Lang J-P and Abrahams B F 2011 *J. Am. Chem. Soc.* **133** 11042-45
[6] Ma X and Tian H 2010 *Chem. Soc. Rev.* **39** 70-80
[7] Sun R and Ma X 2013 *Tetrahedron* **69** 1069-73
[8] Paw W, Cummings S D, Mansour M A, Connick W B, Geiger D K and Eisenberg R 1998 *Coordination Chemistry Reviews* **171** 125
[9] Suzuki S, Sugimura R, Kozaki M, Keyaki K, Nozaki K, Ikeda N, Akiyama K and Okada K 2009 *J. Am. Chem. Soc.* **131** 10374-75
[10] Lai S-W, Chen Y, Kwok W-M, Zhao X-J, To W-P, Fu W-F and Che C-M 2010 *Chem. Asian J.* **5** 60-65
[11] Yu B, Guo S, He L and Bu W 2013 *Chem. Commun.* **49** 3333-35
[12] Pospieszny T 2015 *Mini-Reviews in Organic Chemistry* 12 258
[13] Singh A S and Sun S-S 2012 *J. Org. Chem.* **77** 1880-90
[14] Li G, Still W C 1993 *Tetrahendron Letters* **34** 919-22
[15] Ballardini R, Balzani V, Clemente-León M, Credi A, Gandolfi M T, Ishow E, Perkins J, Stoddart J F, Tseng H-R and Wenger S 2002 *J. Am. Chem. Soc.* **124** 12786-95
[16] Ashton P R, Ballardini R, Balzani V, Fyfe M C T, Gandolfi M T, Martínez-Díaz M-V, Morosini M, Schiavo C, Shibata K, Stoddart J F, White A J P and Williams D J 1998 *Chem. Eur. J.* **4** 2332-41
[17] Ni J, Zhang X, Wu Y-H, Zhang L-Y and Chen Z-N 2011 *Chem. Eur. J.* **17** 1171-83

IOP Conf. Series: Materials Science and Engineering **733** (2020) 012024 doi:10.1088/1757-899X/733/1/012024

Influencing factors and experimental study on performance of composite cement-based grouting

Tianling Du[1], Ying Liu[1]
[1]Research Institute of Highway Ministry of Transport, Beijing 100088, China
*Corresponding author's e-mail:1042736537@qq.com

Abstract: The performance of grouting material directly affects the grouting effect under cement concrete pavement slab. In this paper. the factors affecting the fluidity, strength and bleeding rate of grouting materials are analysed by repeated tests. The test results show that the slurry with good suspension is not easy to be diluted by water and has certain waterproof performance; The water-cement ratio is in the range of 0.35 to 0.5, and the suspension is good and not easily separated; Incorporation of fly ash will cause early strength reduction and increase of slurry bleeding. It is recommended not to exceed 50% of cement. Finally,it is suggested that the strength of the grout should not be as high as possible, and the problem of strength development at the late stage of grouting and matching of the strength of the base layer should be fully considered.

1.Introduction

With the extension of service life, cement concrete pavement appeared different degree of disease, , such as Pumping mud, faults, cracks, broken slab, etc., the occurrence of these diseases are more or less closely related to plate bottom emptying [1]. At present, the grouting technology is one of the most effective and economical measures to solve the problem of plate bottom emptying. In order to reduce the voiding damage of cement concrete pavement slabs, the cement pavement maintenance specifications require that the slabs should be grouted immediately after the voids appear [2].

The grouting stabilized plate technology has been used as a preventive maintenance measure. China and other countries in the world generally adopt the grouting technology to deal with emptying. However, lot of engineer practices demonstrate the grout under slab technology is not so easy to control.The grouting effect of the cement concrete pavement not ideal, and some pavement damage is accelerated after grouting. One of the reasons is that the performance of the grout is not satisfactory. Grouting technology has high requirements on the comprehensive performance of materials. The grouting material with excellent performance can really strengthen, repair and prolong the service life of the concrete pavement. Otherwise, the pavement performance will be further reduced。

Grouting materials for cement concrete pavement are mainly cement slurry, cement fly ash slurry, cement mortar and asphalt perfusion for under-board plugging[3]. Li Zihua [4] applied cement slurry, expansion agent, fly ash, etc. for grouting treatment, and achieved good results. Zhou Zhijian [5] used cement mortar to grout the concrete slab. It was found that the sand can effectively reduce the shrinkage of the slurry, and at the same time save the amount of cement, but the proportion increases the bleeding of the slurry. Li Ye and Gao Wei [6] studied the process of adding emulsified asphalt and chopped polypropylene fiber to the bottom of the cement slurry to improve the compressive/folding ratio and flexibility of the grouting material.

In this paper, the factors that concern suspension, fluidity, strength, expansion rate and bleeding performance of grouting materials were discussed, and a reasonable index range was given. It is further proposed that the strength properties of grouting materials should take into full consideration their compatibility with the existing base strength.

2. Test

2.1. raw materials

The cement used in the test is Portland cement P·O42.5， the mechanical properties of the cement are shown in Table 1; Grade I fly comes from a power plant， its specific surface area of fly ash is 450m2/kg； The natural sand contains less than 1% mud and we just need fine sand below 0.6mm pore size； Other admixtures include high-efficiency water reducing agent, pumping agent, early strength agent, expansion agent, and air entraining agent； Engineering water.

Table1. Physical and mechanical properties test of cement

Type of cement	Fineness (%) (0.08mm)	Setting time (min)		Stability	Flexural Strength (MPa)		Compressive strength (MPa)	
		Initial	Final		3d	28d	3d	28d
P.O42.5	1.78	147	217	qualified	4.8	8.20	21.7	43.0
GB 175-2007	10	≥45min	≤600min	qualified	≥3.5	≥6.5	≥17.0	≥42.5

2.2 Test methods

2.2.1. strength

According to the test method T0506-2005 test method of cement mortar strength in JTG E30, the standard of transportation industry, the size of the specimen is 4cm×4cm×16cm, and non-vibration molding is adopted.

2.2.2. Vertical fluidity

According to the test method T0508-1 cement slurry flow measurement method in JTG E30. During the test, the prepared slurry was poured into the inverted cone. The volume of the inverted cone was 1725 ml ± 5 ml. The time required for the entire slurry to pass through the cone was measured as vertical fluidity (at room temperature, the time pure water flowed out is 8s).

2.2.3. Horizontal spread

The test shall be carried out in accordance with Appendix A of technical code for application of concrete admixtures (GB 50119-2003) for which the truncated cone is round die dimensions: height (60 ± 0.5) mm; inner diameter of the upper opening (70 ± 0.5) mm; an inner diameter of the mouth

(100±0.5)mm; the outer diameter of the lower mouth is 120mm.

2.2.4. Expansion

The test of cement slurry expansion is carried out using self-developed equipment. As shown in Figure 1.

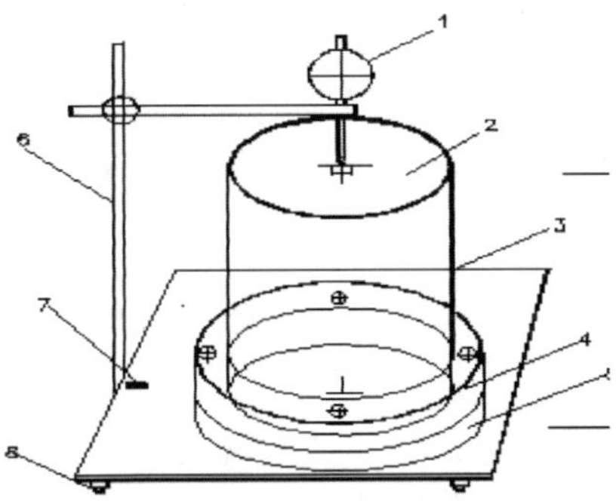

Note:
1 Dial indicator;
2 Test cover;
3 Transparent plexiglass cylinder test mode;;
4 Flange;
5 Cylindrical test base;
6 Support;
7 Level bubble;
8 Leveling screw。

Figure1. Cement (sand) slurry expansion rate test device

2.2.5. Bleeding

The test shall be carried out in accordance with Appendix C4 of Technical Specifications for Construction of Highway Bridges and Culverts (JTG/T F50-2011).

3. Test plan

The mix ratio of grouting material under cement road slab is different from that of concrete material. There is no mature design calculation formula, and it can only be repeatedly tested according to the experience of grouting. The preliminary selection of water-to-binder ratio is between 0.35 and 0.45, the amount of water reducing agent is 0.7% to 1.0%, the amount of pumping agent is 0.5% to 1.0%, and the expansion agent is selected to be 8% to 12%.

4. Results and discussion

4.1. Test results and analysis

4.1.1. Liquidity

The vertical fluidity (s) and horizontal spread (mm) are generally used to comprehensively evaluate the fluidity of the grouting material.

The test results show that the pumping agent can increase the vertical fluidity of the mortar, and the high-efficiency water reducing agent can increase the horizontal spread.

Through further tests, when the vertical fluidity is between 14s and 23s (as is shown in Fig. 2), the workability of the slurry is better, the suspension is good, and the segregation is not separated. When the fluidity is less than 14s, the slurry is relatively thin and easy to be separated. The slurry above 23s is thicker, although it is not easy to isolate, but the horizontal fluidity becomes smaller. Therefore, it is preferable to control the vertical fluidity in the range of 14s to 24s, and the horizontal spread is generally

370 to 390 mm (as is shown in Fig. 3).

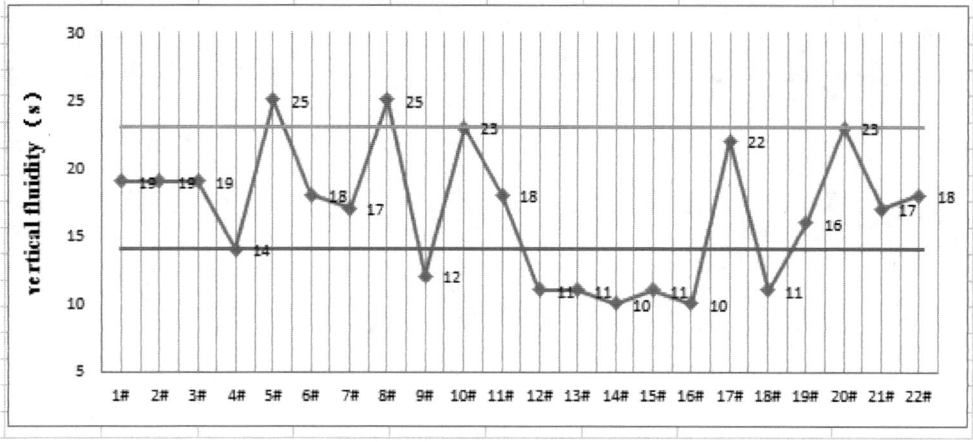

Figure2. Vertical fluidity of grout

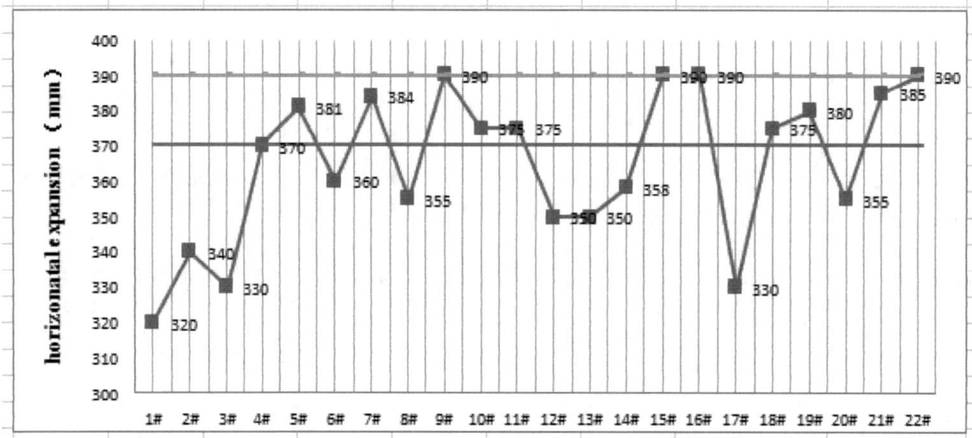

Figure3 Horizontal spread of grout

4.1.2. Bleeding

(1) Water-to-binder ratio

The water-to-binder ratio is large, and the flowability of the slurry is good, but the slurry is easily separated and the bleeding rate is also increased, which affects the strength of the slurry. For the sand-free grouting slurry, the water-to-binder ratio can be controlled from 0.35 to 0.45, and the mortar can be increased increased to 0.6. The final result is based on engineering trials.

(2) Sand effect

The addition of fine sand can increase the strength of the grout, reduce the shrinkage performance of the slurry, and at the same time reduce the amount of cement and other cementing materials, thereby reducing the engineering cost of the grouting plate, but the large-size sand is prone to segregation and bleeding. Therefore, we use fine sand. Of course, the fineness modulus of sand and the amount of sand are related to the situation of voiding. The finer the sand, the higher the amount can be. The larger the amount of voiding of the plate, the larger the amount of sand can be correspondingly larger, according to the construction site. The specific situation to determine the program.

The amount of sand is closely related to the fluidity, expansion property, bleeding property and strength of the slurry. For the grouting material with fine sand, the bleeding of the grouting material is increased, because the fine sand must meet the requirements of the fluidity, and the water must be increased. However, only the hydration reaction of the cement is needed. Adding fine sand grout, the water-to-gel ratio is about 0.4~0.59. The amount of fine sand must be tested to achieve the best ratio of cement particles, fly ash particles and sand particles in the cement slurry. According to the observation and test results in the test, the ratio of sand/cement = 0.7~1.2 is used, the bleeding rate of the slurry is small, and it is close to no water at 24h.

(3) Influence of fly ash

The addition of fly ash can reduce the amount of cement, save costs, improve the workability and shrinkage of the slurry, and facilitate the development of long-term strength. However, it was found in the experiment that the addition of fly ash not only reduces the early strength of the slurry, but also causes the slurry to produce bleeding, especially when it exceeds 50%, it is prone to bleeding. In the test, it was concluded that considering the various factors, the amount of fly ash is recommended to be less than 50% of the cement.

4.1.3 Suspension

Good suspension cement base composite grouting material is the premise of low bleeding rate and good strength performance. We further tested the grout with good fluidity and no segregation during the above test preparation process: leave a small amount of water at the bottom of a large basin, and inject the grout into the pot and find that the slurry can be used at the bottom of the basin. The water is pushed away and suspended in the surface layer of the slurry, and the slurry itself is not easily diluted by water and has waterproof properties. Imagine that if the actual pavement grouting, the well-suspended grout can ischarge the water at the bottom of the board out of the cement road panel under a certain pressure, thus playing the role of grouting stability.

4.1.4 Strength

(1)Impact of bleeding

It is preferred to carry out a strength test in a test ratio in which the slurry has good suspension property, large fluidity, and less bleeding after 24 hours. The test results show that the bleeding rate has a great influence on the strength. As shown in Fig. 5, the strength of the slurry with bleeding after 24 hours is significantly lower than that of the non-bleeding slurry.

It can be seen from Fig. 4 that the strength of 2# slurry is the highest, the quality of the slurry is good, the slurry has no segregation, and the suspension is good, and there is almost no bleeding at 24 hours. 1# and 3# have a high rate of bleeding, and their early and late compressive and flexural strengths are lower, 4# less water, and the strength is centered.

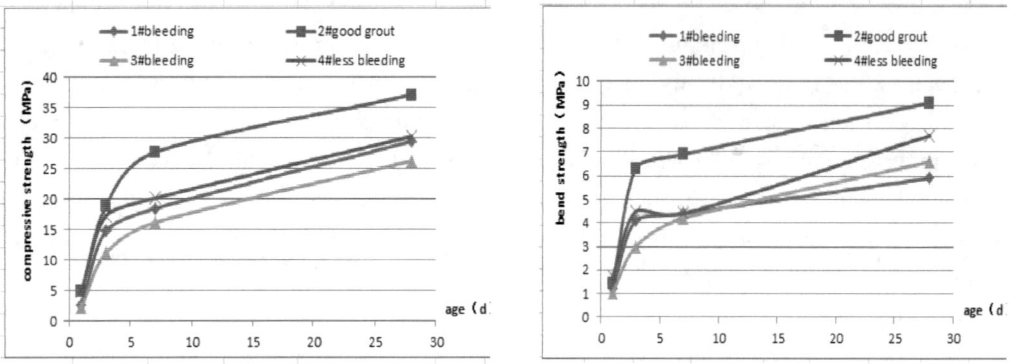

(a) on the compressive strength (b) on the bending strength

Figure4 Effect of bleeding on strength

(2)Impact of expansion

The dosage of UEA in this test is 10%~15%. It can be seen from Fig. 5 that the effect of expansion ratio on the intensity is inversely proportional to the early 1d to 7d. The slurry with a small early expansion rate has a higher growth rate in the later stage. For example, the early expansion rate of the 16# slurry is 4.4%, but the growth rate of the late strength is the lowest. The early expansion rate of the 20# slurry is 0.2%, but the growth rate of the late strength is the highest. It can be seen that the expansion rate of the grouting material is too large to cause a decrease in strength in the later stage.

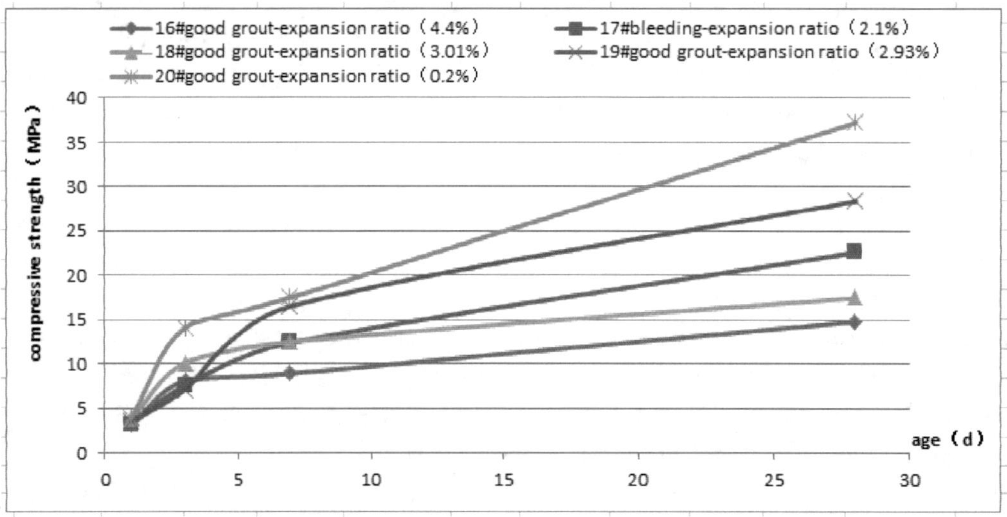

Figure5 Expansion effect on the strength

4.2. Grout strength and base strength matching

For bottom grouting materials, the coordination of the strength of the grouting material with the strength of the substrate is critical. Grouting treatment technology is mainly to restore the uniform support conditions of the base layer under the panel. If the strength of the grouting material is much greater than the strength of the base layer, it will cause uneven support of the bottom of the board, which will lead to further damage of the panel.

5. Conclusion

(1) The pumping agent can increase the vertical fluidity of the mortar, and the high-efficiency water reducing agent can increase the horizontal spread. When the vertical fluidity is 14s~24s and the horizontal spread is generally 370~390mm, the composite slurry has good suspension, no segregation and stratification, and close to no bleeding after 24h. The slurry with good suspension is not easily diluted by water and has certain waterproof performance;

(2) The water-to-binder ratio directly affects the fluidity of the slurry, but the performance of the slurry should be fully considered to control the water-to-binder ratio of 0.35~0.45 (without sand), and the water-cement ratio can be appropriately increased for the sanded grout.

(3) The amount of sand is closely related to the fluidity, swelling property, bleeding property and strength of the slurry. It is recommended to use sand/cement = 0.7~1.2 ratio.

(4) Fly ash has a certain effect on the fluidity of the grout, but it is unfavorable to the early strength, and will also increase the rate of bleeding of the slurry. Therefore, it is recommended that the amount of fly ash is less than 50% of the mass of the cement.

(5) The strength of the grout should be matched with the strength of the base layer. When the strength of the grouting material is much greater than the strength of the base layer, it will cause uneven support of the bottom of the plate, which will lead to further damage of the panel.

References:

[1] Jin Z Q . Maintenance manual for cement concrete pavement maintenance [M]. Beijing: China Communications Press, 2003.

[2] Technical Specification of Cement Concrete Pavement Maintenance for Highway. JTJ073.1-2001

[3] Fu Z, Jin Z Q. Construction and Maintenance Technology of Cement Concrete Pavement [M]. Beijing: China Communications Press, 2003.

[4] Li Z H. Application of bottom grouting technology for cement concrete pavement. Liaoning Transportation Science and Technology, 2004.

[5] Zhou Z J, Experimental Analysis of De-air Grouting Materials at the Bottom of Concrete Road Slabs, Journal of Changchun Institute of Technology (Natural Science Edition) 2009.

[6] Li Y , Gao W, Research on De-air Grouting Materials at the Bottom of Cement Concrete Road Slabs, Low Temperature Building Technology, 2007.

[7] Code for application technique of cementitious grout GB/T50448-2008.

Analysis of crystallization and melting behavior of composites before and after annealing

Z H Zhu [1], N Zhang [1], T Wang [1] and M Y Hao [2]

1 School of advanced manufacturing technology, Guangdong Mechanical & Electrical Polytechnic, Guangzhou, China;

2 The Key Laboratory of Polymer Processing Engineering, Ministry of Education, South China University of Technology, Guangzhou, China

Email: 184823174@qq.com

Abstract

The crystallization and melting behavior of sisal fibers reinforced PLA composites were studied in this work. It can be found that the addition of sisal fiber as reinforcement shortened the nucleation period of PLA and promoted the formation of transverse crystals in PLA composites. Annealing treatment can improve the transition temperature and crystallinity of PLA composites.

1. Introduction

PLA has good physical properties, biocompatibility and biodegradability which has been widely used in various fields. [1] Whereas, the production cost and poor mechanical properties limit the extensive application of PLA. Researchers have done a lot of work to modify PLA and optimize the processing process to promote the industrialization of PLA. [2] Among them, blending modification of PLA with natural fibers is a better method. Natural fibers have many excellent properties over synthetic fibers and can be used to reinforce the polylactide (PLA). They can not only improve the properties of PLA, but also expand the application to achieve the sustainable development. [3]

PLA is a semi-crystalline polymer and natural fiber has a high specific surface area. [4] Therefore, the addition of natural fiber has a great influence on the crystallization behavior of PLA composites. Many tests can be used to study the crystallization behavior of polymers. Differential scanning calorimetry (DSC) has been widely used because of its simplicity and reliability, which was also used to analyze the crystallization and melting behavior of PLA/natural fiber composites in this study.

2. Experiment

2.1. Materials

PLA 3051D, Nature Works ™， was purchased by Shenzhen Bright China Industrial Company

(China). SF bundles (GB/T 15031-94) were provided by Guangxi Sisal Company of Guangxi Province, China. Other chemical reagents were commercial products.

2.2. Samples Preparation

SFs were submerged into 10% NaOH solution (mass concentration) for 3 h. Then, ASFs were washed with distilled water and dried in the open air for 48 h. USFs were mixed with ASFs to fabricate HSFs in a mass ratio of 1:1 and then PLA, ASFs, USFs and HSFs were dried at 100 degrees for 4 h. Then PLA was compounded with SFs with a mass ratio 4:1 using a two-roll plastic mill (SK-160B, Shanghai Rubber Machinery Factory, china) at 180 degrees. Four kinds of samples were made: pure PLA, PLA/USFs, PLA/ASFs, and PLA/HSFs. Then, they were compressed into sheets by a hot press (QBL-350, Wuxi No.1 Rubber Plastics Mechanical Co. Ltd, China). The compression molding temperature and pressure were 190 degrees and 10 MPa, respectively.

2.3. DSC test

About 5 mg of the PLA composites and the pure PLA were placed in the crucible respectively, the scanning temperature range was set as 25-200 ℃, the protective gas was nitrogen atmosphere, and the purge flow rate was set as 20 ml/min.

2.4. Polarizing Microscope

Polarizing microscope (POM) investigation was performed with a POM (Axioskop 40 POL, ZEISS, Germany)

3. Results and discussion

3.1 DSC analysis of pure PLA and PLA composites

DSC curves of PLA and PLA composites during the secondary heating process are shown in figure 1. The thermal properties and the calculated crystallinity of PLA and three composites are also given in table 1. It can be seen that both PLA and composites have a cold crystallization peak on the DSC curve, and the cold crystallization peak temperature of PLA is 128.7℃. It was also found that the crystallization peak of pure PLA was small and wide which fully proved that PLA was a polymer with difficulty in crystallization. However, the crystallization peak of PLA/USFs PLA/ASFs and PLA/HSFs is sharp and narrow and the temperature of the cold crystallization peak of the three composites also decreased to 109.6℃ 110.0℃ and 113.7℃, respectively. This indicates that the addition of sisal fiber can obviously improve the crystallization performance of PLA, which is mainly due to the heterogeneous nucleation of sisal fiber. [5-6] Effect of sisal fiber on the crystallization of PLA can also be seen from the crystallinity data of table 1. The crystallinity of PLA composites was 24.3%, 22.1% and 29.6%, respectively, which were higher than that of pure PLA of 18.0%.

Table 1 Thermal characteristics of PLA and composites

materials	cold crystallization temperature (℃)	melting point (℃)	crystallinity (%)
PLA	128.7	156.5	18.0
PLA/USFs	109.6	151.5/159.8	24.3
PLA/ASFs	110.0	150.0/156.6	22.1
PLA/HSFs	113.7	154.1/161.8	29.6

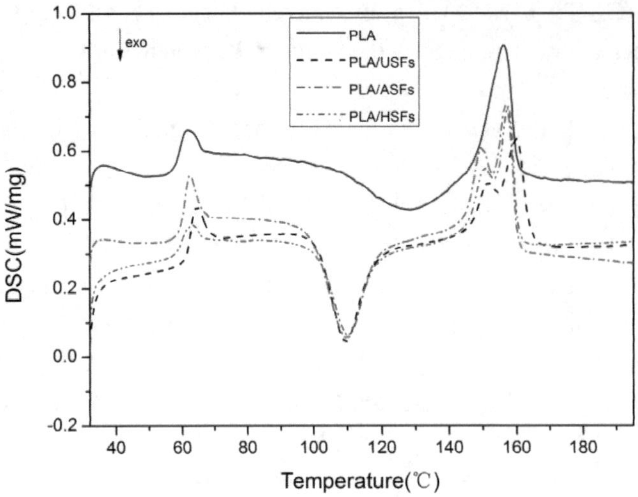

Fig. 1 DSC curves of PLA and composites

In addition, it can be found that PLA has only one melting peak, while the three composites have a small and obvious melting peak near 150℃ and a large melting peak. Due to the complexity of polymer crystallization, multiple melting behavior exists widely in the heating process of crystalline polymer. It has been pointed out in literatures [134-136] that transverse crystals may appear on the surface of fibers when fibers and semi-crystalline polymers are blended. Therefore, it may be speculated that the reason why there is a small melting peak near 150℃ on the DSC curves of the composites may be the occurrence of transverse crystals.

3.2 DSC analysis of pure PLA and PLA composites after annealing

It was well known that crystallization temperature has a very important effect on the structure and morphology of polymer crystallization, such as crystal integrity, wafer thickness, crystal shape and so on. In the process of production and molding of polymer products, the uneven crystal orientation and shrinkage are caused by the inconsistent cooling speed during the molding of the products, which leads to the internal stress of the products. It is necessary to annealing the molded products in order to eliminate or reduce the internal stress in the molded products to avoid the deformation or cracking. Annealing is an effective method for the properties of polymer products, through which the crystallinity, crystal structure and crystal size of the polymer can be adjusted to obtain different physical properties.

Figure 2 shows the DSC curves of PLA and composites before and after annealing treatment. in which AT110 represents the annealing temperature of 110℃. It can be seen from figure 2 that all the melting peaks becomes sharp after annealing treatment, which indicates that the crystal perfection of pure PLA and composites increases after annealing treatment. In addition, the disappearance of cold crystallization peak on DSC curve after annealing in figure 4-5 also indicates that annealing can improve the crystal perfection of PLA and PLA composites. It can also be seen from table 2 that the crystallinity of pure PLA and PLA composites after annealing treatment are higher than that before annealing treatment, which is because the frozen molecular chain of PLA is rearranged in chain segments, and the molecular structure becomes more perfect and regular during annealing treatment and thus the crystallinity is improved.

The glass transition temperatures of pure PLA and PLA composites before annealing treatment were higher than that before annealing treatment. It is because the crystal perfection of the materials is improved , the movement ability of the molecular chain segment is reduced and thus the glass transition temperature is increased. In addition, it can be found that the small melting peaks of the annealed composites near 150℃ do not disappear, which also indicates that a new crystal structure generated because of the addition of sisal fiber.

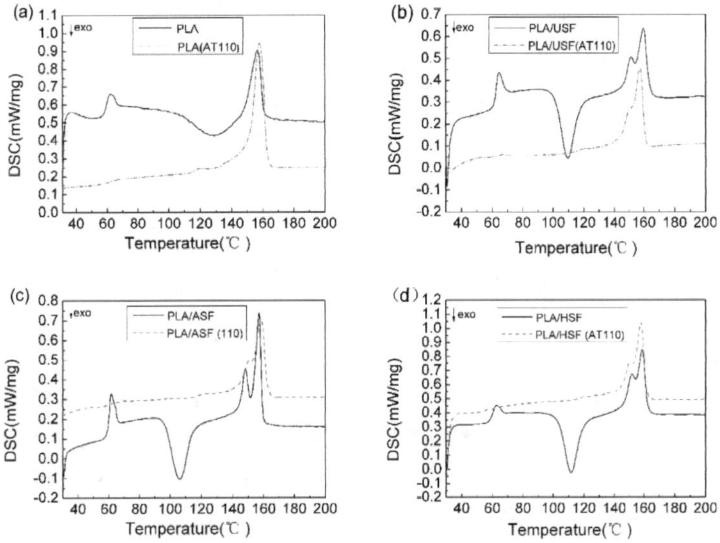

(a) PLA; (b) PLA/USF(AT110); (c) PLA/ASF(AT110); (d) PLA/HSF(AT110)

Fig. 2 DSC curves before and after annealing of PLA and composites

Table 2 Thermal characteristics of PLA and composites after annealing

materials	glass temperature (℃)	cold crystallization temperature (℃)	melting point (℃)	crystallinity (%)
PLA(AT110)	68.7	—	157.9	29.3
PLA/USF(AT110)	66.4	—	151.6/157.6	31.6
PLA/ASF(AT110)	69.6	—	150.7/158.5	30.1
PLA/HSFAT110)	68.2	—	150.7/158.1	39.8

3.3. POM analysis

POM of PLA and PLA composites are presented in figure 3. It can be seen that the crystals of PLA are like branches. However, a layer of crystals perpendicular to the axial direction of sisal fibers appears around SFs which indicated that a new crystal morphology namely transcrystallization appeared because of the addition of SFs in PLA matrix.

The reason is mainly because some factors make the fiber become the nucleating material that induces the crystallization of the polymer matrix and thus a large number of crystal nuclei and crystals produced on the fiber surface which squeeze against each other and grow in a directed way. The appearance of transcrystallization also explained why the DSC curves of PLA composites showed a small melting peak near 150℃.

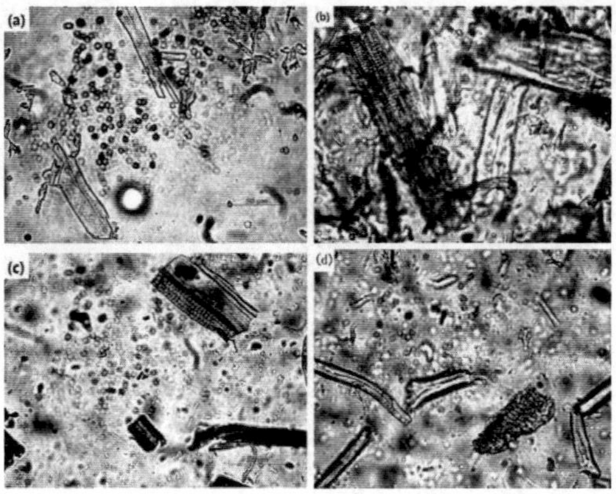

(a) PLA; (b) PLA/USFs; (c) PLA/ASFs; (d) PLA/HSFs

Fig. 3 POM investigation of PLA and composites

4. Conclusion

The crystallinity of the PLA composites was higher than that of pure PLA because of the addition of sisal fiber as reinforcement which also shortened the nucleation period of PLA and promote the formation of transverse crystals in PLA composites. Annealing treatment can improve the transition temperature and crystallinity of PLA composites.

Reference

[1] Auras R, Harte B, Selke S. An overview of polylactides as packaging materials. Macromol. Biosci. 2004, 4, 835-864.

[2] Agopyan V, Savastano H, John V M, et al. Developments on vegetable fibre-cement based materials in Sao Paulo, Brazil: an overview[J]. Cement & Concrete Composites, 2005, 27(5):527-536

[3] John M J, Thomas S. Biofibres and biocomposites. Carbohyd. Polym. 2008, 71, 343-364.

[4] Gottschalk C, Frey H. Hyperbranched polylactide copolymers[J]. Macromolecules, 2006, 39(5):1719-1723

[5] Wang C, Liu C R. Transcrystallization of polypropylene composites: nucleating ability of fibres[J]. Polymer, 1999, 40(2):289-298

[6] Campbell D, Qayyum M M. Melt crystallization of polypropylene: effect of contact with fiber substrates[J]. Journal of Polymer Science, Polymer Physics Edition, 1980, 18(1):83-93

Acknowledge

This research was supported by Key Research Platforms and Projects of Universities of Guangdong Province (grant no. 2018GKTSCX111) and Funding Project for Scientific Research of High-level Talents of Guangdong Mechanical & Electrical Polytechnic (grant no. Gccrcxm-201903).

IOP Conf. Series: Materials Science and Engineering **733** (2020) 012026 doi:10.1088/1757-899X/733/1/012026

Experimental study on the bonding of microfluidic chip

Zhang Jinghui[1], He Jianyun[1]*, Luo Xidan[1], Tong Jinge[1], Ma Xu[1] and Yang Weimin[1]*

[1]Beijing University of Chemical Technology, Beijing 100029, China

Email: 591690385@qq.com

Abstract. In this paper, the ultraviolet radiation curing technology is applied to the processing and manufacturing of microfluidic chips, and a new method of irradiation bonding of photocuring microfluidic chips is studied. Research has shown that by photocuring bonding method, the bonding strength of the photocuring microfluidic chip is high, the bonding speed is fast, the efficiency is high, and since the auxiliary solvent such as adhesive is not needed, the microchannel is not polluted, and the permanent bonding is realized. This new method of photocuring bonding has effectively solved the bonding problem of the microfluidic chip and the photocuring bonding strength satisfies the application requirements.

Introduction

Since the concept of 'micro-total analysis system' was introduced in 1990[1], microfluidic chips have played a significant role in analytical chemistry, medical testing and genetic diagnosis. This technique greatly reduces the analytical reagent loss and greatly increases the analysis speed [2]. The traditional microfluidic chip materials mainly include silicon, glass and polymer [3]. Because of their advantages such as more types, lower cost and simpler forming process, the polymeric materials have become the mainstream material for microfluidic chip manufacturing [4]. The traditional methods for the production of polymer microfluidic chips are mainly hot pressing, injection molding, laser ablation [5-7]. UV radiation curing is an advanced manufacturing technology. It has the characteristics of fast curing, less pollution, energy saving and excellent performance, which meets the requirements of environmental protection and energy saving [5]. The combination of UV curing technology and injection molding technology forms an ultraviolet curing injection molding technology [8]. The dimensional accuracy of the product is superior to that of the thermoplastic injection product [10], and the molding rate can be adjusted according to the process conditions in the experiment. After molding a single microfluidic chip, bonding is required to form a microchannel for detection. The traditional microfluidic chip bonding methods mainly include: hot-pressing method and adhesive method. The hot-pressing method has a higher heating temperature and a higher pressure, which affects the accuracy of the microstructure. For adhesive method, it is necessary to apply adhesive on the surface of the microfluidic chips to bond the two chips together.

In this paper a new bonding method of UV-curing under pressure to microfluidic chips is studied. This new bonding method is different from traditional bonding methods.

1. The experimental part

1.1. Main raw materials

LE-6702, UV resin, bifunctional polyurethane acrylate, DongtouHengli Printing Materials Co., Ltd.;PUA3, UV resin, tri-functional polyurethane acrylate, DongtouHengli Printing Materials Co., Ltd.;LE-6706, UV resin, hexa-functional polyurethane acrylate, DongtouHengli Printing Materials Co., Ltd.; EOEOEA, Monofunctional Monomer, Ethoxyethoxyethylacrylate, Changshu Hengrong Trading Company; TPGDA, bifunctional monomer, tripropylene glycol diacrylate, Changshu Hengrong Trading Company; TMPTA, trifunctional monomers, trimethylolpropane triacrylate, Changshu Hengrong Trading Company;1173 photoinitiator,2-hydroxy-2-me-thyl-1-phenyl-1-propanone, Changshu Hengrong Trading Company.

1.2. Main equipment

UV Curing Injection Molding Machine (Self-made); Laser Microscope, OLS 4100; Microscope, BX51; Digital Electronic Scale, ME104E, Olympus Industrial Company.

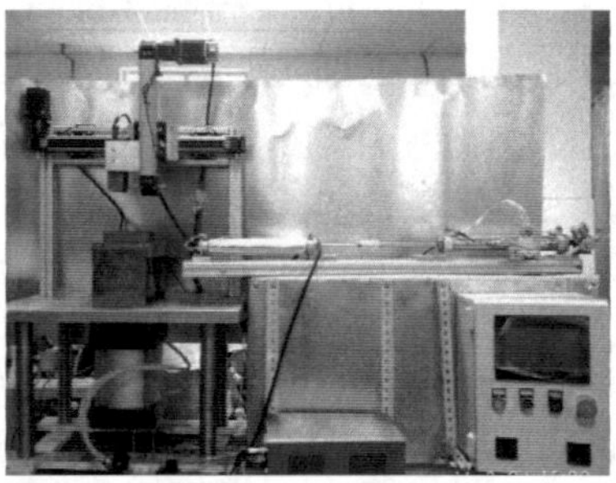

Figure 1.UV Curing injection molding device

1.3. Experimental steps

1.3.1. Mix the UV-curing resin, monomer, photoinitiator thoroughly according to a certain percentage;

1.3.2. Then put it into the vacuum to remove the entrained bubbles;

1.3.3. Using self-made UV curing injection molding device to form the substrates and cover sheets of microfluidic chips;

1.3.4. Using self-made UV-curing device to bond the microfluidic chips.

2. Research on bonding strength of microfluidic chip

Microfluidic chips are usually divided into substrates and cover piece. The process of stably bonding the substrate and the cover piece under certain conditions is the bonding of the microfluidic chip. The bonded microfluidic chip flow channel has laminar flow features, surface capillary effect, as well as rapid heat transfer and diffusion effects [8-9].

2.1. Bonding process

Figure 2 is a microfluidic chip bonding process schematic.

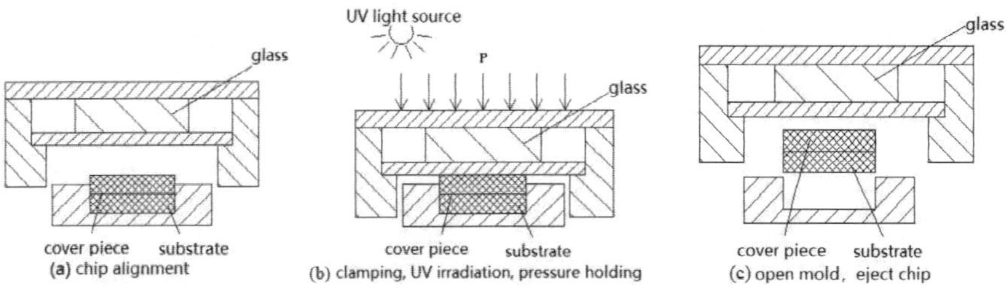

Figure 2 Schema of UV curing microfluidic chip bonding

Two kinds of bonding methods are experimentally studied, which are UV-curing bonding under pressure and bonding just under pressure.

2.1.1. Bonding just under pressure method

First, clean the surface of the UV-cured substrate and the cover chip, and put them align in order in the mold cavity, then apply the bonding pressure. The bonding pressure respectively set to 0.3MPa, 0.4MPa, 0.5MPa.

2.1.2. UV-curing bonding under pressure method

Using UV-curing equipment, first, clean the surface of the UV-cured substrate and the cover chip, and put them align in order in the mold cavity, then apply the bonding pressure at the same time UV irradiation, UV intensity is set to 2500mw/cm^2 and the pressure is set to 0.3MPa, 0.4MPa, 0.5MPa, respectively.

2.2. Bonding strength test

The average bond strength of microfluidic chips obtained by different bonding methods is characterized by the following calculation results:

$$\sigma_b = \frac{F}{A} \tag{1}$$

In formula (1) σ_b is the chip bonding strength; F is the pulling force required when the chip is peeled; A is the bonding area.

With a universal tensile tester, the chip is pulled at a speed of 2 mm / min, the tensile force is measured when the substrate is separated from the cover piece, and the average bond strength is calculated by the formula (1).

The bonding quality of the microfluidic chips has an important influence on the performance of the chip. If bubbles are formed on the bonding surface, it will reduce the quality of the chip and affect the detection accuracy.

3. Results and discussion

3.1. The bonding strength of the microfluidic chip using the in-mold direct pressure bonding method

Figure 3 is a diagram of the relationship between the bonding pressure and the bonding strength of microfluidic chips using in-mold direct pressure bonding method. Under the condition of bonding just under pressure, after the microfluidic chip bonded for a period of time, partial separation was easy to occur at the compacted position.

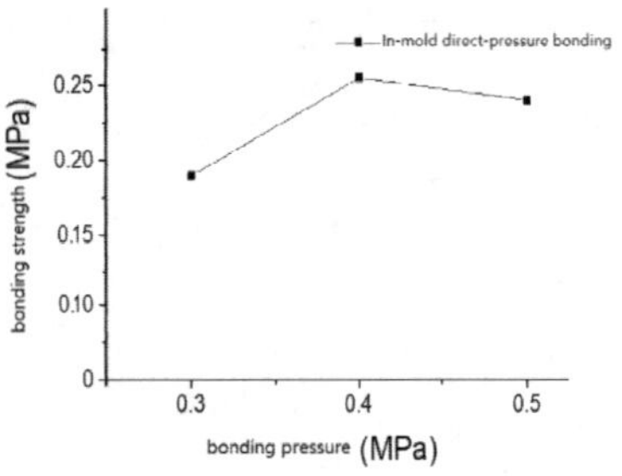

Figure 3. The relationship between the bonding pressure and the bonding strength

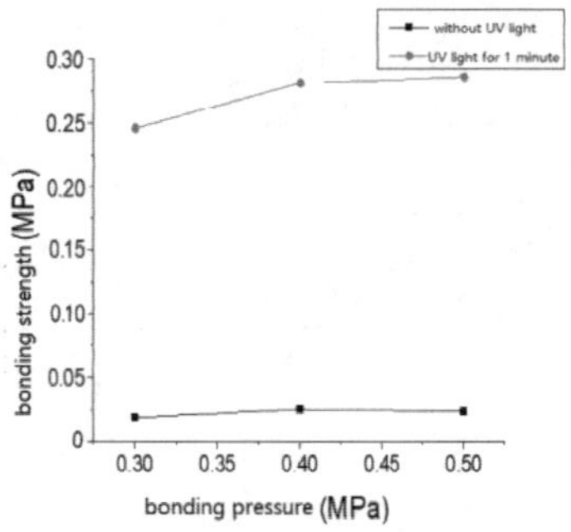

Figure 4. The relationship between the bonding pressure and the bonding strength

As it can be seen from Figure 3, when the microfluidic chip was bonded just under pressure, the bonding strength is low, and the bonding strength does not change significantly with the increase of the bonding pressure during the bonding.

3.2. Bonding strength of microfluidic chips using UV-curing under pressure method

Figure 4 is a diagram showing the relationship between the bonding pressure and the bonding strength of the microfluidic chip bonded by UV irradiation under pressure. It can be seen from Fig.4, the bonding strength of the microfluidic chip bonded by UV-curing under pressure is much greater than that of bonded just under pressure.

For the microfluidic chips bonded by UV curing under pressure, there is no separation at the compacted position of the UV-coupled microfluidic chip, and the bonding bubble defects is very few, which will not block or contaminate the flow channel. The bonding process is relatively quickly.

3.3. The effects of bonding time on the bonding strength of microfluidic chips

Figure 5 is a diagram showing the effects of bonding time on the bonding strength of microfluidic chips.

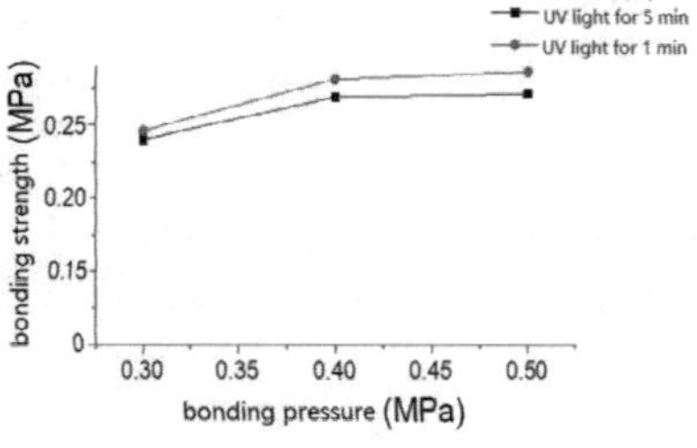

Figure 5. The effect of bonding time on bonding strength

As shown in figure 5, when the bonding pressure is 0.3 MPa, 0.4 MPa and 0.5 MPa respectively, the chip bonding strength under UV irradiation for 5 minutes is 0.240 MPa, 0.269 MPa and 0.271 MPa. Compared to the bonding strength under UV irradiation for 1 minute, the bonding strength changes little. Due to the fast photocuring reaction speed, further improvement of irradiation intensity and irradiation time has little effect on chip bonding strength.

4.Conclusion

In this paper a new bonding method of UV-curing under pressure to manufact microfluidic chips is studied. This new bonding method is different from traditional bonding methods.

For microfluidic chips, the bonding strength of the UV-curing microfluidic chip bonded under pressure is much greater than that bonded just under pressure. The method of adhesive bonding produces a large amount of air bubbles, which can easily contaminate and block the microchannels. Using the method of UV-curing under pressure, the bonding speed is fast, the efficiency is high, and since the auxiliary solvent such as adhesive is not needed, the microchannel is not polluted, and the permanent bonding is realized. This new method effectively solves the bonding problem of the microfluidic chip.

Reference

[1] Manz A, Graber N and Wimder H M Miniaturized total chemical analysis systems: A novel concept for chemical sensing 1990 *J. Sensors and Actuators* B **1(1)** 244-248

[2] Fitzpatrick E, McBride S, Yavelow J, Najmi S, Zanzucchi P and Wieder R Microfluidic techniques for single-cell protein expression analysis 2006 *J. Clin. Chem.* **52** 1080-1088

[3] Wang J and Chen G,Chatrathi M P Capillary yelectrophoresis microchip with a carbon Nanotube modiffied electrochemical detecyor 2004 *J.Anal Chem.* **76** 298-302

[4] Quake S R and Scherer A From micro-to nanofabrication with soft materials 2000 *J. Science.* **290(546)** 1536-1540

[5] Wang Xiaodong, Luo Yi, Liu Chong, Ma Liqun, Wen Min and Wang Liding Process Parameter Determination in Fabrication of Microchannel of Plastic (PMMA) Microfluidic Chips Using Hot-embossing Method 2005 *J. China Mechanical Engineering* **22 （16）** 2061-2063

[6] Song Mancang, Liu Ying, Zhu Tieli, Du Liqun, Wang Minjie and Liu Chong Injection Molding of Plastic Microfluidic Chip 2011 *J. Nanotechnology and Precision Engineering* **9 （4）** 329-334

[7] Wen Min, Wang Xiaodong, Liu Chong and Ma Liqun Research on hot embossing and thermal bonding for fabrication of PMMA microfluidic chips 2004 *J. Optics and Precision Engineering* **12(3)** 272-276

[8] Chang Le, Lin Renyou, Ding Yumei, Yang Weimin and Xie Pengcheng Research of Molding Formability of Products with Microstructure Based on Ultraviolet Radiation 2014 *J. China Plastics* **28 （3）** 71-74

[9] Chang Le，Cai Tianze，Ding Yumei, Yang Weimin, Xie Tongwei and Xie Pengcheng Research on the Monomer for Replication Fidelity of Micro Parts in UV Curing Injection Molding 2014 *J. China Plastics* **28(10)** 61-64

Effect of the in-situ polymerized polyurethane on laminated epoxy adhesive /aramid fabrics composites

GuangLei Lv and He Rui

CNOOC Safety Technology Services Co., Ltd, Tianjin, China
E-mail: 731735101@qq.com

Abstract: In order to avoid the high brittleness of epoxy resin and the low peeling strength of its laminated composite, polyurethane was synthesized by in-situ polymerization in epoxy adhesive system. The results of IR, SEM and mechanical properties test showed that the peel strength and shear strength of laminated epoxy adhesive /aramid fabrics composites reached the optimum value when the molar ratio of NCO/OH was 1.2, the content of MoCA curing agent was 1.5% and the content of polyurethane prepolymer was 20%. The brittleness of epoxy resin was reduced to a certain degree, and the bonding properties of epoxy adhesive/aramid fabrics composites were improved.

1. Introduction

Because of its high strength, high stability, high insulation and low cost, the epoxy resin has been widely used as adhesive in the fields of architecture, aerospace and integrated circuit.[1] However, because the main drawback of epoxy resin is brittle and easy to fault, it limits its application. To improve its peel strength, shear strength, and broaden its application, improving the toughness of epoxy resin is an important way.

At present, the methods for improving the toughness of epoxy resin are divided into chemical methods and physical methods. The physical method is to mix fillers into epoxy resin for modification. Inorganic particles are dispersed in the cracks of the material by blending with epoxy resin, which restricts the further development of cracks. [2] However, due to the energy absorbed by particles is limited, the modification effect of inorganic particles on epoxy resin is limited, which only restrains the cracks to a certain extent. The chemical method is the essential modification of the rigid chain of epoxy resin. Through the combination of chemical bonds, flexible segments are added into the original molecular chain to achieve the purpose of toughening.

Polyurethane (PU) is composed of polyester or polyether in the soft segment and isocyanate and cross-linking agent in the hard segment. The active isocyanate group can react with the hydroxyl group in epoxy resin, which can make the soft segment in polyurethane bonded on the epoxy resin, increasing the flexibility and toughness of the molecule chain. [3] Therefore, many researchers have focused on epoxy resin toughened by polyurethane in recent years. In order to further explore the toughening effect of PU, different proportion of polyurethane prepolymer, IPDI and polyester polyol, and MOCA curing agent are used in synthesis, and the mechanical properties of modified epoxy resin were explored.

2. Experiment Details

2.1. Raw materials

Epoxy resin E-51 (industrial product), Isophorone diisocyanate (IPDI, analytical pure), polyester polyol (PPG-1000), 3,3'-dichloro-4,4'-diaminodiphenylmethane (MoCA), dibutyltin dilaurate (analytical pure).

2.2. Preparation of polyurethane prepolymer

Polyester polyol is added into a three-port flask to raise the temperature to 80 ℃. After melting, it is continuously heated for 30 minutes to remove water from polyester polyol. And then IPDI was added into the flask with a certain proportion(- NCO and - OH molar ratio are 0.8, 1, 1.2, 1.4, 1.6, respectively), which is continuously heated for a period of time. Finally, a certain amount of MOCA is added into the flask and stirred to the reaction finished, then, it is cooled down naturally to room temperature.

2.3. Preparation of polyurethane modified epoxy resin

The different proportion of polyurethane prepolymer was added into the epoxy resin E-51, stirred until uniformly mixed, and added a certain amount of dibutyltin dilaurate, then it was reacted at 80 ℃ for 20 minutes.

2.4. Preparation of epoxy resin adhesive/aramid fiber laminated composite

The prepared modified adhesive and T-31 curing agent (30%, relative to the adhesive) were well mixed, applied on aramid fiber and steel plate. Then the composite was vacuumed and cured at 80 ℃ for about 6 hours. The epoxy resin adhesive/aramid fiber laminated composite sample was cut into a strip adhesive sample with a length of 40 mm and a width of 25 mm and cured at room temperature for 48 hours. The sample preparation was completed.

2.5. Characterization

FTIR of the PU grafted epoxy resin samples were tested by Fourier infrared spectrometer (Nicolet 6700, Thermo Fisher Technology Co., Ltd.), with a wavenumber range of 600-4000cm-1 and resolution of 4cm-1.The SEM of PU grafted epoxy resin samples were tested by scanning electron microscope with the magnification of 2000.The elongation at the break of the composite, the peel strength and shear strength are tested by the standard adhesive property test method with microcomputer controlled electronic universal testing machine (cmt4503, Shenzhen Xinsansi material testing Co., Ltd.).

3. Results and discussion

3.1. Structure of PU grafted epoxy resin

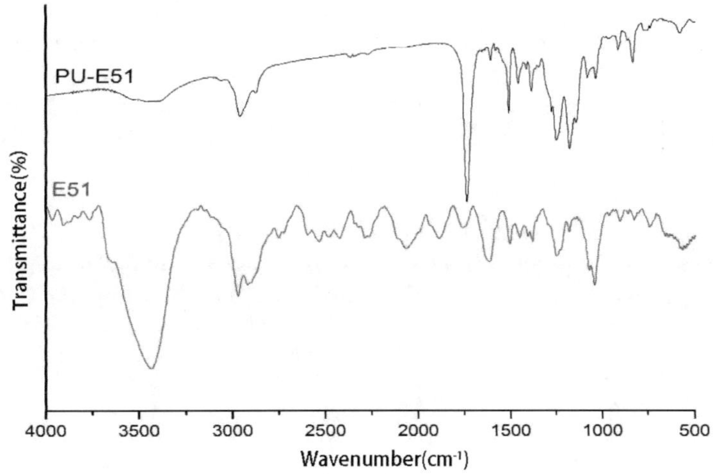

Figure 1. FTIR spectra of epoxy resin and modified epoxy resin

Fig. 1 shows the FTIR spectrum of epoxy resin and modified epoxy resin. From the E-51 spectrum in Fig. 1, it can be seen that the absorption peak at 3450 cm^{-1} is the characteristic absorption peak of - OH, and the characteristic absorption peaks of the benzene ring skeleton are 1606cm^{-1} and 1508cm^{-1}, respectively. The adsorption peak at 914cm^{-1} is the characteristic absorption peak of an epoxy group, which is consistent with the structure of bisphenol-A epoxy resin. Compared with the FTIR spectrum of pure epoxy resin, the characteristic absorption peak of epoxy group at 914cm^{-1} in modified epoxy resin was retained, the characteristic absorption peak at 1730cm^{-1} which was belonged to - C = O , and - NCO group characteristic peak was not found at 2272cm^{-1}, which was due to the reaction between - NCO group in PU prepolymer and - OH in epoxy resin. The results of the FTIR spectra showed that PU prepolymer was grafted to the epoxy resin molecular chain.[4]

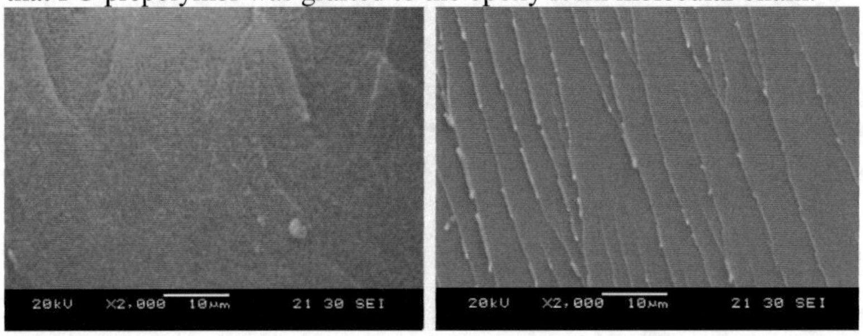

Figure 2. SEM images of epoxy resin（left） and modified epoxy resin（right）

The images in Fig. 2 were the SEM pictures of epoxy resin and modified epoxy resin. It can be seen that the cross-section of epoxy resin without polyurethane particles is smooth and clear, which is a typical brittle fracture; while the cross-section crack distribution of epoxy resin modified with polyurethane particles is relatively dense. The results of SEM analysis indicated that it had more energy absorption after modification and the material toughness is improved. [5]

3.2. Mechanical properties of epoxy resin composite

As shown in Figure 3, the peel strength of modified epoxy resin was increased at first and decreased afterward with the increase of NCO / OH molar ratio. With the increase of the molar ratio, the excessive NCO group becomes the terminal groups at both ends of the polyurethane chain gradually. The -NCO acts as the active group by bonding into the polyurethane elastomer and the epoxy resin. This leads to the toughening effect from two aspects, in which the molecular weight of the modified polymer increased rapidly, and the interpenetrating network structure appeared. In terms of adhesion, epoxy resin has strong adhesion to metal materials, but when a small amount of polyurethane is added, a small amount of hydroxyl reacts with the -NCO group, so the relative density of epoxy group and the adhesion to metal are decreased. With the continued increase of polyurethane content, the ammonia ester group (-NHCO-) which has strong polarity increases more, and it can form the hydrogen bond with metal, which enhances intermolecular force and bonding strength, so the adhesive force increases. [6] When the content of - NCO continues to increase, it reacts with the hydroxyl group in epoxy resin completely, which makes the adhesive force decrease again, and the interpenetrating structure disappears, so the toughening effect weakens. According to the results, the best formula for NCO / OH molar ratio to prepare the polyurethane prepolymer is 1.6.

3.3. Effect of MoCA content on mechanical properties

Figure 4 shows the mechanical properties of modified epoxy resin with the different amount of curing agent MoCA(MoCA wt% = 0.5%, 1%, 1.5%, 2%, 2.5%). It can be seen from the figure that the peel strength, shear strength and elongation at the break all increase at first and decrease

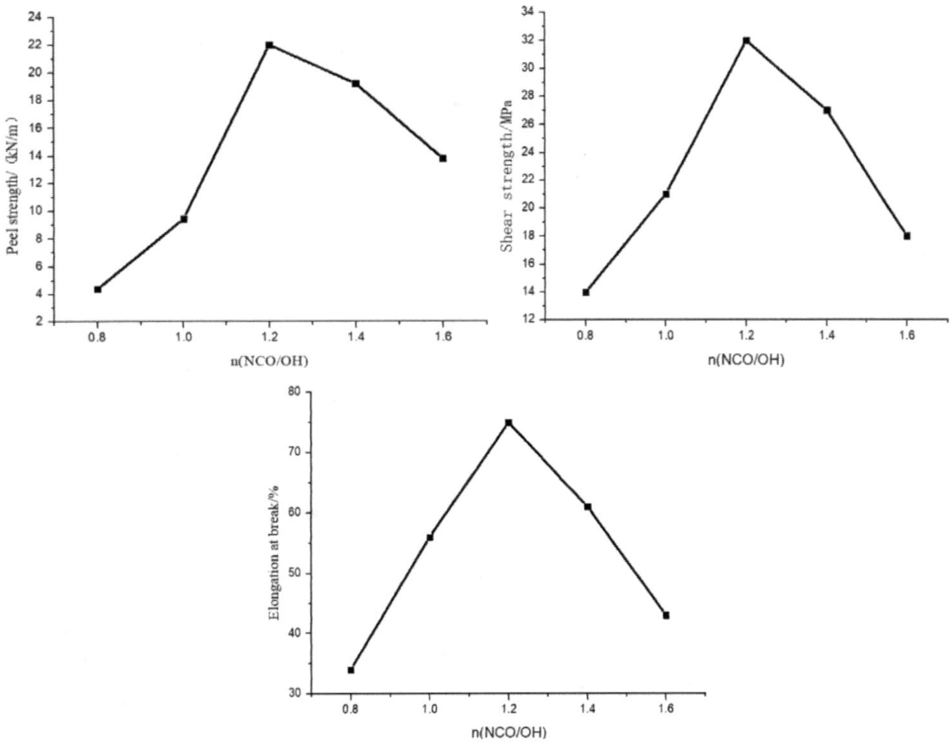

Figure 3. Effect of n (NCO/OH) on mechanical properties of epoxy resin

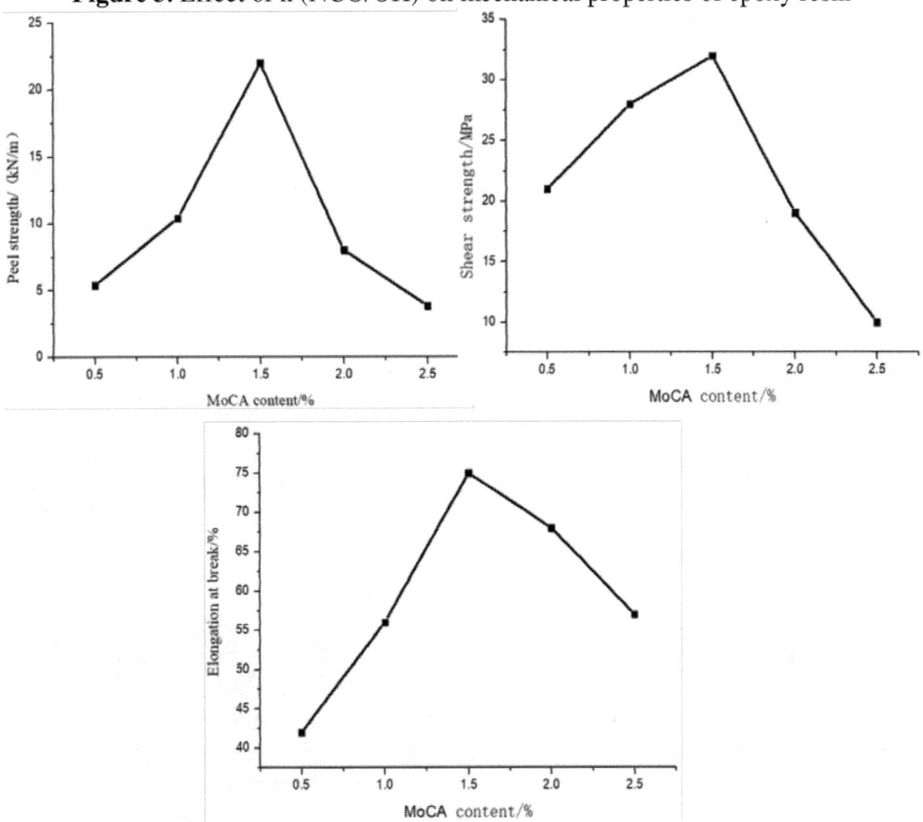

Figure 4. Effect of MoCA content on mechanical properties of epoxy resin

IOP Conf. Series: Materials Science and Engineering **733** (2020) 012027 doi:10.1088/1757-899X/733/1/012027

afterward with the increase of MOCA dosage. This is because the addition of MoCA changes the curing degree of polyurethane prepolymer. With the increase of curing degree, the toughening mode gradually changes from the way of chemical synthesis of the block copolymer to the way of physical mixing epoxy resin with polyurethane. It can be seen that the toughening effect is best when the MoCA addition amount is 1.5%.

3.4. Effect of prepolymer content on mechanical properties

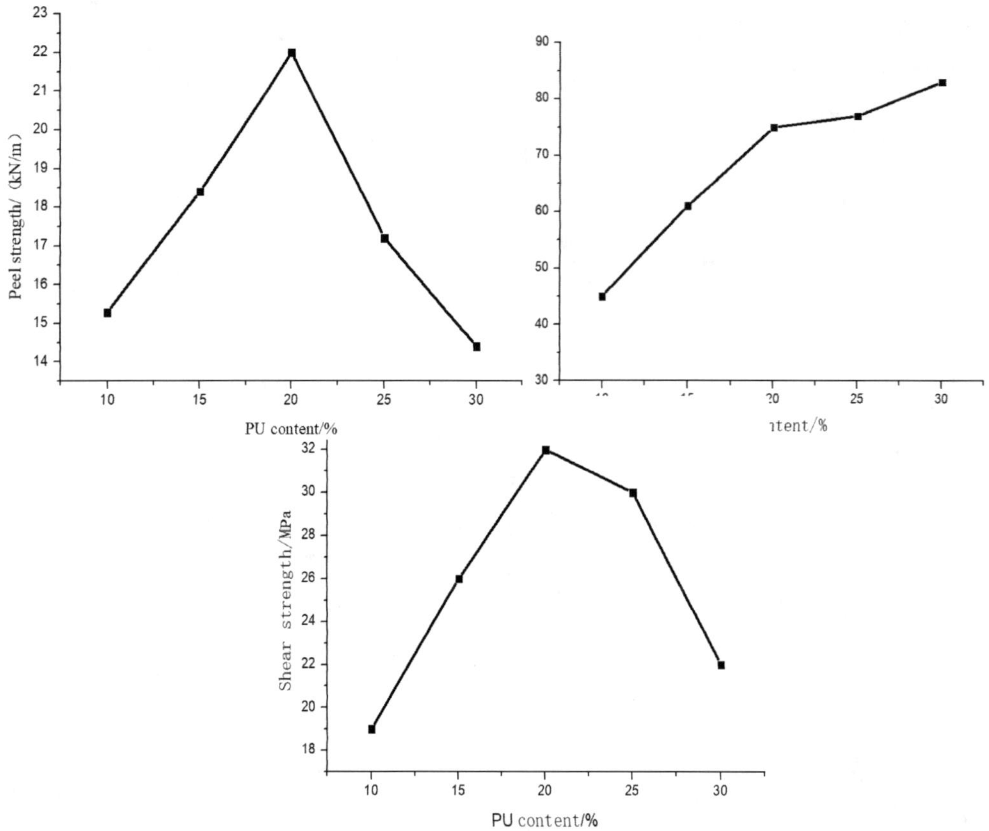

Figure 5. Effect of PU content on mechanical properties of epoxy resin

The mechanical properties of epoxy resin with different mixed content of PU prepolymer were shown in Figure 5. The results show that the break elongation keeps increase, but the peel strength and shear strength of the composite increase first and then decrease afterward with the increase of PU prepolymer. It is due to the more prepolymer is added into the epoxy resin, the more NCO group increase, the more complete reaction with - OH in epoxy resin is, the lower adhesion of epoxy resin is. As results shown in pictures The adhesion effect is best when there are 20% PU prepolymer added into epoxy resin.

4. Conclusion

Polyurethane as a soft segment in-situ polymerized in epoxy resin has obvious toughen the composite. When the molar ratio of -NCO/-OH for the prepolymer synthesized is 1.2, the peel strength of the epoxy resin adhesive/aramid fibers laminated composite reaches the expected goal. Moreover, the shear strength and break elongation are improved, the effect of improvement is significant. The addition of MoCA can control the degree and speed of the pre-polymerization of polyurethane. When the amount of MoCA is 1.5wt%, it will not overheat or have a low polymerization degree, which makes the soft segment in-situ polymerized in epoxy resin have the best toughening effect. This formula retains the excellent adhesive properties of epoxy resin while the defects of low peel strength of epoxy resin in the laminated composite are reduced. It is a kind of modification with practical application significance.

5. References

[1] Meng Z, Xuesong G, Lin W, Gang L, Yifeng Y, Yijun J, Aibing Chen.Study on toughening properties of polyurethane for epoxy resin 2018 *J. Applied Chemical Industry.* 47 pp 1850~53.

[2] Feiming Z, Yunfeng Z, Dan W. Study on epoxy resin adhesive modified with non-isocyanate based polyurethane 2017 *J. Adhesion.* 38 pp 19~24

[3] Yueju W, Jie L, Jianjun Y, Jianan Z, Qingyun W, Mingyuan Wu. Progress in Research and Application of Epoxy Adhesive Modified by Polyurethane 2008 *J. Chemistry and Adhesion.* 30 pp 54-57

[4] Shifu W M 2010 *Fourier transform infrared spectroscopy*(Beijing: Chemical Industry Publisher) pp 291-339

[5] Jieliang W and Guozheng L, Wen Zhao.Interpenetrating polymer networks based on polyurethane 2003 *J. Insulating Materials.* 36 pp 33-37

[6] Ailian L, Dongling Y, Hongxun S, He Yan. Study on the polyurethane-epoxy adhesive 2002 *J. China Adhesives.* 11 pp 8-12

[7] Kehong F, Yundong M, Zizhou P. Study on Toughening Modification of Epoxy Laminate by Polyurethane Particles 2019 *J. Insulating Materials.* 8 pp 11-15.

IOP Conf. Series: Materials Science and Engineering **733** (2020) 012028 doi:10.1088/1757-899X/733/1/012028

Development and Evaluation of a novel High-temperature resistant coating flocculant

Fan Zhang[1], Jinsheng Sun, Kaihe Lv, Xiaofeng Chang

1 School of Petroleum Engineering, China University of Petroleum (East China), Qingdao, Shandong, People's Republic of China
E-mail: zhangfan51@qq.com

Abstract. In the laboratory experiment, 2-Acrylamido-2-methyl-1-propanesulfonic acid (AMPS), N, N-Dimethylacetoacetamide (DMAA), (Z)-Butenedioic anhydrides (MA) and 3-Methacryloxypropyltrimethoxysilane (KH-570) copolymers were used as reaction monomers, potassium persulfate was used as initiator, and a new type of temperature and salt resistant coating flocculant DMA-57 for water-based drilling fluid was synthesized by free radical copolymerization. The synthesis conditions were optimized by orthogonal experiment, and its performance was evaluated. The experimental results show that DMA-57 has good thermal stability, thermal decomposition temperature up to 339 °C. Under the condition of 150 °C, 3.0 MPa and 0.1% dosage, the linear expansion rate of calcium soil can be reduced by 83.83%, and the inhibition rate of calcium soil can be reduced by 67.5%, which has a good effect of flocculation inhibition.

1. Introduction

With the development of drilling technology, various difficult drilling operations such as large angle, horizontal well and directional deviated well are increasing[1, 2]. Therefore, there is a higher standard in the aspects of stabilizing the well wall, preventing the collapse, controlling the hydration and dispersion of drilling cuttings and improving the drilling speed [3-7]. In drilling fluid system, coating agent is an important additive. In the drilling fluid system, coating agent is an important additive, which is used to prevent drilling cuttings and shale from hydration and dispersion, to keep the wellbore stable and prevent collapse. The function of coating agent is to form a protective film through macromolecular polymer on the surface of drilling cuttings and shale, to prevent the cuttings from sticking to each other and increase the screening resistance, so as to form an effective coating on drilling cuttings. The existing flocculants are unstable, easy to deteriorate, large dosage, which cannot meet the needs of drilling construction. Therefore, it is urgent to develop a new strong inhibition flocculant which can meet the needs of high-temperature [8-10].

In this study, a novel inhibiting flocculant (DMA-57) was developed, which has good thermal stability, temperature resistance, good lubrication ability and certain filtration reduction effect, and the effect of inhibiting hydration and dispersion was achieved by adsorption and film formation on clay surface through coating.

2. Materials and methods

2.1. Materials

AMPS (AR), DMAA (AR), and MA (AR) were purchased from Macklin Biochemical Co., Ltd (Shanghai, China). KH-570 (GC), Potassium persulfate (AR), and NaOH (97%) were supplied by Shanghai Wokai Chemical Reagent Co., Ltd. The sodium bentonite used for the drilling fluid cement slurry was purchased from Huawei Bentonite Group Co., Ltd., China. The calcium bentonite used for linear expansion experiment and the drilling cuttings used for rolling recovery experiment were purchased from Bohai Drilling Co., Ltd., China.

2.2 Preparation of polymer

The three monomers of MA, AMPS and DMAA with a certain molar ratio were added into a certain amount of distilled water for complete dissolution (the total monomer concentration of the synthesis reaction was 20 wt-%), and the pH value of the mixture was adjusted to 8.0 with 50 wt-% sodium hydroxide solution. Subsequently, the mixture was placed in a three port flask, heated to 40.0 °C in a constant temperature water bath, and then 0.5 wt-% KH-570 of monomer mass was added. Then, the initiator (potassium persulfate) was added at the rate of 1 drop per 2 seconds. The polymerization reaction should be kept in N2 throughout the whole process. The synthesized product was washed with acetone and dried in a vacuum oven at 60 °C, after grinding, the product of white solid powder is named DMA-57.

2.3 Orthogonal experiment

According to the synthesis principle of polymer fluid loss agent, the main factors that affect the performance of polymer are monomer molar ratio, reaction temperature and initiator dosage. The reaction conditions of synthesis experiment are further optimized by orthogonal experiment. The rolling recovery rate of rock cuttings is taken as the evaluation index of orthogonal experiment. The experimental condition of rolling recovery rate is 150 °C/16h. The orthogonal test of three levels and three factors design is shown in Table 1.

Table 1. Factors and Levels of Polymerization Experiment

Factor level	A	B	C
	Mole ratio	Reaction temperature (°C)	Initiator ratio (%)
1	6:2:2	30	0.3
2	5:3:2	40	0.5
3	4:4:2	50	0.7

2.4 Structure characterisation techniques

The polymer DMA-57 was characterized by IRTRACER-100 FTIR (Shimadzu Switzerland). The spectral range of infrared scanning is 4000-500 cm-1, and the scanning resolution is 4 cm-1. Thermal stability of polymers was measured by Mettler-Toledo thermal analyzer. The heating rate was 10 k/min, the heating temperature was 40-1000 °C, and the protective gas was nitrogen.

2.5 Inhibitive properties

2.5.1 Linear expansion test

In this experiment, using calcium bentonite, 10g calcium bentonite was accurately weighed and placed in a standard mold, and pressurized 10MPa with a manual hydraulic pump for 10min, which is recommended according to the shale expansion test standard. Then, the linear expansion test is carried out with shale dilatometer to measure the expansion amount for 24 hours, the experimental temperature is 150 °C, and the pressure is 3.0 MPa so as to evaluate the inhibition and anti-expansion ability of coating agent on shale [11-13].

2.5.2 Rolling recovery test

Shale samples were crushed on steel plates and core particles were collected between 6 meshes and 10 meshes. The samples were placed in a constant temperature oven and dried at 90 °C for 10 hours. Then, 50 g samples and experimental solution were put into the aging tank and rolled for 16 hours at 150 °C. The remaining shale cuttings were washed with 40 mesh sieve, dried at 103 °C for 4 hours and weighted [14-16].

3. Results and discussion

3.1 Orthogonal Test of DMA-57

It can be seen from the range analysis results of orthogonal experiment that the influence of various factors on the properties of the synthesized products is A > C > B, that is, the mole ratio of monomer is the main factor, followed by the amount of initiator, and then the reaction temperature. It can be seen from the results of orthogonal experiment that the optimal reaction conditions are: mole ratio of DMAA: amps: Ma = 5:3:2, reaction temperature is 40 °C, initiator amount is 0.5% of the total monomer. The results are shown in Table 2.

Table 2. Orthogonal Experiment Results for DMA-57

Sample	A	B	C	Rolling recovery
	Mole ratio of DMAA:AMPS:MA	Reaction temperature/°C	Initiator amount/%	
1	6:2:2	30	0.3	45
2	6:2:2	40	0.5	55
3	6:2:2	50	0.7	40
4	5:3:2	40	0.7	80
5	5:3:2	50	0.3	75
6	5:3:2	30	0.5	78
7	4:4:2	50	0.5	58

IOP Conf. Series: Materials Science and Engineering **733** (2020) 012028 doi:10.1088/1757-899X/733/1/012028

8	4:4:2	30	0.7	51
9	4:4:2	40	0.3	59
K1	130	175	179	
K2	233	194	246	
K3	168	173	171	
R	103	21	75	

3.2 DMA-57 characterization

3.2.1 Structural characterisation of DMA-57

Fig.1 shows the FTIR spectra of DMA-57. The characteristic peaks around 3447 cm^{-1} at DMA-57 were identified as the stretching vibration absorption peaks of –OH in MA. The characteristic peaks around 2983 cm^{-1} at DMA-57 were assigned to the stretching vibration peaks of DMAA. The characteristic peaks around 1292 cm^{-1} were identified as the vibration absorption peaks of the ester group in KH-570. The absorption peaks around 1035 cm^{-1} were identified as the vibration absorption peaks of Si–O in KH-570. The characteristic peaks around 1021 cm^{-1}, which indicated the polymer is partially hydrolyzed to form Si–O–Si. The absorption peak of CO in AMPS was observed at 559 cm^{-1}.The results of FTIR showed that the synthesised product had a target functional group unique to the monomer, all the monomers were fully reacted, and the desired target product was obtained through synthesis [17-20].

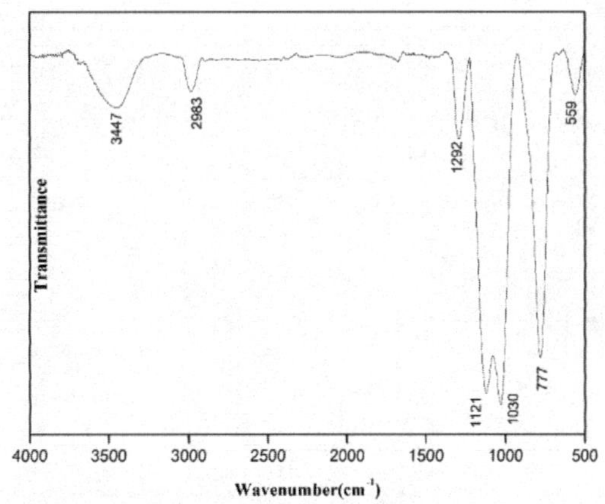

Fig.1 FTIR spectrum of DMA-57

3.2.2 Thermal stability of DMA-57

The TGA curve of DMA-57 is shown in Fig 2. Weightlessness occurring before 185°C is mainly due to the volatilization of water in the sample, the mass of the sample is reduced by 14.1%. The bound water adsorbed by strong hydrophilic groups such as amide group began to volatilize at 185-339°C, the mass of the sample is reduced by 53.3%. The thermogravimetric curve of DMA-57 decreases from 339 to 658 °C. The weight loss is attributed to the breakage of the main chain and the side chain. The thermogravimetric curve of DMA-57 decreases from 658 to 997 °C, the mass of the sample only reduced 2.4%, which proved that DMA-57 had good thermal stability.

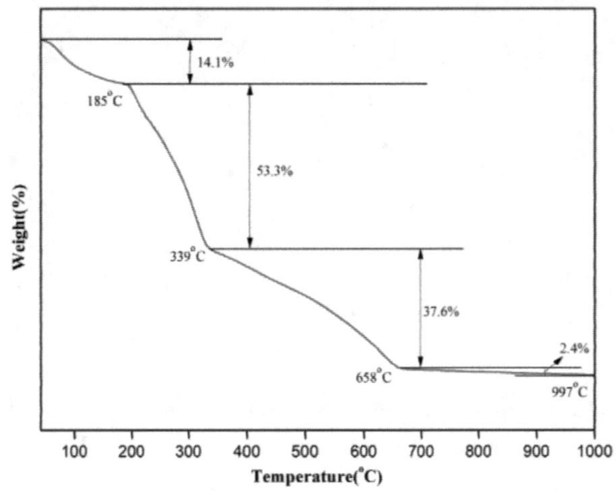

Fig.2 TGA curve of DMA-57

3.3 Inhibition evaluation

3.3.1 Linear expansion test

The linear expansion rate of the coating agent synthesized in this study was measured by experiments, and compared with the coating agent inhibitors commonly used in water-based drilling fluid, in distilled water, the linear expansion rate of calcium soil is 98%. The results are shown in Table 3.

Table 3. Contrast experiment of linear expansion rate at high temperature and high pressure

DMA-57 /%	Expansion rate/%	FA-367 /%	Expansion rate/%	PLUS-L /%	Expansion rate/%	SDJZ-1 /%	Expansion rate/%
0.1	30.5	0.1	35.4	0.1	33.7	0.1	34.3
0.3	27.8	0.3	31.9	0.3	31.4	0.3	32.5
0.5	26.1	0.5	28.4	0.5	29.2	0.5	29.2
0.7	25.2	0.7	28.0	0.7	28.3	0.7	27.8
0.9	24.3	0.9	26.8	0.9	26.7	0.9	26.5

The results of the high temperature and high pressure expansion rate experiment are shown in Table 3. Considering the use cost and the effect of inhibiting expansion comprehensively, comparing with the three commonly used coating agents on the market, the addition of DMA-57 in the range of 0.3% ~ 0.5% has anti-expansion performance.

3.3.2 Rolling recovery test

The recovery of tap water is 15.3%, The rolling recovery is 4.7% in the based slurry system, the results show that the rolling recovery can be increased to 83.83% by adding 0.1% DMA-57, and DMA-57 has a better coating flocculation effect when adding 0.1~0.5%, the rolling recovery can reach 84.26%, which is better than other similar inhibitors.

Table 4. Effect of coating agent dosage on rolling recovery

Coating agent/%	0.1	0.3	0.5	0.7	0.9
DMA-57	83.83	84.26	85.56	87.63	88.62
FA-367	81.41	81.23	82.51	82.96	85.94
PLUS-L	74.51	73.96	76.89	78.19	81.97
SDJZ-1	77.87	75.46	77.19	80.26	82.16

4. Conclusions

(1) The developed coating flocculant DMA -57 has good coating inhibition, thermal stability. The linear expansion test at 3Mpa and 150 °C can verify that the high temperature and high pressure anti expansion ability of DMA-57 is superior to that of similar products.

(2) The amount of coating flocculant DMA-57 is relatively low. Through the rolling recovery experiment, we can see the formation of rock debris, which can effectively flocculate useless solid phase. In terms of adsorption mode, dma-57 can carry out multiple adsorption, which is a strong adsorption of Si-O-Si formed by cation electrostatic adsorption in DMAA and Si-O bond on KH-570 and inorganic silicon on clay.

Acknowledgments

This research was financially supported by Petro China Innovation Foundation (Grants 2018D-5007-0306), Joint Funds of the National Natural Science Foundation of China (No. U1762212), CNPC Science and Technology Project (No. 2018A-3907).

REFERENCES

[1] Pašić B, Gaurina Međimurec N, Matanović D. Wellbore instability: causes and consequences. Rudarsko-geološko-naftni zbornik. 2007;19(1):87-98.

[2] Liu X, Zeng W, Liang L, Lei M. Wellbore stability analysis for horizontal wells in shale formations. J Nat Gas Sci Eng. 2016;31:1-8.

[3] Striolo A, Cole DR. Understanding shale gas: Recent progress and remaining challenges. Energ Fuel. 2017;31(10):10300-10.

[4] Ahmad M. Petrophysical and mineralogical evaluation of shale gas reservoirs: a Cooper Basin case study 2014.

[5] Sondergeld CH, Newsham KE, Comisky JT, Rice MC, Rai CS, editors. Petrophysical considerations in evaluating and producing shale gas resources. SPE Unconventional Gas Conference; 2010: Society of Petroleum Engineers.

[6] Mao H, Guo Y, Wang G, Yang C. Evaluation of impact of clay mineral fabrics on hydration process. Rock and Soil Mechanics. 2010;31(9):2723-8.

[7] Wilson M, Wilson L. Clay mineralogy and shale instability: an alternative conceptual analysis. Clay Minerals. 2014;49(2):127-45.

[8] Balaban RD, Vidal ELF, Borges MR. Design of experiments to evaluate clay swelling inhibition by different combinations of organic compounds and inorganic salts for application in water base drilling fluids. Appl Clay Sci. 2015;105:124-30.

[9] Shadizadeh SR, Moslemizadeh A, Dezaki AS. A novel nonionic surfactant for inhibiting shale hydration. Appl Clay Sci. 2015;118:74-86.

[10] An Y, Jiang G, Ren Y, Zhang L, Qi Y, Ge Q. An environmental friendly and biodegradable shale

inhibitor based on chitosan quaternary ammonium salt. J Petrol Sci Eng. 2015;135:253-60.

[11] Huang X, Shen H, Sun J, Lv K, Liu J, Dong X, et al. Nanoscale Laponite as a Potential Shale Inhibitor in Water-Based Drilling Fluid for Stabilization of Wellbore Stability and Mechanism Study. Acs Appl Mater Inter. 2018;10(39):33252-9.

[12] Jain R, Mahto V, Sharma V. Evaluation of polyacrylamide-grafted-polyethylene glycol/silica nanocomposite as potential additive in water based drilling mud for reactive shale formation. J Nat Gas Sci Eng. 2015;26:526-37.

[13] Yang L, Jiang G, Shi Y, Yang X. Application of ionic liquid and polymeric ionic liquid as shale hydration inhibitors. Energ Fuel. 2017;31(4):4308-17.

[14] Xu JG, Qiu ZS, Zhao X, Zhong HY, Li GR, Huang WA. Synthesis and characterization of shale stabilizer based on polyethylene glycol grafted nano-silica composite in water-based drilling fluids. J Petrol Sci Eng. 2018;163:371-7.

[15] Xiong Z, Tao S, Li X, Fu F, Li Y. Preparation of fluidized catalytic cracking slurry oil-in-water emulsion as anti-collapse agent for drilling fluids. Petroleum. 2016;2(4):361-8.

[16] Patel AD, Stamatakis E, Davis E. Shale hydration inhibition agent and method of use. Google Patents; 2001.

[17] Chen Y, Zhao Y, Zhou S, Chu X, Yang L, Xing W. Preparation and characterization of polyacrylamide/palygorskite. Appl Clay Sci. 2009;46(2):148-52.

[18] Wang Q, Zhao Y, Zhou S, Li M, Xing W, editors. Effect of modification conditions on synthesis polyamide/attapulgite composite material. Computer Distributed Control and Intelligent Environmental Monitoring (CDCIEM), 2011 International Conference on; 2011: IEEE.

[19] Wan T, Li RX, Wu DQ, Hu ZW, Xu M, Cheng WZ, et al. Rheological behaviors and structure of hydrophobically associating AM-SMA copolymers synthesized by microemulsion polymerization. Polym Bull. 2014;71(11):2819-31.

[20] Wu YM, Sun DJ, Zhang BQ, Zhang CG. Properties of high-temperature drilling fluids incorporating disodium itaconate/acrylamide/sodium 2-acrylamido-2-methylpropanesulfonate terpolymers as fluid-loss reducers. J Appl Polym Sci. 2002;83(14):3068-75.

Microstructure analysis of surface coating of graphene modified Cr12MoV die steel by induction cladding

Liu Heping[12], Cheng Shaolei [1], Liu Langlang[1], Sun Fenger[1]

1 School of Materials Science and Engineering, North University of China, Taiyuan 030051, China
2 Chemical and Materials Engineering, University of Alberta, Edmonton, Alberta, Canada

E-mail: peace666@126.com

Abstract In this work, 316 stainless steel powder was used as base powder and graphene was selected as reinforcer. A stainless steel graphene composite coating was successfully prepared by induction cladding on the surface of cold working die steel Cr12MoV. The effects of different coating processes on the microstructure of the coating were investigated by metallographic microscope and scanning electron microscopy. The experimental results show that the bond strength of the coating and the mold steel substrate is perfect. No significant oxides are present in the grain boundaries. With the raise of induced current, the surface hardness of the composite coating increases gradually. In addition, with the increase of graphene content, the surface hardness of the coating also shows the same upward trend.

1. Introduction

The structure of graphene is a planar hexagonal honeycomb composed of a layer of carbon atoms, which has strong corrosion resistance. The corrosion resistance and wear resistance of metals can be improved by using graphene as metal protective layer. Graphene is one of the materials with the highest known strength. It also has good toughness and flexibility. The theoretical Young's modulus of graphene is 1.0 TPa and the intrinsic tensile strength is 130 GPa. Graphene is also increasingly used in surface strengthening of steel materials [1,2].

Generally speaking, wear, corrosion and fracture are the three main forms of material failure. The direct and indirect economic losses caused by corrosion, fracture and wear account for about 4% of the world's annual average GDP. The surface strength and corrosion resistance of steel are always contradictory. At present, the high strength steel materials used are generally high carbon steel. But it also brings an unavoidable problem, that is, the corrosion resistance of steel. But with the discovery of graphene, this contradiction has been well resolved.

Mold is the main processing tool for manufacturing industrial components. The quality of die steel directly affects the quality of pressure processing technology, precision production and production cost of products. The main series of cold-rolled die steel is high hardness and cold work, which is mainly used for dies requiring high compression resistance and wear resistance. The Cr12MoV used in the present work is a cold working die steel, which can reduce the carbon content of high carbon Cr12 steel. Cr12MoV is mainly used to manufacture various dies and tools with large cross-section, complex shape and heavy working load [3-5].

In this paper, a 316L/graphene composite coating was added on the surface of Cr12MoV die steel by induction cladding. The purpose of improving corrosion resistance and surface strength of die steel surface is realized, and the service life and application field of die steel are improved.

2. Experimental

2.1 Matrix Treatment

The Cr12MoV was chosen as the matrix material in the experiment. The die steel is cut into small discs with a diameter of 20 mm and a thickness of 3 mm. Before induction cladding, the two sides of die steel were polished with 600-1200 mesh sandpaper to remove the oxide on the surface of die steel. This is conducive to the adhesion of the coating material to the surface of the sample and to the enhancement of the bonding strength after cladding.

2.2 Mixing of Coated Powder Materials

In this work, the mixed powder of graphene and 316L stainless steel was used as coating material. The mass fractions of graphene were 0.2%, 0.3%, 0.4% and 0.5% respectively. In order to make the mixture of graphene and stainless steel powder uniform, dry ball milling was used in the experiment. The rotational speed of the ball mill is 200 r/min, the milling time 2 h, and the ball-to-material ratio 10:1.

2.3 Induction Cladding

The powder is coated on the surface of the sample which has been pretreated separately and compacted to make the surface smooth. Then the coated sample is placed in the coil of the induction cladding instrument. The high frequency induction heater is produced by Zhengzhou Kechuang Electronics Co., Ltd. The inverted frequency is 20-60 KHz, and the maximum heating current is 630 A. The induction heating current used in the experiment is 360A, and the induction heating time is 20s.The effects of different induction heating currents on the properties of the coating were also studied. The induction heating currents were 280 A, 320 A, 360 A and 400A, respectively. The surface of die steel was pretreated before induction heating, and the thickness of powder laying was 0.5-1 mm.

2.4. Detection

The morphology and structure of the samples before and after electrochemical measurement were characterized by scanning electron microscopy (SEM, SU-5000), metallographic microscope (MDS 400). The surface hardness of the samples was tested by micro Vickers hardness tester (HVS-1000A).

3. Results and Discussion

3.1 Effect of different graphene content on cladding layer

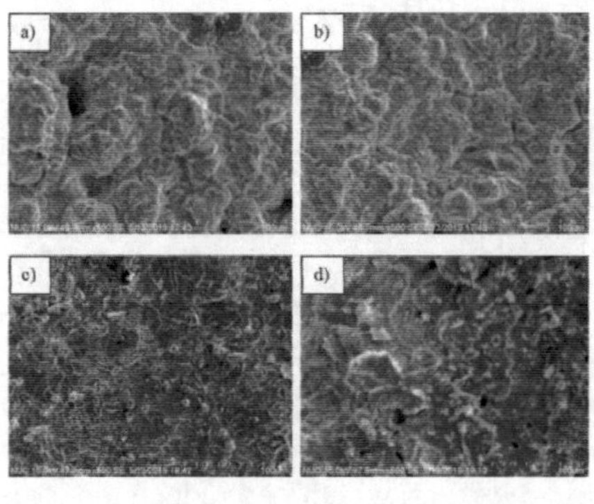

Figure 1. Surface morphology of cladding layer with different graphene content
(a) 0.2% (b)0.3% (c) 0.4% (d) 0.5%

In Figure 1, the cladding effect is good at different graphene content, and there are no large pore and particulate matter on the surface of the cladding layer. Figure 1 (a) and (b) have almost the same effect as the cladding layer, but the void content in Figure 1 (b) decreases and the surface becomes smoother. When the content of graphene reaches 0.4%, the surface properties of the cladding layer are the best, and no obvious voids are produced. However, with the increase of graphene content, many small holes appear in Figure 1 (d), and the smoothness of the cladding layer surface decreases.

With the increase of graphene content, graphene exists at the grain boundary of the cladding layer. When the content of graphene is small, it has little effect on the cladding layer. However, as the content increases, graphene increases the risk of oxidation of grain boundaries at high temperatures, resulting in a large number of carbides and oxides at the grain boundaries. As a result, a small number of micro-pores appear on the surface of the cladding layer in Figure 1 (d).

3.2 Effect of Different Induction Currents on Cladding Layer

In this experiment, the output induction current intensity is proportional to the output power. The larger the induced current, the greater the output power and the more heat energy the sample receives. Four samples with graphene content of 0.3% were prepared by different induced currents. After induction cladding, the original structure of the coating surface was observed by scanning electron microscopy (SEM). The surface structure of the coating is shown in Figure 2.

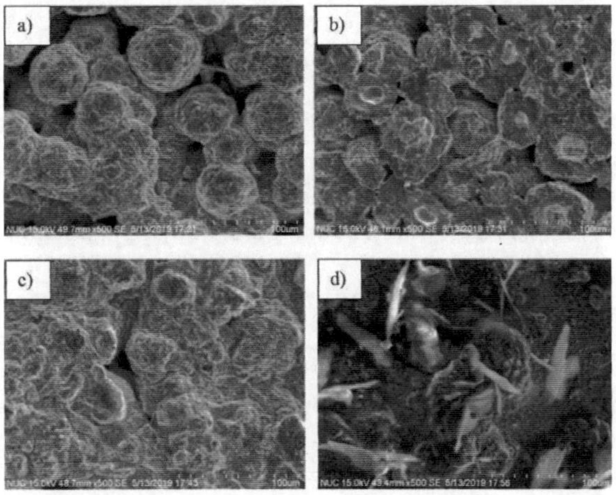

Figure 2 Surface morphology of cladding layers with different induced currents
(a) 280A (b) 320A (c) 360A (d) 400A

It is observed that when the induced current is 280A, a large number of unmelted particles can be seen on the surface of the cladding layer from Figure 2 (a). In Figure 2 (b), when the current rises to 320A, the unmelted stainless steel powder particles on the surface of the cladding layer decrease. It can be seen from Figure 2 (c) that there are no obvious unmelted powder particles on the surface of the cladding layer, and that the surface of the cladding layer is relatively flat, but there are still a few holes, which are caused by insufficient induction heating temperature; while Figure 2 (d) shows that there are a few over-burning phenomena on the surface, but the over-burning is not obvious. Induction cladding current is large and the temperature of induction heating is too high.

3.3 Metallographic analysis of cladding layer with different graphene content

Figure 3 is a photographed cross-section metallographic photomicrograph of the different graphene contents at an induced current of 360A. Compared to Figure3 (a) and (b), the cladding layers in Figure3 (c) and (d) are more dense and the grains are relatively small. The effect of graphene on the structure of cladding layer can be seen from Figure 3. In Figure 3, when graphene content is 0.2% and 0.3%, graphene migrates to the bonding zone and agglomerates during induction cladding. A small amount of oxides appeared in the cladding layer. In Figure 3 (c) Graphene content is 0.4%, the oxide content in the cladding layer increases, and the agglomeration of graphene is more obvious, which leads to the increase of defects in the cladding layer. However, graphene, originally near the bonding zone, also appears in the cladding layer. The graphene content is 0.5% in Figure 3 (d). Because of the high content of graphene, the agglomeration of graphene near the bonding zone is intensified, and the oxide in the cladding layer is also more. At the same time, graphene diffuses to the matrix at high temperature, and the phenomenon of grain boundary oxidation appears in the matrix

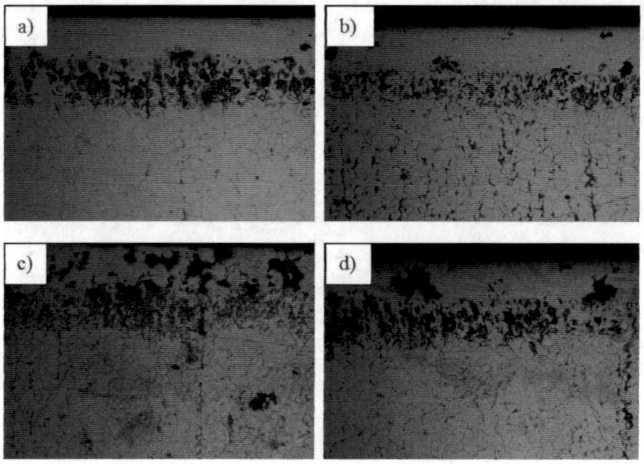

Figure 3 Cross-sectional metallographic photos of samples with different graphene content (200X)
(a) 0.2% (b) 0.3% (c) 0.4% (d) 0.5%

3.4 Metallographic Analysis of Cladding Layer under Different Induction Heating Currents

Figure 4 is a metallographic photograph of different induced current intensities. The graphene content is 0.3%. The cross-sectional structure of the sample was divided into matrix and coating, and no obvious transition zone was observed. Graphene in the cladding layer migrates and agglomerates to the bonding zone.

In induction cladding, the surface temperature of the sample is very high. First, the coating melts by heat transfer. Because the melting point of Cr12MoV is lower than 316L stainless steel, the surface of the sample dissolves preferentially to form a molten pool, and there is oxidation more or less. The surface of the coating contacts with the external environment, which has the fastest heat dissipation, a large temperature gradient, and forms equiaxed crystals. The grain size becomes smaller and the structure is more compact.

Contrast diagrams Figure 4(a), (b), (c), (d) show that with the increase of induction current, the oxidation of the sample surface intensifies. The best cladding effect is that the induction current is 360A. At this time, there are fewer oxides in the cladding layer and fewer defects in the cladding layer.

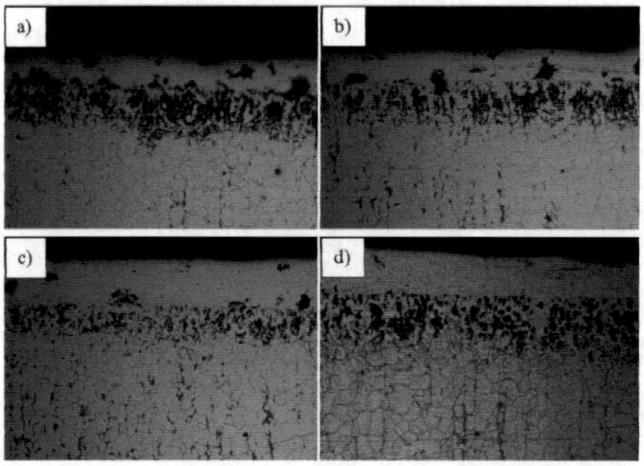

Figure 4. Metallographic micrographs of different induced current intensities (200X)
(a) 280A, (b) 320A, (c) 360A, (d) 400A

3.5 Effect of graphene content on mechanical properties

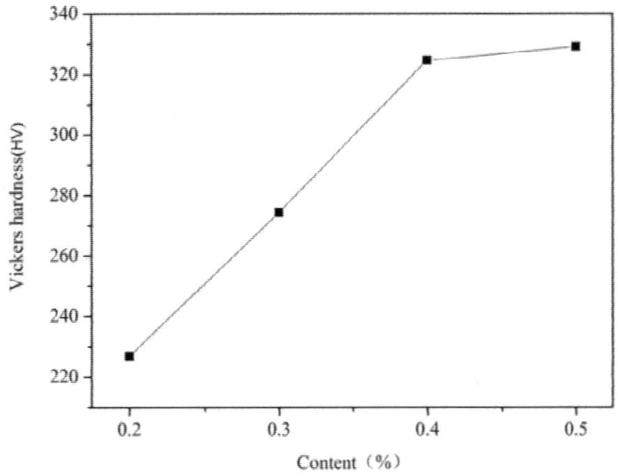

Figure 5 Vickers hardness curve with different graphene content

The purpose of this experiment is to study the effect of different graphene content on the hardness of metal coatings. The test parameters are set as follows: Four groups of samples are selected, and the induced current intensity is 360A. The graphene content was 0.2%, 0.3%, 0.4% and 0.5% respectively. Figure5 shows Vickers hardness curve of graphene coatings with different content by induction cladding. The surface hardness of the cladding layer increased with the increase of graphene content. However, when graphene content reaches 0.5%, the slope of Vickers hardness curve decreases.

In the process of induction cladding, the hard phase with more graphene content precipitates, and the graphene in the coating distributes evenly. As the second phase, the hard phase also prevents the movement of dislocations and plays the role of pinning dislocations. Therefore, the hardness of the coating increases with the increase of graphene content. It is found that the increase of graphene content can significantly improve the microhardness of the coating. However, the agglomeration of graphene in the matrix leads to the increase of oxides in the cladding layer and the increase of defects in the cladding layer. The enhancement effect of graphene on the cladding layer is reduced. So when the graphene content is 0.5%, the slope of the curve in Vickers hardness decreases.

3.6 Effect of Different Induction Heating Currents on Mechanical Properties of Coatings

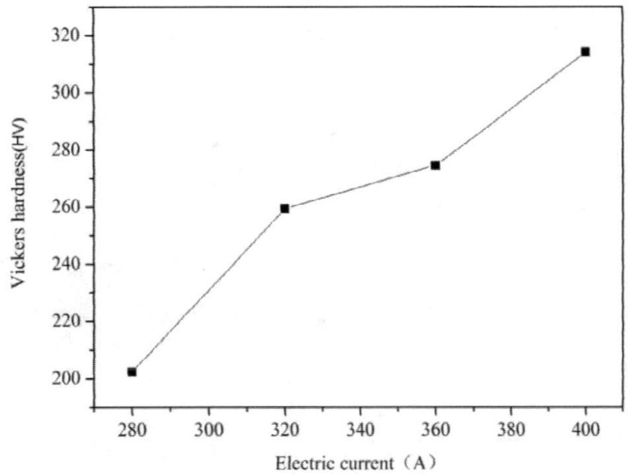

Figure 6 Hardness curves at different induced currents

The experimental scheme is as follows: using four groups of samples, the induced current intensity is 280 A, 320 A, 360 A and 400A respectively, and the content of graphene is 0.3%. Figure 6 shows the microhardness curves at different induction currents.

Combined with metallographic images, when the current of induction cladding is 280-400A, the surface hardness of the cladding layer increases with the increase of current intensity. At the beginning, with the increase of induction current, the unmelted powder in the cladding layer decreases, and the bonding strength of the cladding layer surface increases. Therefore, the surface hardness of the cladding layer will be improved. However, when the induction current is 360A, graphene agglomeration occurs in the bonding zone of the cladding layer, and the defects in the cladding layer decrease, so the hardness of the cladding layer surface increases. When the induced current reaches 400A, a small amount of overburning occurs in the cladding layer, and obvious grain boundary oxidation occurs in the cladding layer. These oxides act as hard particles in the cladding layer, so the hardness of the cladding layer will continue to increase[6]. However, the existence of these oxides also leads to the decrease of the toughness of the cladding layer, which makes the material prone to fracture.

4 Conclusion

1. Graphene has obvious strengthening effect on the surface of cladding layer.
2. Graphene agglomerates at the bonding zone of the cladding layer. With the increase of graphene content, the defects such as oxides on the surface of the cladding layer also increase.
3. Different induction heating currents have great influence on the cladding effect. When the induction heating current is large, the surface of the cladding layer will appear obvious overburning phenomenon.

Acknowledgments

This research work was financially funded by the China Scholarship Council (No.201908140060), the Postdoctoral Science Foundation of China (No.2016M590214), the Key Research and Development (R&D) Projects of Shanxi Province (No.201803D121028).

References

[1] Zhou H, Jiang M, Xin Y, et al. 2016, *Surface deposition of graphene layer for bioactivity improvement of biomedical 316 stainless steel*. Materials Letters, 192: p 123-127.

[2] Baig Z, Mamat O, Mustapha M. 2018, *Recent Progress on the Dispersion and the Strengthening Effect of Carbon Nanotubes and Graphene-Reinforced Metal Nanocomposites: A Review*. Critical Reviews in Solid State & Materials Sciences, 43(6): p 1-46.

[3] Pang G X , Chen Z Y , Li Z L , et al. 2011,*Research on Ion Nitriding Temperature Effect on Wear Resistance of 35CrMo Steel*. Advanced Materials Research, 341-342:p 128-132.

[4] Wu B, Peng L, Fei Z, et al. 2018, *Effect of parameters on picosecond laser ablation of Cr12MoV cold work mold steel*. Applied Physics A, 124(1): p 11.

[5] Sun F, Zhang D Q, Cheng L, et al. 2019,Microstructure Evolution Modeling and Simulation for Dynamic Recrystallization of Cr12MoV Die Steel During Hot Compression Based on Real Metallographic Image[J]. Metals and Materials International,p 1-16.

[6] Zhao Y, Jie S, Li J. 2014, *Effect of rare earth oxide on the properties of laser cladding layer and machining vibration suppressing in side milling*. Applied Surface Science, 321(3):p 387-395.

First-principles investigations of structural, elastic, thermodynamic and electronic properties of Pt_3Hf compound under pressure

Zongbo Li[1, 2], Kai Xiong[1, 2*], Yingjie Sun[1], Xiuhua Chen[1], Junjie He[1], Shunmeng Zhang[1], Yingjie Fu[1], Yong Mao[1, 2*]

1 School of Materials Science and Engineering, Yunnan Provincial Engineering Laboratory of Copper-based and Advanced Conductive Materials, Yunnan University, Kunming 650091, China.
2 Materials Genome Institute of Yunnan University, Yunnan University, Kunming 650091, China.
E-mail: xiongkai@ynu.edu.cn, maoyong@ynu.edu.cn

Abstract. The pressure effects on the structure, elasticity, thermodynamics and electronic properties of Pt_3Hf compounds were investigated by first-principles calculations based on density functional theory. The obtained equilibrium lattice parameter is consistent with the experimental and theoretical values. The calculation results show that the elastic constants, bulk modulus, shear modulus, Young modulus, anisotropy factor and Debye temperature, sound velocities increase with the increase of pressure. The B/G ratio and Poisson's ratio results show an upward trend with increasing pressure, which means that increasing external pressure can enhance the ductility of Pt_3Hf compound. Analyzing the charge density difference and the density of states reveal that charge accumulation gradually increases with the increase of pressure between Pt and Hf atoms.

1. Introduction

Platinum group metal alloys (PGMs) have attract great attentions because of their have high melting point, superior oxidation high specific strength, and corrosion resistance [1], which means that it important in high temperature structural materials applications field. Especially, the high-strength properties of Pt-based intermetallic compounds is far beyond the traditional Ni-based super-alloys [2]. Among Pt-based alloys, Pt_3Hf intermetallic compound has attract physical properties on both on industrial and high-temperature field. Among them, Pt_3Hf compound is an $L1_2$ structure with good mechanical properties in high temperature structural materials.

It is well known that pressure plays an important character in the mechanical, elastic, thermodynamic and electronic properties of material performances[3]. Therefore, studying the elastic properties of Pt_3Hf intermetallic compounds under extreme high pressures conditions is of great significance for Pt-based alloys. However, there are few reports on the experimental and electronic calculations of the structural properties of Pt_3Hf intermetallic compounds at high pressure. Investigations on the structure, elasticity, thermodynamic and electronic properties of Pt_3Hf compound under high pressure are valuable and urgently needed for its high-temperature applications. For Pt_3Hf, the previous studies mainly investigated the mechanical properties elastic constant elastic modulus at zero GPa. POPOOLA et al. [4] And the investigations of high temperature thermodynamic properties in Pt-Hf systems for published paper [5]. Because it is difficult to achieve and control under the experimental conditions under high pressure, but can easily achieved under simulated conditions to provide some guidance for the experiment. However, there are no literature reports on experimental or theoretical studies on the structure, elasticity, thermodynamics and electronic properties of Pt_3Hf intermetallic compounds under high pressure.

Thus, it is very urgent and meaningful to investigate the mechanical and thermodynamic properties of Pt_3Hf intermetallic compound with pressure. In this paper, we focus on the structure, elastic, electronic and thermodynamic properties of Pt_3Hf intermetallic compound under high pressure (from 0~100 GPa).

2. Computational method

The calculations were performed within the frame of density function theory (DFT) by using the Vienna plane wave ab initio simulation package (VASP) [6]. Using the projector augmented wave method, (PAW) [7] exchange-correlation functional was treated by the generalized gradient approximation (GGA) with the Perdew–Burke–Ernzerhof (PBE) [8]. The atoms basis PAW was selected on $5d^9 6s^1$ for Pt and $5d^2 6s^2$ for Hf. The kinetic cut-off energy of 500 eV was shown an excellent convergence of the total energies and structural parameters. The Brillouin zones of the unit cells were represented by the 9×9×9, 15×15×15 Monkhorst-Pack grid meshes. The energy change was converged to lower than 1.0×10^{-5} eV, and the maximum force less than 0.01 eV/Å.

3. Results and discussion

3.1 Structural properties

The Pt_3Hf intermetallic compound belongs to the Cu_3Au - type structure. The Hf and Pt atoms locate at the (0, 0, 0) and (0.5, 0.5, 0) sites, respectively. The Pt_3Hf intermetallic compound is shown in Fig. 1. The Birch-Murnaghan (BM) equation of state (EOS) expressed as [9]:

$$E(V)=E_0+\frac{9V_0B_0}{16}\left([(\frac{V_0}{V})^{\frac{2}{3}}-1]^3B_0'+[(\frac{V_0}{V})^{\frac{2}{3}}-1]^2[6-4(\frac{V_0}{V})^{\frac{2}{3}}]\right) \quad (1)$$

Where V_0 represent equilibrium volume, E_0 represent total energy, B_0 repersent equilibrium bulk modulus, and B'_0 represent pressure derivative.

Figure 2 show the relationship between total energy and volume change of Pt_3Hf intermetallic compounds. Obviously, the equilibrium volume at the lowest energy is 65.13 (Å^3). The Table 1 shown the equilibrium lattice parameters a (Å), equilibrium volume V (Å^3), bulk modulus B_0 (GPa), its pressure

IOP Conf. Series: Materials Science and Engineering **733** (2020) 012030 doi:10.1088/1757-899X/733/1/012030

derivative B'_0 and density ρ (g/cm^{-3}) of Pt_3Hf intermetallic compound. The present results are good consistent with the available experimental data [10] and the reference theoretical calculations results [11], which improve the data is reliable. The calculated density ρ is about 0.8% lower than the result in Ref [12]. The lattice parameters ratios a/a_0 and lattice volume ratios V/V_0 as functions of pressure in the range from 0 to 100 GPa are shown in Fig. 3, explicit that the a/a_0 and V/V_0 decrease gradually with the increase pressure.

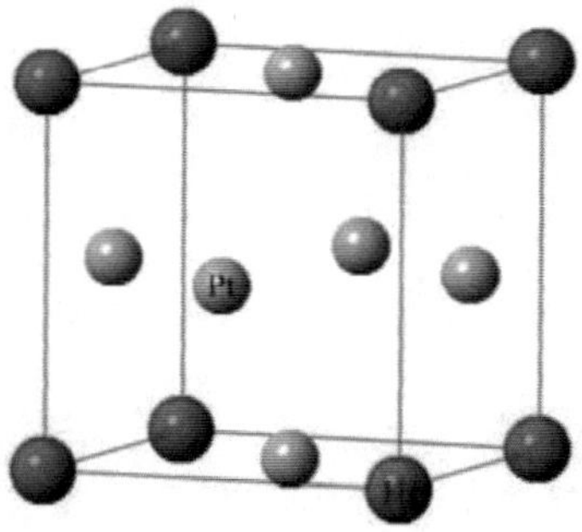

Fig. 1. The crystal structure of $L1_2$ Pt_3Hf. The light gray atoms represent Pt, and the red atoms are Hf.

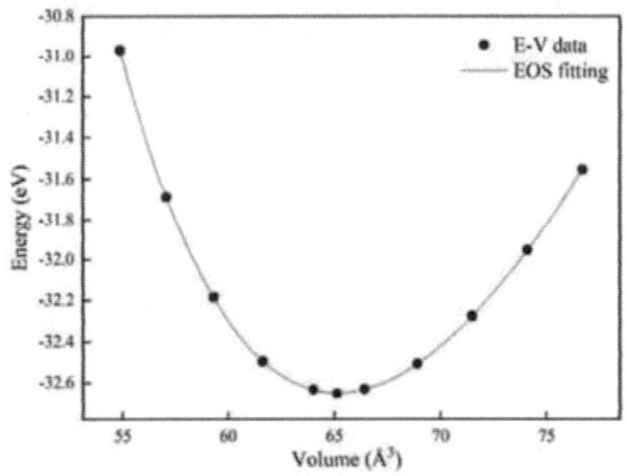

Fig. 2. The E-V data and EOS fitting curve of Pt_3Hf.

Table 1. The equilibrium lattice parameters a (Å), equilibrium volume V (Å^3), density ρ (g/cm^{-3}), bulk modulus B (GPa), and its pressure derivative B' of Pt_3Hf calculated at zero pressure, compared with available experimental and theoretical results from references.

alloy		a [Å]	V [Å^3]	B_0 [GPa]	B'_0	ρ [g/cm^{-3}]
Pt_3Hf	This work	4.023	65.13	236.99	4.9	19.470
	Exp	4.020 [a]				
	Cals	4.029 [b]		238 [c]		19.626 [d]

a [13], b [11], c [14], d [12]

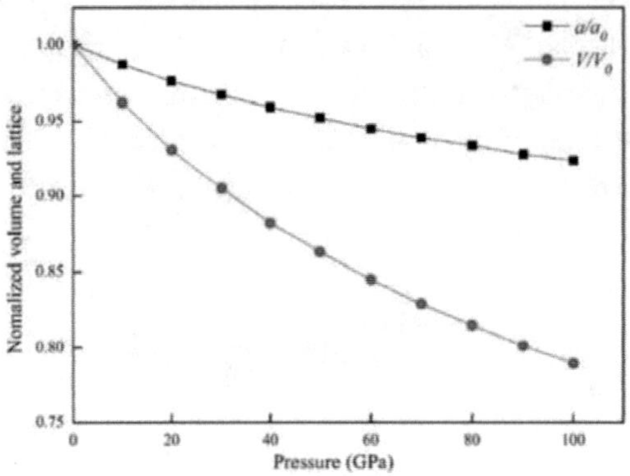

Fig. 3. The variation of the lattice parameter ratio a/a_0 and the lattice volume ratio V/V_0 of Pt_3Hf compound with increasing pressure

3.2 Elastic properties

The elastic constants make great important parameters to reflecting the relation between stress and strain in a crystal. It is used to describe the mechanical crystals, dynamic and thermodynamic properties. There are three independent elastic constants used to describe the structure of cubic crystal system (C_{11}, C_{12} and C_{44}) [15]. The bulk modulus and shear modulus of Pt_3Hf intermetallic compound are calculated by using the Voigt-Reuss-Hill (VRH) approximation [16]. The calculation criteria are defined as [17]:

$$B=(C_{11}+2C_{12})/3 \tag{2}$$

$$G_V=(C_{11}-C_{12}+3C_{44})/5 \tag{3}$$

$$G_R=5(C_{11}-C_{12})C_{44}[4C_{44}+3(C_{11}-C_{12})] \tag{4}$$

$$G=(G_V+G_R)/2 \tag{5}$$

$$E=9GB/3B+G \tag{6}$$

The calculated elastic constants and elastic modulus, B/G ratio, Poisson's ratio (v) and anisotropy index (A^Z) are shown in Table 2. Obviously, the values are good consistent with available theoretical data at zero pressure. The calculated elastic constants satisfy the mechanical stability criterion of cubic crystal $C_{11} > 0$, $C_{44} > 0$, $C_{11} - C_{12} > 0$, $C_{11} + 2C_{12} > 0$ under different pressures [18], indicating that Pt_3Hf intermetallic compound is satisfy mechanical stability within the scope of 0 to 100 GPa.

Table 2. The calculated elastic constants and elastic modulus, B/G ratio, Poisson's ratio (v) and anisotropy index (A^Z) of Pt_3Hf under different pressure (in GPa).

	p	C_{11}	C_{12}	C_{44}	B	G	E	v	B/G	A^Z
Pt_3Hf	0 [a]	351	181	126	238	120	309	0.280	1.98	
	0	353	178	129	236	110	287	0.298	2.13	1.48
	10	403	206	153	272	128	333	0.296	2.11	1.56
	20	450	232	176	305	145	377	0.294	2.09	1.61
	30	496	261	197	340	160	416	0.296	2.11	1.68
	40	542	291	219	374	175	454	0.298	2.13	1.74
	50	583	318	238	406	188	490	0.299	2.15	1.79

60	625	347	257	440	201	523	0.302	2.18	1.84
70	667	376	276	473	213	557	0.304	2.21	1.89
80	706	403	294	504	225	589	0.305	2.23	1.94
90	746	429	313	535	238	622	0.306	2.24	1.97
100	783	452	329	562	249	652	0.307	2.25	1.99

a [14]

The bulk modulus B measures the resistance to volume changes. The shear modulus G measures the resistance to shape change. The Young's modulus E defined the stiffness of the material. The larger the result of Young's modulus E, the harder is the material. As shown the figure 4 shows the calculated elastic modulus of Pt_3Hf intermetallic compound under pressure, the bulk modulus B shear modulus G Young's modulus E increase linearly with the increase the pressure. Thus, the addition pressure contributes to an increase hardness and stiffness of the Pt_3Hf intermetallic compound.

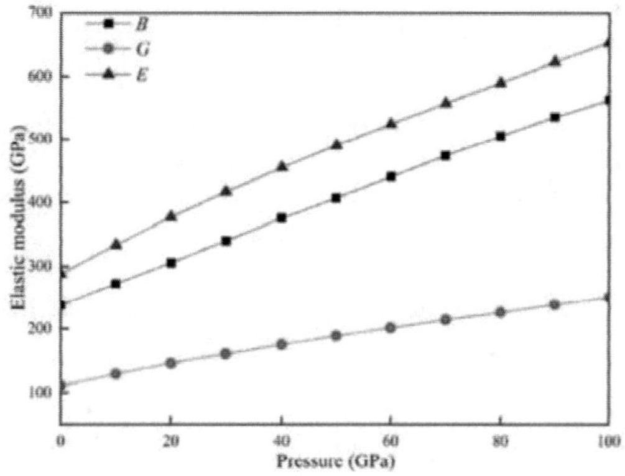

Fig. 4. The calculated the elastic modulus of Pt_3Hf intermetallic compound with the different pressures.

The research of the mechanical properties of materials usually considers the effects of micro cracks and lattice distortion on its mechanical properties and the elastic anisotropy of elastic properties is usually a necessary condition for the two. Therefore, it is necessary to study the elastic anisotropy to improve the mechanical durability of the material. Figure 5, the calculated 3D Young's modulus graph with the different pressure. For cubic structure Pt_3Hf, the Young's modulus E calculation criteria are defined as [19]:

$$\frac{1}{E}=S_{11}-2(S_{11}-S_{12}-\frac{S_{44}}{2})(l_1^2l_2^2+l_2^2l_3^2+l_1^2l_3^2) \tag{7}$$

The l_1, l_2 and l_3 represents the direction cosines in the sphere coordination, S_{ij} represent the elastic compliance constant. It can be obtained from the inverse of the elastic constant matrix. As is shown the Fig. 5, the Pt_3Hf intermetallic compound three-dimensional stereograms of Young's modulus increase with the increasing pressure, which means higher pressure can improve the elastic anisotropy values and good to the hardness of the Pt_3Hf intermetallic compound.

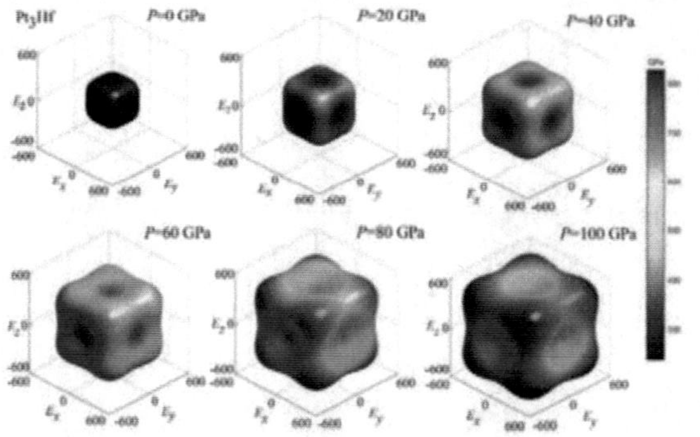

Fig. 5. The evolution of Young's modulus for Pt_3Hf as the pressure changes.

Elastic anisotropy play an important role in physical and mechanical properties of material [20]. The anisotropy factor (A^Z) measures the degree of anisotropy of the materials. If the A^Z value equal to one, means the material completely isotropic. While the value departed from the value of one, means the elastic anisotropy of this material. Form Table 2 and Fig. 6 (a), obviously that the all results of A^Z no equal to one and larger than one and increase with increasing the pressure for Pt_3Hf intermetallic compound, which means that elastic anisotropy can be improved.

$$A^Z = \frac{2C_{44}}{C_{11}-C_{12}} \tag{8}$$

It is well known that Poisson's ratio v (-1 < v < 0.5) is used to measure the material's ability to resistance shear. The larger the value of Poisson's ratio, the stronger the resistance to shearing, the better the plasticity of the material [21]. If the value of v is less than 0.26, indicate the material it a brittle manner. Form Fig. 6 (b), it can observe all the Poisson's ratio values of Pt_3Hf intermetallic compound are larger than 0.26 and increase with the pressure increases, it indicates that the increase in pressure can improve the plasticity.

$$v=(E-2G)/2 \tag{9}$$

Pugh [22] proposes the value of the ratio of B/G to judge the plasticity and brittleness of the material behaviors. The larger value of the B/G ratio related to the ductility, otherwise the low B/G ratio value related to brittleness. If $B/G > 1.75$, the material behavior in a ductile nature, otherwise the material behavior in a brittle manner. As it can see form the Fig. 6 (c), the B/G ratio values performed an upward trend with raise the pressure, and the B/G ratio all values larger than 1.75, which indicates that Pt_3Hf intermetallic compound has a good ductility at higher pressures.

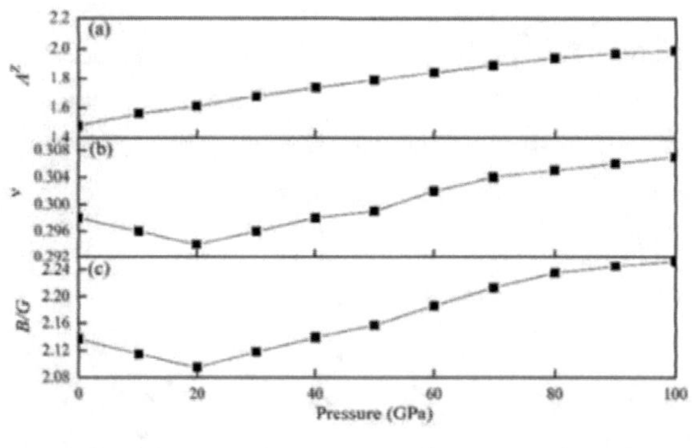

Fig. 6. The calculated the (a) B/G ratio, (b) Poisson's ratio (v), and (c) the anisotropic index (A^Z) of Pt_3Hf

with the increase of pressures.

3.3 Debye temperature

The Debye temperature is a crucial role parameter describing the material. To further understand the thermodynamic properties and lattice stability under pressure, it is meaningful to calculate the Θ_D values at different pressures. We calculate the Deby temperatures for Pt_3Hf intermetallic compound calculated by using the average sound velocity and given by reference [23]:

$$\Theta_D=\frac{h}{k}[\frac{3n}{4k}(\frac{\rho N_A}{M})]^{\frac{1}{3}}v_m \tag{10}$$

Where h is the Planck's constant, k_B is the Boltzmann's constant, N_A is the Avogadro number, ρ is the mass density of the material, M is the molecular weight, n is the number of atoms per formula unit. And the average elastic wave velocity V_m is given by the equation:

$$v_m=\left[\frac{1}{3}\left(\frac{2}{v_t^3}+\frac{1}{v_l^3}\right)\right]^{-1/3} \tag{11}$$

Where V_l and V_t are longitudinal and transverse elastic sound velocities, respectively, which given by Navier's equations [24].

$$v_l=\sqrt{\left(B+\frac{4G}{3}\right)\Big/\rho} \tag{12}$$

$$v_t=\sqrt{G/\rho} \tag{13}$$

Form the table 3, Fig. 7 and Table. 3 display the pressure dependence of longitudinal (v_l), transverse (v_t), average sound velocities (v_m), and Debye temperature (Θ_D) of Pt_3Hf intermetallic compound increasing at the pressure increase. As the pressure increases, the covalent bond between the Hf atom and the Pt atom of the Pt_3Hf intermetallic compound increases. The present calculated results could be provided an effective prediction for the future experiment.

Table 3. The calculated longitudinal (v_l), transverse (v_t), average sound velocities (v_m) (m/s) and Debye temperature Θ_D (K) of Pt_3Hf compoundwith different pressure P (GPa).

Alloy	P	v_l	v_t	v_m	Θ_D
Pt_3Hf	0	4445.58	2386.23	2664.57	312.99
	10	4684.58	2522.53	2816.12	335.01
	20	4888.30	2640.17	2946.81	354.42
	30	5077.73	2733.25	3051.45	370.46
	40	5251.98	2818.43	3147.24	385.41
	50	5402.37	2891.86	3229.81	398.43
	60	5547.54	2956.97	3303.52	410.31
	70	5685.32	3018.98	3373.69	421.75
	80	5809.34	3075.53	3437.61	432.21
	90	5925.78	3132.82	3501.98	442.82
	100	6026.24	3182.57	3557.85	452.26

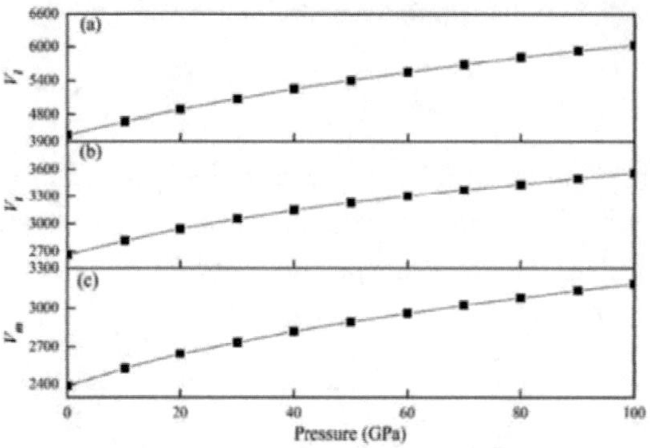

Fig. 7. The calculated the v_m (m/s) (a), v_t (m/s) (b) and v_l (m/s) (c) of Pt_3Hf compound with the different pressure.

3.4.Electronic properties

Electronic density of states is used to understand the mechanical and thermodynamic properties of materials. In order to further understand the electronic structure and chemical bonding strength character of Pt_3Hf compound, the total density of states (TDOS) and the density of partial states (PDOS) are calculated by first-principles based on the density function theory. Form the Fig. 8, there is no energy gap near the Fermi level, so the Pt_3Hf intermetallic compound exhibits metallic properties. The main bonding peaks of Pt_3Hf is predominantly derived from the Pt-5*d* and Hf-5*d* states in energy range, and the energy region from -10.2 eV to 7.4 eV. Meanwhile, we find the pseudogap occurs above Fermi level, shown the covalent bonding in Pt_3Hf compound.

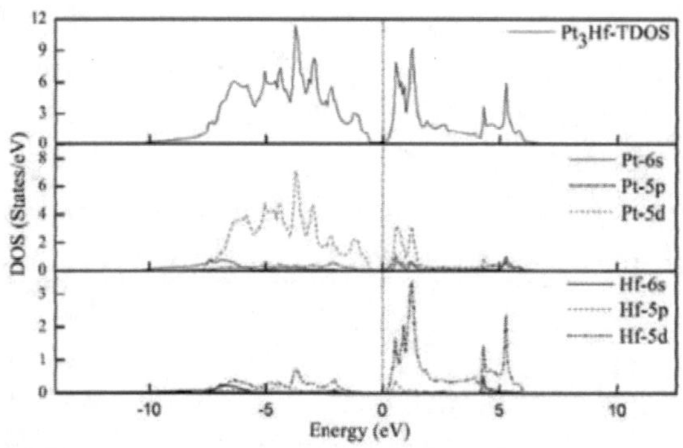

Fig. 8. The total density of states (TDOS) and partial density of states (PDOS) for Pt_3Hf at 0 GPa.

The charge density difference distribution directly reflects the physical image of charge distribution in the system. The redistribution of charge density caused by pressure can be clearly shown, so that it has a more intuitive understanding of the bonding characteristics between atoms. The charge density difference is calculated as:

$$\Delta\rho=\rho(Pt_3Hf)-\rho(Pt_3)-\rho(Hf) \quad (14)$$

Where $\rho(Pt_3Hf)$, $\rho(Pt_3)$ and $\rho(Hf)$ are the charge densities for the corresponding species with the same atoms coordinate. The charge density difference distribution is visualized by VESTA [25]. The effect of pressure on the charge density difference distribution has been investigated. From the Fig. 9, it is shown the charge density difference in the (001) plane of Pt_3Hf at 0 GPa to 100 GPa. And we can know the lower density (blue) region higher density (red) region. And it is clear that with the pressure increase the charge is transfer Pt and Hf atoms and the covalent bonding between Pt and Pt is enhanced. The ionic bonding between Pt and Hf becomes stronger in pressure increase.

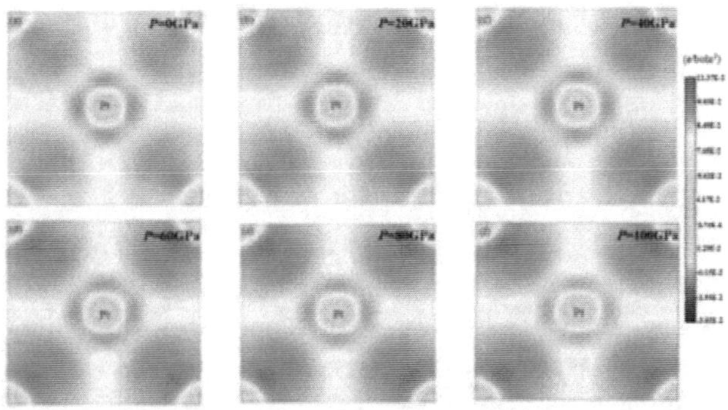

Fig. 9. Two-dimensional plane view of charge density difference distribution for Pt_3Hf intermetallic compound. The red color region indicates the positive charge density and blue one indicates the negative charge density.

4. Conclusions

To summarize, the structural, elastic, thermodynamic and electronic properties of Pt_3Hf compound under high pressure are studied by using the generalized approximation (GGA) calculations. The relevant conclusions are listed as follows:

(1) The calculated lattice parameter and volume are consistent with the experimental and theoretical results at zero pressure. The obtain elastic constant C_{ij}, bulk modulus B, shear modulus G, Young's modulus E, B/G ratio, Poisson's ratio (ν), are calculated. The results show that the mechanical more stable, has a high hardness and stiffness, ductility and plasticity all enhance under high pressure. The calculated anisotropic A^Z and 3D curved surface of Young's modulus relevant are all anisotropic at zero pressure and are more obviously anisotropic under high pressure.

(2) The Density of states (DOS) shows that the Pt_3Hf compound exhibits metallic characteristics. The difference charge density reveals the change of charge transfer under high pressure. It can conclude that the bonding strength between the atoms of the Pt_3Hf compound is enhanced under high pressure.

Acknowledgments

This work was financially supported by the National Natural Science Foundation of China (Nos. 51801179 and U1502272), the Yunnan Science and Technology Projects (Nos. 2018ZE001, 2018ZE006, 2018ZE016, 2018ZE021 and 2018ZE023), the Yunnan Applied Basic Research Projects (Nos. 2018FB083 and 2018FD011), the Yunnan Education Department Projects No. 2017ZZX232.

References

[1] L. A. Cornish, R. Süss, A. Watson, S. N. Prins, Building a Thermodynamic Database for Platinum-Based Superalloys: Part I. Platinum Metals Review 51, 104-115.

[2] Yanjun Liu, Huawei Huang, Yong Pan, Guanghui Zhao, Zheng Liang, First-principles study on the phase transition, elastic properties and electronic structure of Pt_3Al alloys under high pressure. Journal of Alloys and Compounds 597 (2014), 200-204.

[3] Xiao-Li Yuan, Mi-An Xue, Wen Chen, Tian-Qing An, Yan Cheng, Investigations on the structural, elastic and electronic properties of the orthorhombic Zirconium–Nickel alloy under different pressure. Computational Materials Science 65, 127-132.

[4] Adewumi Isaac Popoola, Lesley Heath Chown, Lesley Alison Cornish, Theoretical investigations of Pt3X (X = Al, Sc, Hf, Zr) ground state. Turkish Journal of Physics 38 (2014), 10-16.

[5] P. J. Meschter, W. L. %J Metallurgical Transactions A Worrell, An investigation of high temperature thermodynamic properties in the Pt-Zr and Pt-Hf systems. 8 (1977), 503-509.

[6] Z. H. Sun, X. X. Wang, A. K. Soh, H. A. Wu, On stress calculations in atomistic simulations. Modell. Simul. Mater. Sci. Eng. 14 (2006), 423-431.

[7] J. Cormier, J. M. Rickman, T. J. Delph, Stress calculation in atomistic simulations of perfect and imperfect solids. J. Appl. Phys. 89 (2001), 99.

[8] J. P. Perdew, K Burke, M Ernzerhof, Generalized Gradient Approximation Made Simple. Physical Review Letters 77 (1998), 3865-3868.

[9] Francis Birch, Finite Elastic Strain of Cubic Crystals. Physical Review 71 (1947), 809-824.

[10] R. J. Wagner, L. Ma, F. Tavazza, L. E. Levine, Dislocation nucleation during nanoindentation of aluminum. Journal of Applied Physics 104 (2008), 601.

[11] Yuhua Zhang, Zhiyao Duan, Chan Xiao, Guofeng Wang, Density functional theory calculation of platinum surface segregation energy in Pt3Ni (111) surface doped with a third transition metal. Surf. Sci. 605 (2011), 1577-1582.

[12] Weiwei Xing, Xing-Qiu Chen, Dianzhong Li, Yiyi Li, C. L. Fu, S. V. Meschel, Xueyong Ding, First-principles studies of structural stabilities and enthalpies of formation of refractory intermetallics: TM and TM3 (T=Ti, Zr, Hf; M=Ru, Rh, Pd, Os, Ir, Pt). Intermetallics 28 (2012), 16-24.

[13] L. M. Pecora, P. J. Ficalora, Some bulk and thin film properties of ZrPt3 and HfPt3. Journal of Electronic Materials 6 (1977), 531-540.

[14] ADEWUMI ISAAC POPOOLA, COMPUTATIONAL STUDY OF NOBLE METAL ALLOYS. (2013), pp. 143-156.

[15] J. Wang, S. Yip, S. R. Phillpot, D. Wolf, Crystal instabilities at finite strain. Phys Rev Lett 71 (1993), 4182-4185.

[16] V. Paidar, Yi-Shen Lin, M. Cak, V. Vitek, Why is the slip direction in CuZn and FeAl different than in CoTi? Intermetallics 18 (2010), 1285-1287.

[17] Liu Na, Xueye Wang, Yali %J Materials Chemistry Wan, Physics, First principles calculations of structural, elastic, electronic properties of Ir 3 Zr with L1 2 structure under high pressure. 162 (2015), 807-812.

[18] Dongmin Shi, Bin Wen, Roderick Melnik, Shan Yao, Tingju Li, First-principles studies of Al–Ni intermetallic compounds. Journal of Solid State Chemistry 182 (2009), 2664-2669.

[19] Jinghan Wang, Sidney Yip, S. R. Phillpot, Dieter Wolf, Crystal instabilities at finite strain.

Physical Review Letters 71, 4182-4185.

[20] Martin Ostoja-Starzewski, Luis Costa, Shivakumar I. Ranganathan, Scale-Dependent Homogenization of Random Hyperbolic Thermoelastic Solids. Journal of Elasticity 118 (2014), 243-250.

[21] B. Mayer, H. Anton, E. Bott, M. Methfessel, J. Sticht, J. Harris, P. Schmidt, Intermetallics 11 (2003) 23e32.,

[22] S. F. Pugh, XCII. Relations between the elastic moduli and the plastic properties of polycrystalline pure metals. The London, Edinburgh, and Dublin Philosophical Magazine and Journal of Science 45 (2009), 823-843.

[23] Orson L. Anderson, A Simplified Method for Calculating the Debye Temperature From Elastic Constants. Journal of Physics & Chemistry of Solids 24 (1963), 909-917.

[24] P Wachter, M Filzmoser, Rebizant Jean, Electronic and Elastic Properties of the Light Actinides Tellurides. 293 (2001), 199-223.

[25] K Momma, F Izumi, VESTA 3 for three-dimensional visualization of crystal, volumetric and morphology data. Journal of Applied Crystallography 44 (2011), 1272-1276.

IOP Conf. Series: Materials Science and Engineering **733** (2020) 012031 doi:10.1088/1757-899X/733/1/012031

Investigating the effect of the conductivity of the electron transportation layer on the organic light-emitting diodes

Jinxiang chen[1 *], Shipan Wang[1], Hang Zhou[1]
1.Peking University ShenZhen Graduate School University Town, Nanshan District, Shenzhen 518055, P.R.China

Email: chenjinxiang1@tcl.com; Email: zhouh81@pkusz.edu.cn

Abstract. In this paper, we choose three types of n-doped electron transport layer (ETL) materials with different chemical structures to fabricate only-electron carrier device for studying the conductivity of them, namely, 100 nm ETL1, 40 nm ETL2, and 23 nm ETL3. It is also analyzed that the effect of the potential barrier height on the carrier injection efficiency of the single-carrier device when they directly contact with the electrode having the different work function ITO (indium tin oxide film) electrode. When the thickness of the organic electron transport layer is thin， the current density is mainly limited by the injection barrier in the interface, However, When the thickness of the organic electron transport layer is thick, the organic electron transport layer with very low carrier mobility is a dominant factor in limiting the current through the device. If ETL1 is in contact with a lower work function aluminum electrode relative to the ITO electrode, the effect of the injection limit on the current is weaker. Considering Poole-Frenkel effect and space charge limited current effect, the field strength-dependent carrier mobility of ETL1 material is estimated by linear-fitting the current-electric field characteristic curve of only-electron device. Organic red light-emitting diode is fabricated with ETL1 thicknesses of 10 nm and 100 nm, respectively. For the organic light-emitting device with 100 nm ETL1, it reaches 14.9 cd/A at a current density of 30 mA/cm^2 , and the color coordinates is (0.658, 0.341).

1. Introduction

Organic light-emitting diodes(OLED) have been widely used in the fields of illumination, panel display, etc., and organic semiconductors(OS) are sandwiched between the different work function electrode to form the basic device frame of OLED.[1] In order to improve the turn-on voltage and luminous efficiency,except for the proper selection of the metal electrode, an interface modifying layer is introduced between the electrode and the organic layer.[2] The later one is a common method that works well. It is well known that the closer the work function of the electrode is to the Fermi level of the semiconductor, the lower the injection barrier can is, so the turn-on voltage of the OLED will be decreased and the luminous efficiency could be increased. Here, a single carrier device are fabricated using these ETL materials having different chemical structures. The effect of the potential barrier from the direct contact between the high work function electrode and the organic semiconductor electron transport layer(ETL) on the carrier transportation will be analyzed from the measured current-electric field (J-E) characteristic curve.

By comparing the *J-E* curves at forward voltages bias with that at negative voltage bias, we can get some information about the interface barrier height. The thickness of ETL is comparable to that of the real OLED device. Actually, the thickness of The thickness of ETL is intensively related to the current density through the whole device. If the contact interface is an ideal ohmic contact, the material bulk-resistance is dominant in limiting current density.[3] In this paper, the luminous efficiency of the OLED is experimentally proved to be improved by adjusting the thickness of the ETL.

The ETL material widely used is usually disordered amorphos hydrogen-carbon materials, and the electronic state of such materials is the localized state disorderly.[4] Almost, all carriers are localized, there are no real free carriers, the migration of most of free electrons in organic semiconductor materials is limited by high-density trap state, and the energy levels of the trap for the organic semiconductor are similar, so most of which are essentially a p-type semiconductor with the hole mobility greater than the electron mobility.[5] However, although the carrier concentration of the ETL is generally relatively low, the conductivity can be significantly improved by doping a certain amount of electron donor material as an additive. [6]

Firstly, it is well known that the carrier mobility of most OS exhibits an electric field dependency. But, here, we introduce a constant factor H, that can be expressed by the Poole-Frenkel(P-F) equation

$$\mu(E) = \mathrm{H}\mu_0 e^{\beta\sqrt{E}} \tag{1}$$

μ_0 is the zero field mobility and β is the poole-frenkel factor.

The no-trap space charge limited current

$$J = \frac{9}{8}\,\varepsilon_r\varepsilon_0\,\frac{E^2}{L}\mu \tag{2}$$

J is the current density, E is the electric field, L is the thickness of the film, ε_r is relative dielectric constant and ε_0 is vacuum dielectric constant.

and so

$$J = \frac{9}{8}\,\varepsilon_r\varepsilon_0\,\frac{E^2}{L}\mathrm{H}\mu_0 e^{\beta\sqrt{E}} \tag{3}$$

It can also be rewritten as

$$\ln\frac{J}{E^2} = \ln\left(\frac{9}{8}\,\varepsilon_r\varepsilon_0\,\frac{\mathrm{H}\mu_0}{L}\right) + \beta\sqrt{E} \tag{4}$$

The deduced formula includes the information that the carrier mobility of the material is a function of the electric field strength.[7]

The conductivity of metal materials well obeys Ohm's law. However, for insulators, low carrier mobility semiconductors, and organic materials with generally high trap densities, current-voltage characteristics is often studied using the P-F effect. The P-F emission effectively explains the phenomenon that as the electric field strength increases, the conductivity of such materials increases. For the low carrier mobility for organic semiconductor materials, under the high-intensity electric field, the internal potential well is lowered to cause the restricted carriers to escape, increasing its conductivity. The formula is as follow.

$$J = \mathrm{C}\,E\,e^{\left[\frac{-\mathrm{q}\left(\varphi_B - \sqrt{\frac{\mathrm{q}E}{\pi\varepsilon_r\varepsilon_0}}\right)}{\mathrm{kT}}\right]} \tag{5}$$

Where φ_B is the barrier height, q is the unit charge, and ε_r is the relative dielectric constant. C is constant factor, k is Boltzmann constant.

When the Poole-Frenkel(P-F) effect becomes the dominant mechanism of conduction current, ln(J/E) and $\sqrt{E}$ should be linear.[8]

2. Experiment

In order to investigate the conductivity of the n-doped electron transport layer, a single electron device was prepared. The structure of a single carrier device is shown in Figure 1. Patterned indium tin oxide (ITO)-coated glass with a sheet resistance is cleaned by ultrasonic in deionized water. The thicknesses of ETL1, ETL2, and ETL3 are 100 nm, 40 nm, and 23 nm, respectively. The devices in Figure 1 (a), (b), (c) are respectively named device A, device B and device C. The ETL by the thermal evaporation covers the ITO anode or the ITO/Al anode. The current density–voltage (J–V) characteristics is recorded with a Keithley 2400 source unit. This paper analyzes the J-E curve under the external electric field by simulation in order to extract the electrical properties of the ETL such as mobility, dielectric constant and so on.

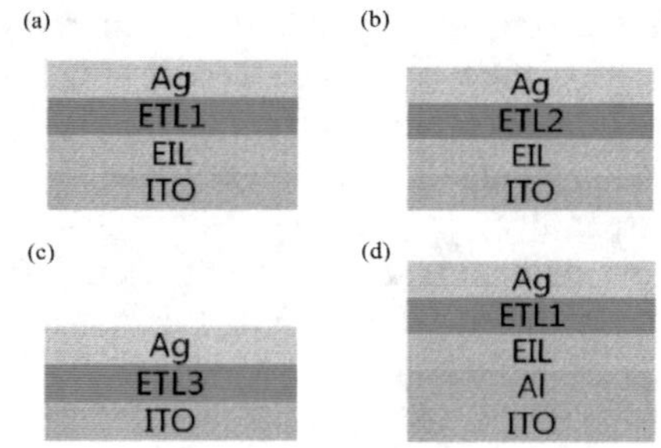

Figure 1.the device structure of the diode based on the ETL doping with Liq_3. (a)、 (b)、 (c) respectively represents the device A、 the device B and the device C.(d) represents the improved structure of ETL1 device ,named as device D.

3. References

Figure 2 shows the current density–field(*J-E*) intensity characteristics of a single-carrier device made by different ETL materials under the forward and negative electric field. The ITO electrode is regard as the positive electrode, reversely, the Ag electrode is the negative electrode. Both device A and device B show bidirectional conductivity. Likely, there is an electron injection layer (EIL) between the ITO electrode and the ETL. The device C is only positively conductive. To some extent, this indicates that large contact potential barrier is formed between ETL3 and the ITO electrode. In general, its Fermi level of the n-type doped ETL can rise above the forbidden band, near the Lowest Unoccupied Molecular Orbital (LUMO) conduction band. In result, there is a very large contact barrier when it directly contacts with the high work function ITO electrode. However, the potential barrier varies with the applied external electric field, the potential barrier will be weakened or enhanced.The applied forward voltage will lower the potential barrier, and the reverse voltage will make the barrier increases, which is how the rectifier diode works. In the ETL3 device structure, the *J-E* characteristic clearly shows the rectification characteristics that the forward current is significantly larger than the reverse current.

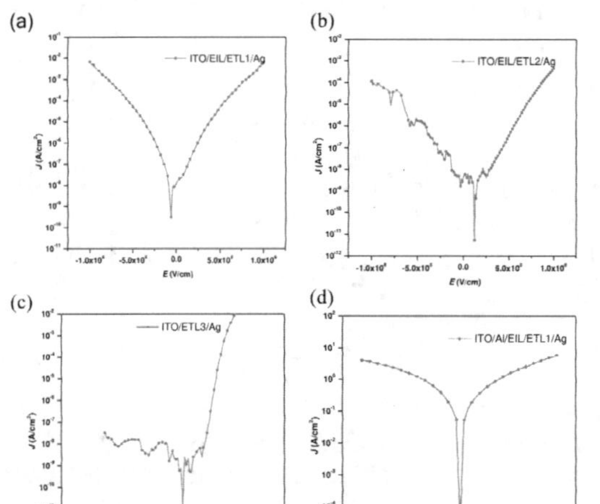

Figure 2. the *J-E* characteristic curve of the diode device with ITO electrode or ITO/Al electrode.

For the device A, the ETL reaches 100 nm, the electric field intensity scanned from -1 MV/cm to +1 MV/cm. When the electric field strength is +1 MV/cm, the current density is 6.4×10^{-3} A/cm^2, then, the electric field strength in the negative direction is -1 MV/cm, and the current density is close to the forward current density, which is 6.8×10^{-3} A/cm^2. For the device A, the thickness of the ETL2 material is only 40 nm. When the electric field strength reaches +1 MV/cm, the current density is only 4.64×10^{-4} A/cm^2. Such low current density means that the limiting-current factor is not only the material bulk-resistence but the interface potential barrier.

We further apply the space charge-limited current model (SCLC)and the P-Femission model to analyze the current density of the device A.

The scan voltage applied to the device A changes from negative bias to positive bias, we get the current density-field intensity curve. Interesting, it is noticeable that the current density is not minimum under the zero-electric filed. Importantly, there is already a certain amount of space charge accumulation inside the device, so, if the current under the forward voltage is analyzed, the effect of space charge on conductivity should be considered. At the same time, for a 100 nm thick ETL material, the field-dependent mobility is not negligible.

Figure 3(a) shows the calculated data used for linearly fitting with Equation (3).The result is as follow.

$$Y=0.0114*X-43.677 \tag{6}$$

Y is the vertical coordinate, X is the horizontal coordinate.

Here, the intercept is -43.677. Based on the latter information in this paper ,we order that the relative dielectric constant is 3.6, so, the $H\mu_0$ is 2.99×10^{-12} cm^2/V^{-1}s^{-1}, where the Poole-Frenkel factor is 0.0114. Substituting them into Equation (1), Figure 3 (b) shows the carrier mobility as a function of electric field strength.

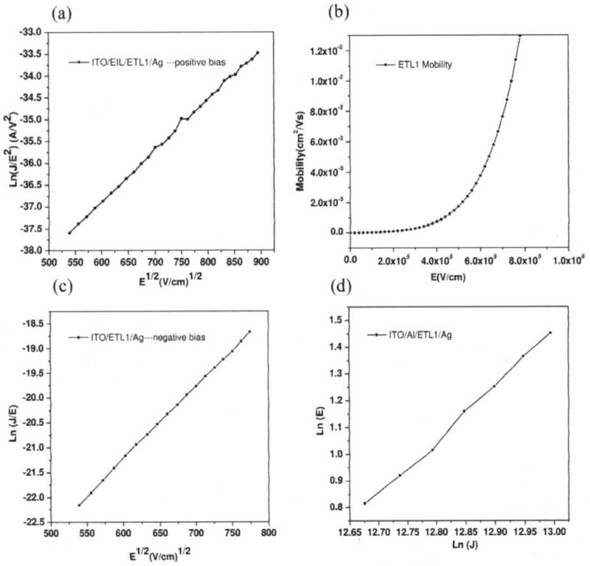

Figure 3. (a) The $\ln(J/(E^2))$ versus $E^{1/2}$ curve characteristic of ITO/ETL1/Ag device applied with positive voltage bias. (b) Carrier mobility μ versus electric field intensity E curve characteristic of ITO/ETL1/Ag device. (c) $\ln(J/E)$ versus $E^{1/2}$ curve characteristic of ITO/ETL1/Ag device applied with negative voltage bias.(d) $\ln(E)$ as a function of $\ln(J)$ for the analysis of the SCLC.

At the start of scanning, the negative electric field is applied to the fresh device A with no current through it ever, the current-voltage characteristic curve is obtained. It is necessary to consider that both the interface contact potential barrier and the P-F effect influence the conductivity. Based on the previous report, The current density as function of the field intensity is fitted using Equation (5).[9] The result is as follow.

IOP Conf. Series: Materials Science and Engineering **733** (2020) 012031 doi:10.1088/1757-899X/733/1/012031

$$Y=0.0146*X-30.019 \quad (7)$$

It is approximately inferred that the relative dielectric constant of the ETL1 material is 3.63,which is in the range of the dielectric constant of the common ETL.It can also be calculated that the height of the barrier is 0.818 V, which is lower than the expected value. It may be that the EIL layer lowers the barrier potential height.

In order to further explore the influence of the interface barrier height on the current-voltage characteristics, the structure of the ETL1 device is improved, and the lower work function metal Aluminium(Al) was vacuum thermal-evaporation deposited on the ITO electrode. According to the literature, the work function of the Al is 4.2 eV, which is lower than that of ITO.The work function of the ITO is 4.7 eV, and the *J-E* characteristic curve is obtained as shown in Fig. 2 (d). Compared with the pure ITO electrode, the current density is greatly improved under the same electric field strength. When the electric field intensity reaches 0.5 MV/cm, the current density reaches 5.92 A/cm^2, much larger than that of pure ITO electrode which is 4.729×10^{-5} A/cm^2. This means that the lower work function metal Al electrode is more conductive.

The double-logarithm of the *J-E* characteristic curves under the forward electric field is shown in Fig. 3(d), we find that when the electric field strength increases to 0.32 MV/cm, the space charge limiting current (SCLC) region gradually appears.

Using the formula (2) to analyze the SCLC from the *J-E.* we get the following linear formula

$$Y=2.0439*X-25.109 \quad (8)$$

From the above formula, the carrier mobility is 3.47×10^{-4} cm^2/V^{-1}s^{-1}, and it can be seen that the electron mobility is greatly improved after ETL1 n-type doping. So, when the electric field intensity is 0.32 MV/cm, the carrier mobility 3.47×10^{-4} cm^2/V^{-1}s^{-1}.During the SCLC process, the mobility is seen as be constant. Lastly ,we calculate the H value is 5.46×10^{-6}.The reason why the value is greatly low is unkown.The zero-field mobility is 5.49×10^{-7} cm^2/V^{-1}s^{-1}.

Intriguingly, red-light OLED devices are prepared using a 10 nm ETL1 film and a 100 nm ETL1 thick film. At around 1000 nits, the color coordinates, the applied voltage, the current density and the current efficiency and luminescence peaks are listed in table 1. The luminous efficiency of thin ETL1 OLED is apparently higher than that of thick ETL1 material. So,the thickness of ETL is one of the parameters which affect the efficiency of the OLED.

Table 1. the characteristic of the organic light-emitting device with different ETL1 thickness

Thickness (nm)	Luminance (nits)	CIE x	CIE y	Voltage (V)	J (mA/cm^2)	CE (cd/A)	Peak (nm)
10	1380	0.656	0.344	3.5	5.60	24.6	616
100	1090	0.659	0.341	4.0	6.19	17.6	616

The corresponding curve is shown in Fig.4. The red-lighting OLED with 100 nm ETL1 has a current efficiency of 14.9 cd/A at a current density of 30 mA/cm^2, a colour coordinate (0.658, 0.341), and a power efficiency of 7.82 lm/W. Unfortunately, the current efficiency commonly decreases with the drop of the current density. It can be seen that the peak of the spectrum is at 616 nm, and there is shoulder peak at near 680 nm, it means that there are more than one kind of photoexcitons. As can be seen from Fig. 4(d), the colour coordinates hardly change with the current density.

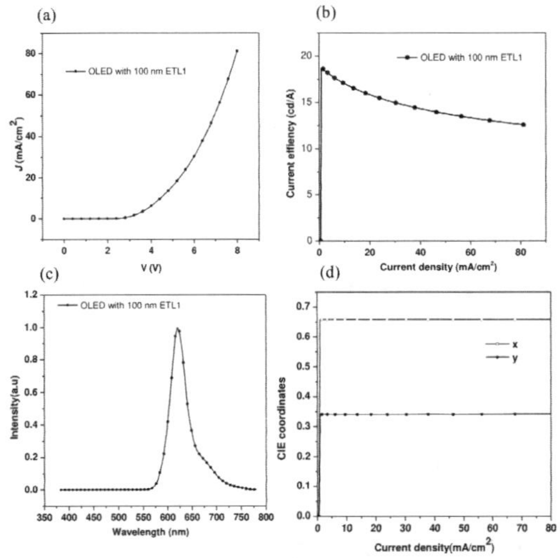

Figure 4. (a) *J-V* characteristic of the conventional structure OLED with 100 nm ETL1.(b) the current efficiency (CE) versus the current density. (c) the wavelength spectrum of the red-emitting OLED. (d) the CIE color coordinates versus the current density .

4. Conclusion

In this paper, the effects of interface contact potential barrier and organic semiconductor bulk material mobility on J-E characteristics of single-carrier devices were studied. Here, we reveal the competition correlation between the contact potential barrier and the bulk-resistance.When the organic film is thin, the current-limiting factor is the interface property. When the ETL material reaches a certain thickness, the limitation of the material bulk-resistance becomes further stronger. Then, comparing the performance of organic light-emitting devices made with two ETL1 thickness, the ETL1 thin film benefits higher luminous efficiency. Single-carrier devices were prepared using n type-doped ETL. The J-E characteristic curves were measured. Considering the Poole-Frenkel effect and the space-charge limiting current effect, the electric field-dependent electron mobility was obtained. The space charge-limited current model was used to fit the J-E curves from the ITO/Al/ETL1/Ag device, and the SCLC electron mobility was determined to be 3.47×10-4 $cm^2/V^{-1}s^{-1}$. So ,this is a useful method to evaluate the carrier mobility of the film.

References

[1] K. Walzer, B. Maennig, M. Pfeiffer, and K. Leo 2007 *Chem. Rev.* 107 1233
[2] G. W. Kim, Y. H. Son, H. I. Yang 2017 *Chem. Mater.* 29 8299
[3] P. N. Murgatroyd 1970 *J. Phys. D: Appl. Phys.* 3 151
[4] W. Vollmann 1974 *Phys. Stat. Sol. (a)* 22 195
[5] A. Kohler *Nat.Mater.* 2012 11 836
[6] A. G. Werner, F. Li, K. Harada, M. Pfeiffer et al 2003 *Appl. Phys. Lett.* 82 4495
[7] C.T. Sun, I.H. Chan, P.C. Kao et al 2011 *Journal of The Electrochemical Society* 158 H1284
[8] K. H. Allers 2004 *Microelectronics Reliability* 44 411
[9] K. Y. Yiang, W. J. Yoo, Q. Guo, A. Krishnamoorthy 2003 *Applied Physics Letters* 83 524

Acknowledgments

The authors thank Mr. Shipan Wang, all the colleagues for the discussion and advice. The author thank Prof. Hang Zhou very much.

IOP Conf. Series: Materials Science and Engineering **733** (2020) 012032 doi:10.1088/1757-899X/733/1/012032

Elastic Modulus Prediction of Particle Reinforced Composites Based on Sphere Model

Yanru Li[1] and Haibo Jiang
1 Associate Professor, Naval Logistical Academy, Tianjin, China
2 Doctor, Naval Logistical Academy, Tianjin, China
E-mail: jianghaibo022@126.com

Abstract. The elastic modulus calculation of particle-reinforced composites is a classic but rather complex issue. By adopting analytic method, this paper is intended to seek solutions and calculations with the help of the symbolic integral operation function of the software Mathematica to derive the analytical expressions of the elastic modulus of particle-reinforced composites. Study shows that the representative volume unit of particle-reinforced composites may be expressed with spherical model. Two calculation methods which feature "first parallel and then serial" and "first serial and then parallel" respectively deliver two slightly different elastic modulus values, and the intermediate value may be deemed as the predicted elastic

1. Introduction

The elastic modulus calculation of particle- reinforced composites is a classic issue. Earlier studies adopted relatively simple models which can be calculated with analytic methods. These studies culminated in 50s and 60s in the last century, and representative studies include the corrected mixture ratio formula of Halpain-Tsai[1], Eshelby elastic inclusion model[2,3], Mori-Tanaka mean field model[4,5], Chamis model[6], Hill-Hashin-Christensen-Lo model[7-9] and bridging model[10], etc.

This paper is intended to attempt to resolve complex and vivid models by using analytic method in order to prevent the considerable error generated by simple models adopted by analytical method and meanwhile eliminate the consumption of excessive computer resource in the case of numerical method which provides non-universal calculation results among other deficiencies.

2. Spherical model

According to the idea of statistical averaging, it is feasible to assume that the particles used as reinforcement are evenly distributed in the space and surrounded by matrix material to form a series of small cubes; taking one of these small cubes as a representative volume unit while the particle in it is a sphere. Let the side length of this representative volume unit (calculating element) is 1, the radius of the particle is r, and then the volume fraction of the particle is

$$V_p = \frac{4\pi}{3} r^3 \tag{1}$$

So the relationship between the radius and volume fraction of particle is

$$r = \left(\frac{3}{4\pi} V_p\right)^{1/3} \tag{2}$$

3. Derivation of elastic modulus formula

Taking the center of the sphere as the origin of coordinates, establish the horizontal rightward *x*-axis and the vertical upward *y*-axis. Assuming that a horizontal force is imposed on this representative volume unit.

3.1 The "series after parallel" calculation mode

Divide the portion falling within $-r \le x \le r$ (the area between two dashed lines in the Figure 1) into *n* equal parts vertically to get *n* small cuboids (micro segments) which have an uniform width of Δx and an uniform height of 1; the middle part of all micro segments is particle material while the upper and lower parts are matrix material; the cross section at *o-xy* plane is shown in Figure 1.

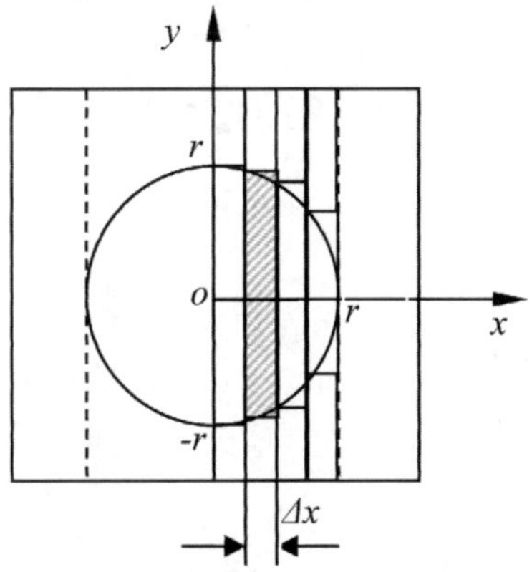

Figure 1. Sphere model for calculation of series after parallel

Let the abscissa of one micro segment is *x* and the particle material is represented by shadow (the three-dimensional image is a disc which has a thickness of Δx), then the volume fraction of the particle material in the micro segment (including the matrix material) is

$$V_{ps} = \pi\left(r^2 - x^2\right) \tag{3}$$

Under the action of the horizontal force, the shaded portion is arranged in parallel with the other portions in the cuboid, so the elastic modulus of this small cuboid is

$$E_x = V_{ps}E_f + \left(1 - V_{ps}\right)E_m = \pi\left(r^2 - x^2\right)E_p + \left(1 - \pi r^2 + \pi x^2\right)E_m$$
$$= \pi\left[\left(\frac{3}{4\pi}V_p\right)^{2/3} - x^2\right]E_p + \left[1 - \pi\left(\frac{3}{4\pi}V_p\right)^{2/3} + \pi x^2\right]E_m \tag{4}$$

Where: E_p and E_m represent the elastic modulus of the particle and matrix material respectively.

IOP Conf. Series: Materials Science and Engineering **733** (2020) 012032 doi:10.1088/1757-899X/733/1/012032

Under the action of the horizontal force, all these n micro segments are arranged in series within the range of $-r \le x \le r$ (the portion between dashed lines in the figure), then according to series formula the reciprocal of the elastic modulus E_{rx} of this portion is

$$\begin{aligned}\frac{1}{E_{rx}} &= \frac{V_{x1}}{E(x_1)} + \frac{V_{x2}}{E(x_2)} + \frac{V_{x3}}{E(x_3)} + \cdots + \frac{V_{xn}}{E(x_n)} \\ &= \frac{1}{E(x_1)}\frac{\Delta x_1}{2r} + \frac{1}{E(x_2)}\frac{\Delta x_2}{2r} + \frac{1}{E(x_3)}\frac{\Delta x_3}{2r} + \cdots + \frac{1}{E(x_n)}\frac{\Delta x_n}{2r}\end{aligned} \tag{5}$$

Where: $V_{xi}(i=1,2,\ldots,n)$ is the volume fraction of the particle material of the ith micro segment, $E(x_i)(i=1,2,\ldots,n)$ is the elastic modulus of the particle of the ith micro segment, and $\Delta x_i(i=1,2,\ldots,n)$ is the width (uniformly the height is 1) of the ith micro segment. When n approaches infinity, the above formula can be calculated with the integral method:

$$\begin{aligned}\frac{1}{E_{rx}} &= \sum_{n=1}^{\infty}\frac{1}{E(x_n)}\frac{\Delta x_n}{2r} = \int_{-r}^{r}\frac{1}{E_x}\frac{\mathrm{d}x}{2r} = \int_{-\left(\frac{3}{4\pi}V_p\right)^{1/3}}^{\left(\frac{3}{4\pi}V_p\right)^{1/3}} \frac{1}{\pi\left[\left(\frac{3}{4\pi}V_p\right)^{2/3} - x^2\right]E_p + \left[1-\pi\left(\frac{3}{4\pi}V_p\right)^{2/3} + \pi x^2\right]E_m}\cdot\frac{\mathrm{d}x}{2\left(\frac{3}{4\pi}V_p\right)^{1/3}} \\ &= \frac{\sqrt[3]{32}\arctan\dfrac{\left(6\pi^{1/2}V_p\right)^{1/3}\left(E_p-E_m\right)^{1/2}}{\left\{\left[4-\left(36\pi V_p^2\right)^{1/3}\right]E_m + \left(36\pi V_p^2\right)^{1/3}E_p\right\}^{1/2}}}{\left(3\pi^{1/2}V_p\right)^{1/3}\left(E_p-E_m\right)^{1/2}\left\{\left[4-\left(36\pi V_p^2\right)^{1/3}\right]E_m + \left(36\pi V_p^2\right)^{1/3}E_p\right\}^{1/2}}\end{aligned} \tag{6}$$

The above formula expresses the elastic modulus of all materials within the range of $-r \le x \le r$ (the portion between the dashed lines in the figure), and the portion of pure matrix material on left and right side is also connected in series with it, so the reciprocal of the total elastic modulus E_1 is

$$\frac{1}{E_1} = \frac{2r}{E_{rx}} + \frac{1-2r}{E_m} = \frac{4\arctan\dfrac{\left(6\pi^{1/2}V_p\right)^{1/3}\left(E_p-E_m\right)^{1/2}}{\left\{\left[4-\left(36\pi V_p^2\right)^{1/3}\right]E_m + \left(36\pi V_p^2\right)^{1/3}E_p\right\}^{1/2}}}{\sqrt{\pi}\left(E_p-E_m\right)^{1/2}\left\{\left[4-\left(36\pi V_p^2\right)^{1/3}\right]E_m + \left(36\pi V_p^2\right)^{1/3}E_p\right\}^{1/2}} + \frac{1-\left(\frac{6}{\pi}V_p\right)^{1/3}}{E_m} \tag{7}$$

3.2 The "parallel after series" calculation mode

Similarly, we derived E_2

$$\begin{aligned}E_2 &= 2rE_{ry} + (1-2r)E_m = 2\left(\frac{3}{4\pi}V_p\right)^{1/3}E_r + \left[1-2\left(\frac{3}{4\pi}V_p\right)^{1/3}\right]E_m \\ &= \frac{4E_mE_p\arctan\dfrac{\left(6\pi^{1/2}V_p\right)^{1/3}\left(E_p-E_m\right)^{1/2}}{\left\{\left[4-\left(36\pi V_p^2\right)^{1/3}\right]E_p + \left(36\pi V_p^2\right)^{1/3}E_m\right\}^{1/2}}}{\sqrt{\pi}\left(E_p-E_m\right)^{1/2}\left\{\left[4-\left(36\pi V_p^2\right)^{1/3}\right]E_p + \left(36\pi V_p^2\right)^{1/3}E_m\right\}^{1/2}} + \left[1-\left(\frac{6}{\pi}V_p\right)^{1/3}\right]E_m\end{aligned} \tag{8}$$

3.3 Prediction of elastic modulus

The calculations with two means - "series after parallel" and "parallel after series" deliver two different elastic modulus E_1 and E_2, and the former is larger than the latter. It is expected that the actual elastic modulus falls between E_1 and E_2, and the intermediate value may be deemed as the predicted elastic modulus value of particle-reinforced composites..

4. Conclusions

(1) The representative volume unit of particle- reinforced composites can be expressed with spherical model and sphere represents the shape of particle in a much better way than that in cube and other shapes; the problem of difficult integral computation can be solved by using symbolic computation software.

(2) Two calculation methods which feature "series after parallel" and "parallel after series" respectively deliver two slightly different elastic modulus values, and the intermediate value may be deemed as the predicted elastic modulus value of particle-reinforced composites.

References

[1] Ganesh V V and Chawla N 2005 Effect of particle orientation anisotropy on the tensile behavior of metal matrix composites: experiments and micro structure-based simulation *Mater. Sci. Eng.* A, **391** 342~353

[2] Eshelby J D 1957 The determination of the elastic field of an ellipsoidal inclusion and related problems *Proceedings of the Royal Society* **241** 376-396

[3] ESHELBY J D 1959 The elastic field outside an ellipsoidal inclusion *Proceedings of the Royal Society* **243** 561-569

[4] Mori T and Tanaka K 1973 Average stress in matrix and average elastic energy of materials with misfitting inclusions *Acta Metall* **21** 571-576

[5] Benveniste Y 1987 A new approach to the application of Mori-Tanaka's theory in composite materials *Mech Mater* **6** 147-157

[6] Chamis C C 1989 Mechanics of composite materials: Past, present and future *J. Comp. Technol* **ASTM 11 No 1** 3-14

[7] Hill R 1964 Theory of mechanical properties of fiber- strengthened materials. Elastic Behavior *J. Mech. & Phys* **Solid 12 No 12** 199-212

[8] Hashin Z and Rosen B W 1964 The elastic moduli of fiber-reinforced materials, *J. Appl. Mech.* **31 No 2** 223-232

[9] Hashin Z 1965 On elastic behavior of fiber-reinforced materials *J. Mech. & Phys*. **Solid 13 No 3** 119-134

[10] Huang Z M 2001 Simulation of the mechanical properties of fi- brous composites by the bridging micromechanics model *Composites* Part A **32 No 2** 143-172

IOP Conf. Series: Materials Science and Engineering **733** (2020) 012033 doi:10.1088/1757-899X/733/1/012033

Simulation of Giesekus fluid flow in extruder using helical coordinate system

E K Vachagina[1], A I Kadyirov[2] and J V Karaeva[2]

1 Leading Researcher, IPEAT FRC Kazan Scientific Center of RAS, Kazan, Russia
2 Senior Researcher, IPEAT FRC Kazan Scientific Center of RAS, Kazan, Russia
E-mail: aidarik@rambler.ru

Abstract. In the present work, it is proposed to use a helical coordinate system for mathematical modeling of non-isothermal Giesekus fluid flow in the extruder in the metric zone. The convenience of using a helical coordinate system is to reduce the three-dimensional problem to two-dimensional without the use of additional assumptions. The helical coordinate system is applicable only for steady flow conditions. Particular results of the numerical implementation of the use of a helical coordinate system are presented on the example of a Giesekus fluid flow in an extruder with a single flight, the results are compared with the flow of Newtonian fluid.

1. Introduction

The screw is a key element of the extruder and its most important part, therefore its design is crucial for the successful operation of any extrusion system [1]. The optimal design of the screw will provide a homogeneous melt with desired characteristics [2]. A detailed study of the hydrodynamics of the polymer flow in the extruder channel is the basis for modeling and understanding the extrusion process as a whole [3]. Experimental studies of liquid polymers showed a nonlinear viscoelastic nature of their behavior; therefore, almost at the same time, analytical solutions to the same problem for non-Newtonian fluids appeared. Analytical solutions of the Couette – Poiseuille flow of a viscoelastic fluid for the helical flow are known, as well as for Poiseuille flow (simpler solution for helical flow) [4]. In a number of works, for the rheological description of polymer flows, models were also used that take into account their viscoelastic properties, for example, FENE-P [5], Pom-Pom [6], Giesekus [7], PTT [8], the Leonov model [9], etc. In [10], it was proved that the PTT and Giesekus models can capture the complex nature of the polymer relaxation time at high Weissenberg numbers.

Despite the complexity of the problem being solved, the use of numerical methods is important for analyzing the flow of polymers in the channels of extrusion equipment. For example, based on numerical results [11] the influence of the screw characteristics (length, diameter, pitch, normal pitch, clearance, twist angle and number of turns of the screw) and physical properties of the material (density, viscosity) on the main process were analyzed.

In the general case, the velocity field of a moving fluid in a channel with a screw insert depends on three independent variables. Typically, mathematical models are used to calculate hydrodynamic fields, in which the basic system of equations is written in a Cartesian coordinate system, sometimes in a cylindrical one, whose axis coincides with the axis of symmetry of the channel. This approach is not based on helical symmetry, which is possessed by both a system of equations together with uniqueness conditions and a geometric region — a helical channel, and requires a large amount of computer memory and time for calculations in a three-dimensional setting. Using a helical coordinate system allows you to get the distribution of hydrodynamic fields, depending not on three, but on two variables. This significantly reduces both the memory costs and the time needed to obtain results. In this work, we used a helical coordinate system to study the processes of the Giesekus fluid flow in a channel with a non-rotating extruder.

2. Problem statement

2.1 Helical coordinate system

In this paper, we consider the flow of an incompressible viscoelastic fluid in a channel of a single-screw extruder (non-rotating) in the metric zone (fig. 1). It is assumed that the fluid motion is stationary, laminar, with a formed velocity profile at the inlet to the pipe. The temperature of the liquid does not change during its movement.

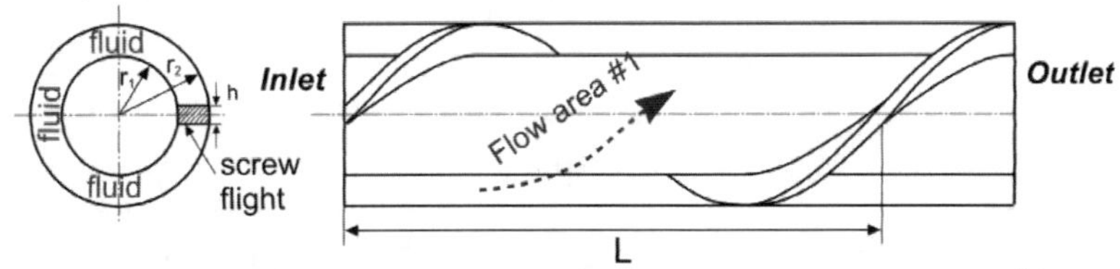

(a) (b)

Fig. 1 Transverse (a) and longitudinal (b) channel cross-sections: L is the lead of helical channel (length of a part of the channel which corresponds to the angular rotation of screw flight by 360°), r_2 is the barrel radius; r_1 is the shaft radius; h – screw thickness

It is proposed to use the following helical coordinate system ξ^1, ξ^2, ξ^3

$$\begin{cases} x = \xi^1 \cos\left(\pm\omega\xi^3\right) - \xi^2 \sin\left(\pm\omega\xi^3\right) \\ y = \xi^1 \sin\left(\pm\omega\xi^3\right) + \xi^2 \cos\left(\pm\omega\xi^3\right), \\ z = \xi^3 \end{cases} \quad (1)$$

where x, y, z - Cartesian coordinate system, $\omega = (2\pi)/S$, R - shaft radius, L - is the lead of helical channel. In eq. (1), the "+" sign is selected in the case when the fluid flow swirling clockwise, and the "-" sign is selected in the case of the flow swirling counterclockwise.

Note that the coordinate system (1) has no singular points, is non-orthogonal, and compared with the coordinate system presented in previous work [12] has a larger number of nonzero components of the metric tensor. It is advisable to use it in the case where the use of the system presented in the work [12] complicated by the presence of a singular point near the flow region, which leads to difficulties in numerical calculations.

2.2. Giesekus model

To describe the rheological behaviour of viscoelastic fluid it is used the Giesekus model [13]

$$\begin{cases} \boldsymbol{\sigma} = \boldsymbol{\sigma}_V + \boldsymbol{\sigma}_N, \boldsymbol{\sigma}_N = 2\mu_N \mathbf{D}, \\ \boldsymbol{\sigma}_V + \lambda \overset{\nabla}{\boldsymbol{\sigma}}_V + \dfrac{\alpha_G \lambda}{\mu_V} \boldsymbol{\sigma}_V \cdot \boldsymbol{\sigma}_V = 2\mu_V \mathbf{D}, \end{cases} \quad (2)$$

where $\overset{\nabla}{\boldsymbol{\sigma}} = \dfrac{d\boldsymbol{\sigma}}{dt} - \boldsymbol{\sigma} \cdot \nabla \mathbf{V}^T - \nabla \mathbf{V} \cdot \boldsymbol{\sigma} = \dfrac{\partial \boldsymbol{\sigma}}{\partial t} + \nabla \boldsymbol{\sigma} \cdot \mathbf{V} - \boldsymbol{\sigma} \cdot \nabla \mathbf{V}^T - \nabla \mathbf{V} \cdot \boldsymbol{\sigma}$ - upper convective derivative of the tensor; $\boldsymbol{\sigma}$; $\boldsymbol{\sigma}$ - stress tensor deviator; $\boldsymbol{\sigma}_N, \boldsymbol{\sigma}_V$ - viscous and elastic components of the tensor $\boldsymbol{\sigma}$; $\mathbf{D}$ - strain rate tensor; μ_N, μ_V - viscosity; λ - relaxation time; α_G - rheological parameter; $\mathbf{V}$ - velocity vector.

2.3. Governing equation

Based on assumption, the system of equations for the transfer of momentum and continuity in the coordinate system (1) can be written as follows

$$\mathrm{Re}^*\left(\left(\frac{\partial v^1}{\partial \eta^1} \mp \kappa\eta^2 \frac{\partial v^3}{\partial \eta^1}\right)v^1 + \left(\frac{\partial v^1}{\partial \eta^2} \mp \kappa\eta^2 \frac{\partial v^3}{\partial \eta^2} \mp 2\kappa v^3\right)v^2 - \kappa^2\eta^1\left(v^3\right)^2\right) =$$

$$-\frac{\partial p^*}{\partial \eta^1} + \frac{\partial s_1{}^1}{\partial \eta^1} + \frac{\partial s_1{}^2}{\partial \eta^2} \mp \kappa s_2{}^3, \quad (3)$$

$$\mathrm{Re}^*\left(\left(\frac{\partial v^2}{\partial \eta^1} \mp \kappa\eta^1 \frac{\partial v^3}{\partial \eta^1} \pm 2\kappa v^3\right)v^1 + \left(\frac{\partial v^2}{\partial \eta^2} \mp \kappa\eta^1 \frac{\partial v^3}{\partial \eta^2}\right)v^2 - \kappa^2\eta^2\left(v^3\right)^2\right) =$$

$$-\frac{\partial p^*}{\partial \eta^2} + \frac{\partial s_2{}^1}{\partial \eta^1} + \frac{\partial s_2{}^2}{\partial \eta^2} \pm \kappa s_1{}^3, \quad (4)$$

$$\mathrm{Re}^*\left(\mp\kappa\eta^2 \frac{\partial v^1}{\partial \eta^1} \pm \kappa\eta^1 \frac{\partial v^2}{\partial \eta^1} + \left(1+\kappa^2\left(\left(\eta^1\right)^2 + \left(\eta^2\right)^2\right)\right)\frac{\partial v^3}{\partial \eta^1} + 2\kappa^2\eta^1 v^3\right)v^1 +$$

$$\mathrm{Re}^*\left(\mp\kappa\eta^2 \frac{\partial v^1}{\partial \eta^2} \pm \kappa\eta^1 \frac{\partial v^2}{\partial \eta^2} + \left(1+\kappa^2\left(\left(\eta^1\right)^2 + \left(\eta^2\right)^2\right)\right)\frac{\partial v^3}{\partial \eta^2} + 2\kappa^2\eta^2 v^3\right)v^2 =$$

$$-\frac{\partial p^*}{\partial \eta^3} + \frac{\partial s_3{}^1}{\partial \eta^1} + \frac{\partial s_3{}^2}{\partial \eta^2} \mp \kappa s_2{}^1 \pm \kappa s_1{}^2 + \kappa^2\eta^1 s_1{}^3 + \kappa^2\eta^2 s_2{}^3, \quad (5)$$

$$\frac{\partial v^1}{\partial \eta^1} + \frac{\partial v^2}{\partial \eta^2} = 0, \quad (6)$$

here $\kappa = \omega R$, $\frac{\partial p^*}{\partial \eta^3} = const$, $\eta^i = \xi^i / R(i=1,2,3)$ - dimensionless variables; и $v^i = V^i / V^*(i=1,2,3)$ dimensionless contravariant velocity components; $p^* = PR/((\mu_N + \mu_V)V^*)$ - dimensionless pressure; $\mathrm{Re}_H^* = V^* R/(\mu_N + \mu_V)$ - modified Reynolds number; V^* - specified velocity.

Mixed dimensionless components of the stress tensor deviator $\mathbf{s} = \frac{R}{(\mu_N + \mu_V)V^*}\boldsymbol{\sigma}$ in the coordinate system (1) have the form

$$s_i^{\ j} = 2(1-\beta)d_i^{\ j} + \beta s_{Vi}^{\ j}, \quad (7)$$

here $\beta = \frac{\mu_V}{\mu_N + \mu_V}$; $d_i^{\ j}$ - mixed components of the dimensionless strain rate tensor, $\mathbf{d} = \frac{R}{(\mu_N + \mu_V)V^*}\mathbf{D}$; $s_{Vi}^{\ j}$ mixed components of dimensionless elastic part, $\mathbf{s}_V = \frac{R}{(\mu_N + \mu_V)V^*}\boldsymbol{\sigma}_V$ stress tensor deviator.

3. Results and discussion

The numerical implementation of the problem was carry out in the «Comsol Multiphysics».

The package allows you to rewrite the governing equations in the helical coordinate system (1). The numerical results were made for the case $\mathrm{Re}_H \to 0$ and $\mathrm{Re}_H = 150$, $We_H = 0.04$, $\alpha = 0.7$, $\beta = 0.3$, $\delta = 0.66$ $(\delta = r_1/r_2)$, $L = 3D$ $(We^* = \lambda V^* / R$ - Weissenberg number). The Reynolds and Weissenberg numbers are calculated using the equivalent diameter, calculated from the channel cross-section.

Fig. 2 shows the distribution of the contours of the dimensionless axial velocity for two Reynolds numbers $\mathrm{Re}_H \to 0$ and $\mathrm{Re}_H = 150$. For comparison, the same figure shows the distribution of the same component for the case of a Newtonian fluid flow at $\mathrm{Re}_H = 150$ and $We_H = 0$. As can be seen from the figure, with increasing the Reynolds number, the maximum value of the dimensionless axial velocity shifts toward the screw wall under the action of centrifugal forces. The distributions of the axial velocity for the Giesekus fluid and the Newtonian fluid has qualitative coincidence, but for the same Reynolds number it can be seen that the contours for the Giesekus fluid flow are more elongated in the tangential direction and narrow in radial direction.

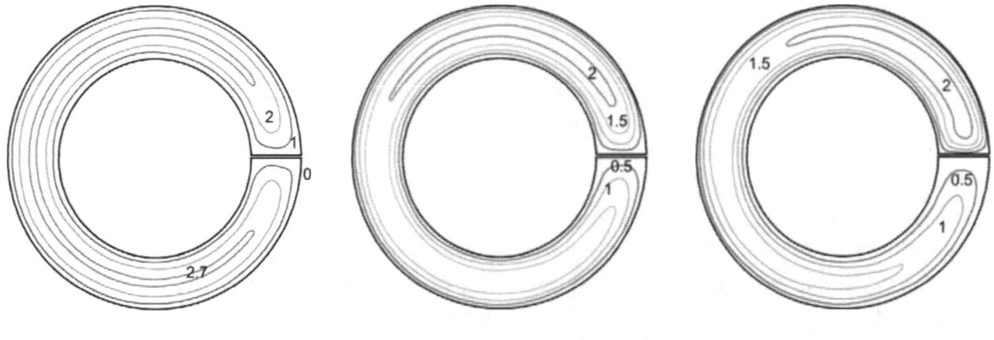

(a) (b) (c)

Fig.2. Contours of the dimensionless axial velocity for two Reynolds numbers in the cross-section of the channel: (a) - $\mathrm{Re}_H \to 0(We_H = 0.04)$; (b) - $\mathrm{Re}_H = 150(We_H = 0.04)$; (c) - $Re_H = 150(We_H = 0)$ – Newtonian fluid.

Fig.3 and 4 shows contours of the dimensionless component of the elastic part of the stress tensor s^{13} and s^{23}. At small Reynolds numbers these distributions are symmetric, then with increasing Reynolds numbers it becomes asymmetric under the influence of a similar change in the velocity field. The maximum values of these components, as expected, are located near the walls. Especially strongly with increasing Reynolds number, the modulus of the component in the region of the screw flight on the side of the incoming flow increases.

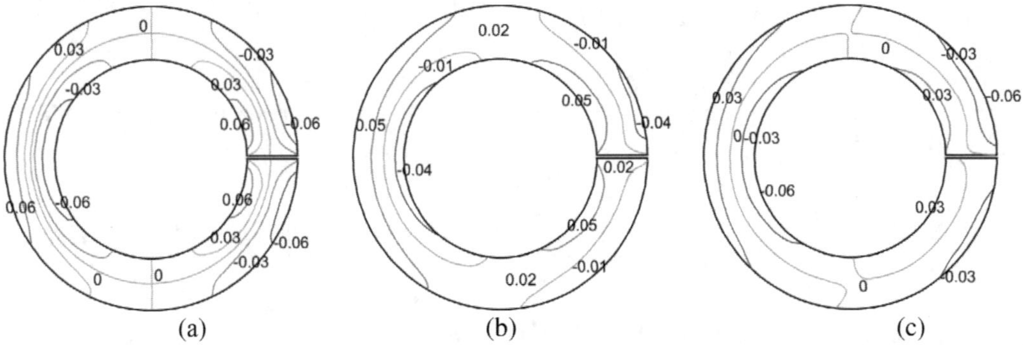

(a) (b) (c)

Fig.3. The contours of the dimensionless component of the elastic part of the stress tensor (s^{13}). Designations are similar fig. 2.

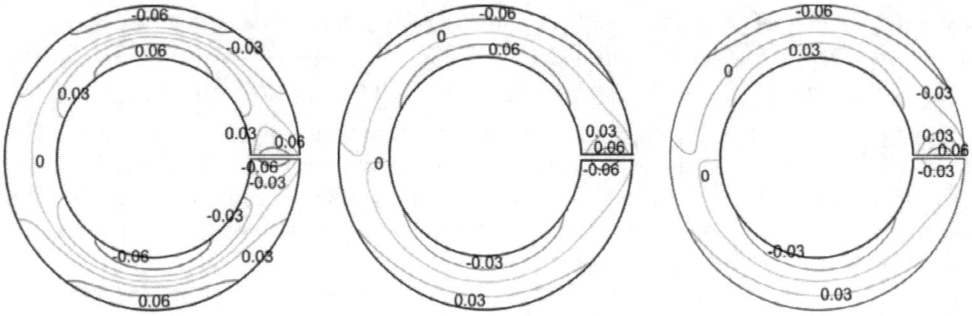

(a) (b) (c)

Fig.4. The contours of the dimensionless component of the elastic part of the stress tensor (s^{23}). Designations are similar fig. 2.

Reference

[1] Kim VS 2005 *Theory and practice of polymer extrusion* (Moscow: Chemistry, KolosS)

[2] Roland W, Kommenda M, Marschik C and Miethlinger J 2019 *Polymers* **11** (2) 334

[3] Wilczynski K, Buziak K, Wilczynski KJ, Lewandowski A and Nastaj A 2018 *Polymers* **10** (3) 295

[4] Cruz DOA and Pinho FT 1983 *J. Non-Newt. Fluid Mech.* **13** 109-143.

[5] Barrett J and Boyaval S 2017 Finite Element Approximation of the FENE-P Model hal-01501197

[6] Hawke LDG, Huang Q, Hassager O and Read DJ 2015 *Journal of Rheology* **59** 995-1017

[7] Mokarizadeh H, Asgharian M and Raisi A 2013 *J. Non-Newt. Fluid Mech.* **196** 95–101

[8] Cruz DOA and Pinho FT 2012 *J. Non-Newt. Fluid Mech.* **167** 95-105

[9] Leonov AI 1976 *Rheol. Acta* **15** 85–98

[10] Holmes LT 2010 Modeling *3D viscoelastic secondary flows in extrusion* (Madison: The University of Wisconsin)

[11] Siregar AN, Mahmod WMFW, Ghani JA, Haron CHC, and Rizal M 2014 *Res. J. Appl. Sci. Eng. Tech.* **7** (10) 2098-2105

[12] Vachagina EK and Kadyirov AI 2014 *Q. J. Mech. Appl. Math.* **67** (4) 553–566

[13] Giesekus H 1982 *J. Non-Newt. Fluid Mech.* **11** 69–109

Acknowledgments

The research was conducted with financial support by RSF (Russian Science Foundation), project number 19-11-00220.

IOP Conf. Series: Materials Science and Engineering **733** (2020) 012034 doi:10.1088/1757-899X/733/1/012034

Temperature effect on the Production of Highly charged Fe Ion beams at 14.5 GHz ECR Ion source

H. P. Liu, J. Y. Li, L. Kang, T. M. Zhang, and J. Li

Institute of Modern Physics, Chinese Academy of Sciences,
Lanzhou 730000, China
E-mail: liuhp2008@impcas.ac.cn

Abstract. Temperature is one of the important parameters that effect the production of metal ion using MIVOC method. Beams of highly charged iron ions are produced with 14.5 GHz all-permanent-magnet electron cyclotron resonance ion source at three temperatures at 320 kV high voltage multi-discipline platforms for highly charged ions at IMP, CAS. At 18 ℃, Fe^{13+} and Fe^{15+} of more than 15.9 eμA and 11.7 eμA have been extracted with an extraction voltage of 15 kV and a microwave power of 300 W. When increasing temperature to 25 ℃ and 30 ℃, more main substance arrives at the plasma chamber of ion source and the outputs of Fe^{15+} and Fe^{16+} decreases, but those of Fe^{13+}, Fe^{12+} and Fe^{11+} increase, the distribution of Fe ions shifts towards low charge state. The conductance of the coupled MIVOC system is estimated at different temperatures. The ionization efficiencies and the effective consumptions rate of ferrocene are also figured out from the vapour pressure and the intensities of iron ion beams. With the increase of the temperature, the conductance and the consumptions increases while the dependence of ionization efficiency on the temperature is different.

1. Introduction

Electron cyclotron resonance (ECR) ion sources have been invented by R. Geller and his coworkers from open-ended fusion plasma decades ago [1]. Now, due to their high performance ECR ion sources become the most powerful devices that produce intense beams of multiply charged heavy ions for accelerators and atomic physics experimental apparatus.

In recent years, demands for the intense and stable metal ion beams are increasing for different research area of materials science at 320 kV high voltage multi-discipline platform for highly charged ions (320 kV HV platform, for short), IMP, CAS. Every year, the platform ran with Fe ion beams no less than 3 months. In order to produce the Fe ions beams easily and low costly, the MIVOC method is adopted. It is a very simple and inexpensive technique widely used all over the world in the generation of Ti, Fe, Ni, Co, Ru, Os and W ion beams at room temperature or slightly elevated temperature. Detailed description about this method can be found in [2-4] and the references therein.

This paper describes the production of iron ion beam at 320 kV high voltage platform, IMP, CAS. The intensities of highly charged Fe ions are measured at three room temperatures. The parameter of the ECR ion source and the coupled MIVOC system are presented. Then, using the spectrum and pressures at both sides of the ECR ion source we deduce the conductance of the MIVOC system, ionization efficiency and effective consumption rate of $Fe(C_5H_5)_2$.

2. Experimental set-up

The experimental set-up consists of a 14.5 GHz all-permanent- magnet electron cyclotron resonance ion source (LAPECR2) and a MIVOC system. LAPECR2 has been designed and fabricated to provide intense multiple charge state ion beams at 320 kV HV platform to meet the demands in the fields of atomic physics, material physics, biological physics and low-energy nuclear astrophysics. The axial magnetic field configuration of the LAPECR2 is: $B_{inj} < 4B_{ecr}$ (Becr is the resonance magnetic field determined by the rf frequency) at injection, $B_{ext} \approx B_{rad}$ (Brad is the radial magnetic field at the inner wall of the plasma chamber) at extraction. For LAPECR2, the optimum operation frequency is 14.5 GHz. More detailed description of LAPECR2 can be found in references [5]. The extraction system consists of a plasma electrode and a conical shape puller electrode at ground potential. The diameters of the plasma electrode and the puller electrode are 10 mm and 14 mm. The ion beams are analyzed with a bending magnet. The plasma chamber is made of stainless steel, the inner diameter of which was 67 mm. The MIVOC chamber is a stainless-steel container of about 50-cm3 capacity. It is attached to the gas feed tube of the 14.5 GHz ECR ion source via a rough, large conductance regulation valve to control the flow rate of the compound vapour (shown in Figure 1). Ferrocene powder is loaded in the MIVOC chamber where the compound vaporizes and diffuses into the ECR ion source through the gas feed tube. The diameters D and the length L of it are 4 mm and 270 mm, respectively.

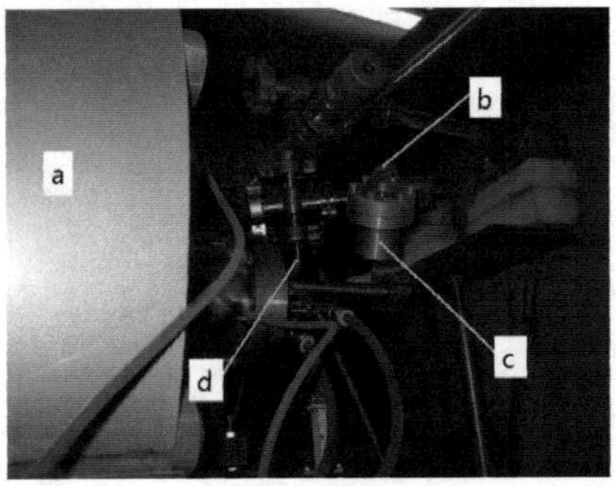

Figure. 1 The MIVOC system coupled to injection side of LAPECR2. (a) APECR2, (b) regulation valve, (c) the MIVOC chamber, (d) gas feed-tube.

At the beginning of this experiment, the inner wall of the stainless-steel plasma chamber was cleaned up by the oxygen plasma for 3 days. A stable vacuum, 3.2×10^{-7} mbar at the injection side (P_{inj0}) and 1.7×10^{-7} mbar at the extraction side (P_{ext0}), achieved with two 700 l/s molecular pumps at the injection and extraction regions respectively before we released the vapor of the compound. Oxygen, which is considered to be the better gas mixing in the generation of metal ions using the MIVOC method [6, 7], is the support gas. The feed rate was 0.08cm3/h at 18°C and 25°C. The suppressor electrode on the Faraday cup was biased to -160 V to suppress the secondary electrons.

Before our measurement, we record the pressures on both sides of the ECR ion source under different microwave power P_w and minus bias voltage.

3. Experimental results

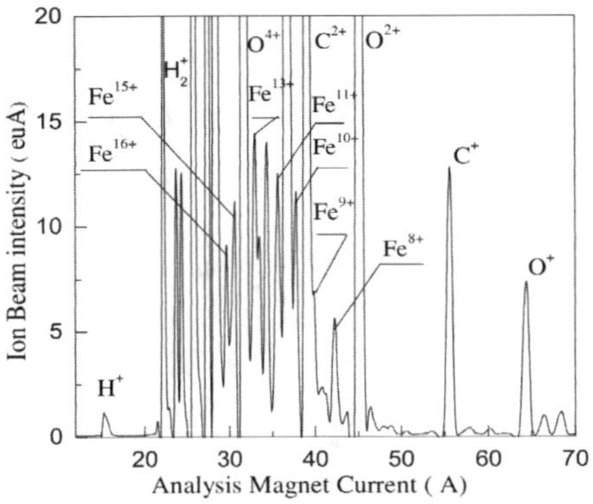

Figure 2. Charge state distribution spectrumfor iron ion at 18°C. The ion source parameters were optimized for $^{56}Fe^{13+}$.

Figure 2 shows the charge state distribution spectrum of Fe ion beams produced with LAPECR2 at room temperature (T) of 18 °C. The P_w is 300 W and an extraction voltage of 15 kV is used. The disc voltage applied in the injection side is -50 V. The regulation valve is fully opened. The ion source is optimized to produce $^{56}Fe^{13+}$ and achieved a beam current of 14.5 eμA of it at the first Faraday cup (FC1) of the platform. At this moment, P_{inj1}=6.7×10^{-7} mbar, P_{ext1} =3.6×10^{-7} mbar, and the ratio of P_{vap} to P_{ox} is about 1.1. The currents of $^{56}Fe^{15+}$ and $^{56}Fe^{16+}$ are 11.2 eμA and 9.7 eμA respectively. The intensity of $^{56}Fe^{13+}$ increases slowly. Three hours later, the intensity of $^{56}Fe^{13+}$ grows up to 16.2 eμA and that of the $^{56}Fe^{15+}$ is 12.3 eμA.

Keeping the microwave power and extraction voltage unchanged, increasing the room temperature T to 25 °C, the currents of $^{56}Fe^{13+}$ and $^{56}Fe^{11+}$ increase to 17.6 eμA and 19.2 eμA, while those of $^{56}Fe^{15+}$ and Fe^{16+} decrease to7.5 eμA and 5.3 eμA. Then, closing support gas inlet valve and raising the room temperature to 30°C, the ion source is optimized to produce $^{56}Fe^{11+}$. The currents of $^{56}Fe^{15+}$ of 5.2 eμA, $^{56}Fe^{13+}$of 18.5 eμA and $^{56}Fe^{11+}$ of 37.0 eμA were read as P_w = 300 W. The analyzing magnet current and the voltage loading on bias disc were fine-tuned to maximize the output of $^{56}Fe^{13+}$. The intensities of Fe ion beams at different temperature are listed in Table 1. From these data, we found that temperature has a big impact on the intensities of highly charged ion beams.

Table 1. The currents of highly charged Fe ions at different temperatures.

T (°C)	$^{56}Fe^{16+}$ (eμA)	$^{56}Fe^{15+}$ (eμA)	$^{56}Fe^{13+}$ (eμA)	$^{56}Fe^{12+}$ (eμA)	$^{56}Fe^{11+}$ (eμA)	$^{56}Fe^{10+}$ (eμA)
18	9.8	11.7	15.9	14.1	12.8	11.9
25	5.3	7.5	16.8	17.4	19.2	17.6
30	1.9	5.2	18.5	32.1	37.0	45.1

When the room temperature enhances, more ferrocene vapor generated and flows into the plasma. With the increase of the main substance in plasma, the currents of Fe^{15+} and Fe^{16+} decrease while those of Fe^{13+}, Fe^{12+}, Fe^{11+} increase. The distribution of Fe ions shifts towards lower charge states. The vapor pressure P_{vap} of ferrocene is calculated basing on the Clausius–Clapeyrou equation [8, 9, 10].

When we raise the room temperature, ferrocene vapor pressure P_{vap} as well as the average pressure *P* in the gas feed-tube increases, leading to the big changes in the conductance of the coupled MIVOC system. The conductance C of the coupled MIVOC system is taken to be

$$C = 2.45 \times 10^{-2} \times D^4 (\upsilon L)^{-1} P \quad (1)$$

where *D* and *L* are the diameter and the length of gas feed-tube mentioned in section 2, *P* is the average pressure in the gas feed-tube, and υ is the internal friction coefficients. For ferrocene υ is estimated to be 1.17 $\times 10^{-2}$ mbar/s [11]. The conductance of the MIVOC system are deduced and displayed in table 2.

Then, according to the spectrum gotten in this experiment, we calculate the ^{56}Fe ion flow rate N_i at the FC1 as

$$N_i = \sum_q (I_q q^{-1}) \quad (2)$$

where q is iron ion charge state and I_q is the iron ion beam intensity of charge q. The number of iron ion produced per second in the plasma is computed by

$$N_r = 2N_i / f_{eff} K\varepsilon \quad (3)$$

where f_{eff} is the efficiency of the vapor flowing from the main plasma to the extraction hole, K is the transmission efficiency between the ion source and the FC1 and ε is the isotopic abundance of ^{56}Fe in nature iron. Depending on the change of the vacuum in the plasma, we can estimate iron particle (neutral atom and ion) flow rate N_t in the main plasma as follow

$$N_t = N_a \Delta P_{ECR} \times 10^{-3} \times 2 \times \upsilon_m^{eff} / 22.4 \quad (4)$$

where $N_a = 6.02 \times 10^{23}$ is Avogadro constant, ΔP_{ECR} is pressure difference caused by vapor at the centre of the resonance zone, $\Delta P_{ECR} \approx (P_{inj1} - P_{inj0} + P_{ext1} - P_{ext0})/2$, υ_m^{eff} is the effective pumping speed of the pumping system,

$$\upsilon_m^{eff} \approx 700 \times \left(\frac{M_N}{M_F}\right)^{1/2} l/s \quad (5)$$

Where M_N is the molecular weight of nitrogen and M_F is the molecular weight of $Fe(C_5H_5)_2$ [12,13]. The ionization efficiency of ferrocene $\varepsilon = N_r / N_t \times 100\%$. The calculated N_r, N_t and ionization efficiency are shown in table 2.

The effective consumption rate of ferrocene R_{eff} is 1.31 mg/h, corresponding to a iron consumption of 0.39 mg/h at 18 ℃. Most of the $Fe(C_5H_5)_2$ vapor condenses on the tube surface of the MIVOC system during it diffuses from the chamber to the plasma. Then, more than 90% of the vapor flowing into the plasma is pumped out of the ECR ion source.

The data in table 2 involved the effect of microwave power and minus bias voltage on plasma pressure is taken into accounted.

Table 2. The conductance C, Fe ion production rate N_r, total Fe particle flow rate N_t, ionization efficiency ε and effective consumption rate R_{eff} at different temperatures.

T (°C)	C (l/s)	N_r (ion/s)	N_t (particle/s)	ε (%)	R_{eff} (mg/h)
18	0.01	3.24×10^{14}	3.87×10^{15}	8.37	1.31
25	0.02	3.40×10^{14}	8.03×10^{15}	4.23	1.56
30	0.04	7.03×10^{14}	1.12×10^{16}	6.28	3.36

4. Conclusions

Stable and intense highly charged iron ion beams have been produced with all permanent magnet ECR ion source using in the MIVOC method experiment at three temperature at 320 kV high voltage platforms, IMP, CAS. The temperature has an impact on the output of iron ion beams, resulting in the distribution of iron ion beams shifted towards lower charge state. The ionization efficiency and the effective consumption rate of ferrocene are calculated and the temperature has quite an influence on them.

The MIVOC method has turned out to be well suited to the production of ion beams with different charge state from high melting point materials, which were composed of the volatile compounds. The iron ion beams extracted are strong and stable. Currents of 2.0 eμA of $^{56}Fe^{13+}$ beam and 0.8 eμA of $^{56}Fe^{15+}$on target maintained in the material irradiation experiment at 320 kV high voltage platforms.

5. References

[1] Geller R 1996 Electron Cyclotron Resonance Ion Sources and ECR Plasmas, Institute of Physics, Bristol.

[2] Koivisto H, Ärje J and Nurmia M 1994 Nucl. Instr. Methods. B 94 291.

[3] Ärje J, Koivisto H and Nurmia M 1993 Operation of the JYFL-ECR ion source, ed. by A.G. Drentje, Proceedings of the 11th International Workshop on ECR Ion Sources, GrÖningen, p.27.

[4] Waldmann H and Matin B 1995 Nucl. Instr. Methods. B 98 532.

[5] L.T.Sun H.W.Zhao J.Y.Li H.Wang B.H.Ma Z.M.Zhang X.Z.Zhang X.H.Guo Y.Shang X.X.Li Y.C.Feng Y.H.Zhu P.Z.Wang H.P.Liu M.T.Song X.W.Ma W.L.Zhan 2007 Nucl. Instr. Meth. B 263 503.

[6] Shirkov G D 1997 Phys. Scr. T73 384.

[7] Tarvainen O, Suominen P and Koivisto H 2004 Nucl. Instr. Meth. B 217 136.

[8] Weast R C 1978 *Handbook of Chemistry and Physics,* 58th ed. CRC, Boca Raton, FL.

[9] Lyman W J, Reehl W F and Rosenblatt D H 1990 Handbook of Chemical property Estimation Methods (American chemical Society, Washington, DC, McGraw-Hill)

[10] Ullrich A, Grübling P, and Zschornack G 1998 Rev. Sci. Instrum. 69 813.

[11] Mansoori Oghaz N, Ghiamati E, Haghighi B, Eskandari Nasrabad A, Papari M M and Bamdad M 2012 Journal of Mol. Liq. 165, 55.

[12] Guo N, Qiu J W, Jiang H C 2007 Xe Vaccum 44 15.

[13] Eckle F J, Bickert P, Lachenmann R, Wortamann B 1996 Vacuum 47 799.

Colorless, heat resistant polyimide films derived from 2,3,3',4'-biphenyltetracarboxylic dianhydride

G Song[1] and C Chen[2]

1 Center for Advanced Low-Dimension Materials, Donghua University, Shanghai, China

2 Suzhou Polytri Material Technology Co.,Ltd, Changshu, China

E-mail: sgl_tong@dhu.edu.cn

Abstract. The development of high performance colorless polyimide (PI) films is one of the targets of PI chemistry, since heat resistant colorless PI films are key materials for reliable plastic substrates in image display devices. In the present work, 2,3,3',4'-biphenyltetracarboxylic dianhydride (a-BPDA) was used to polymerize with diamines to afford a series of PI, as shown in Scheme 1. Due to the fully aromatic composition, the films displayed high thermal decomposition temperatures ($T_{5\%}$). PI-FDA and PI-FFDA had similar glass transition temperatures (T_g) around 425°C, and much higher than that of PI-TFMB. PI-TFMB and PI-FFDA showed a transparent and colorless feature, with high transparency at 400nm (T_{400}). Coefficients of thermal expansion (CTE) were found around 55ppm.°C^{-1} for the series of PI films, larger than the requirement of flexible substrate applications, which could be attributed to the distorted chain structure derived from the a-BPDA component.

1. Introduction

Colorless, heat-resistant polymer films are key materials in the applications of display devices, liquid crystal alignment layers, color filters and optical compensation films. In addition, plastic substrates alternative to the current inorganic glass substrates are urgently required in the field of image display devices [1,2].

Polyimides (PIs) have satisfactory heat resistance, and have been used as electrical insulation materials in a variety of electronic devices. Besides, PI films are known to have good dimensional stability (low-CTE) characteristics. Therefore, PI films should be good candidate for the heat resistant, colorless polymer films [3.4].

The key point to develop colorless PI films is to weaken the charge transfer complex of the polymer chains [5,6], therefore, the monomers with distorted and twisted conformation should be employed, to disrupt the electronic conjugation between the polyimide chains [7].

In the present article, dianhydride monomers with distorted biphenyl linkage: 2,3,3',4'-biphenyltetracarboxylic dianhydride (a-BPDA) was used to prepare colorless PI films, and the properties were discussed from the viewpoint of structure-property relationship.

2. Experimental

2.1 Reactants and reagents

2,2'-Bis(trifluoromethyl)benzidine (TFMB) and a-BPDA was supplied by Sunlight Pharmaceutical Co., Ltd. (Jiangsu, China), 9,9-bis(4-aminophenyl)fluorene (FDA) and 9,9-bis(4-amino-3-fluorophenyl)fluorene (FFDA) were provided by Tokyo Chemical Industry

Co., Ltd. (Tokyo, Japan), N,N-dimethylacetamide (DMAc) and isoquinoline were purchased from Yonghua Chemical Technology Co., Ltd. (Jiangsu, China) and Meryer Chemical Technology Co., Ltd. (Shanghai, China), respectively, m-cresol was provided by Meryer Chemical Technology Co., Ltd. (Shanghai, China) and distilled under reduced pressure prior to use. All the other commercially available reagents were analytical grade and used without purification.

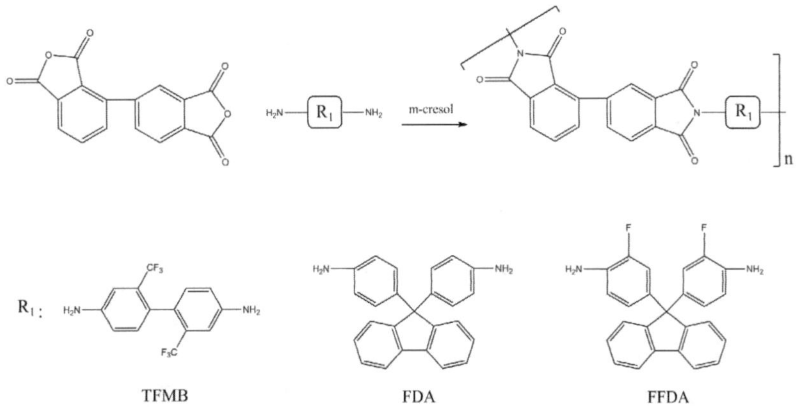

Scheme 1. Synthesis of the polyimides.

2.2 Characterizations

FT-IR spectra were obtained on a Bruker Vertex 70 spectrometer at a resolution of 4cm^{-1} in the range of 400 cm^{-1}-4000 cm^{-1}. Mechanical properties were examined at room temperature on a Instron 5966 universal testing apparatus with crosshead speed of 5mm/min. Thermogravimetric analysis (TGA) was performed using a TA TGA 550 with a temperature ramp of 10°C/min in nitrogen atmosphere. Temperatures at which 5% weight loss occurred (T_d^5) were taken as parameters to evaluate thermal stability. Differential scanning calorimetric (DSC) analysis were conducted with a TA DSC 250 with a heating rate of 20°C/min in nitrogen purge. Thermal dimensional stabilities of the PI films were examined by TA TMA Q400 at a heating rate of 10°C/min in film tension mode and a constant load of 2g was applied to the sample in nitrogen flow (the initial dimension of film samples: 15mm long, 4mm wide, 20μm thick). Dynamic mechanical analysis (DMA) was carried out with a TA DMA 850 at a heating rate of 5°C/min and a load frequency of 1Hz in film tension geometry, glass transition temperature (*Tg*) was determined from the peak point of the loss factor curve. Solubility of the PI resins was qualitatively tested with 10mg powder in 1mL of solvent. Ultraviolet-visible (UV-Vis) spectra of the PI films were recorded with a Shimadzu UV-Vis 2550 spectrometer in the range of 800nm-200nm in transmittance mode.

2.3 Polymerization and film preparation

The polyimides were prepared by equimolar polyaddition of tetracarboxylic dianhydride and diamine, via one-step method in m-cresol, as shown in Scheme 1. A typical procedure was as follows. a-BPDA (2.1185g, 7.20mmol), TFMB (2.3071g, 7.20mmol), isoquinoline (0.45g, 3.5mmol) and m-cresol (25g) was added into a three-necked 100ml flask, which was equipped with mechanical stirrer, condenser, and nitrogen inlet/outlet. The mixture was stirred at 160°C for 1h to completely dissolve the reactants, afterwards, the mixture was heated to 200 °C and kept for 10h. The resulting viscous solution was cooled and precipitated to excessive amount of ethanol. The fibrous product was washed with ethanol for several times, and dried at 120°C under vacuum for 24h to afford polyimide a-BPDA/TFMB.

Polyimide solution was obtained by dissolving PI powder in DMAc with solid concentration of 10%wt. Prior to film casting, the solution was filtered to remove insoluble impurities and degassed by standing still overnight. Afterwards, the solution was cast onto flat glass substrate and dried in an air convection oven at 50°C for 2h, 80 °C for 1h, 150 °C for 1h, followed by 250 °C under vacuum for 1h. After cooling down, the film was peeled off from the substrate by immersion in tap water for a few minutes. The thickness of PI films were controlled to be 20μm approx..

3. Results and discussion

3.1 Preparation of the PI films

Polyimides were prepared based on a-BPDA and several aromatic diamines. One-step polymerization was employed in m-cresol in the presence of isoquinoline, which acted as the catalyst. The polymerization proceeded smoothly and viscous homogenous solution was obtained after several hours of reaction. Figure 1 shows the FTIR spectra of the PI films, the absorptions of imide group were observed at 1780 cm^{-1}(C=O asymmetric), 1720 cm^{-1}(C=O symmetric) and 1370 cm^{-1}(C-N). The characteristic amide (1660 cm^{-1}) and -COOH (1710 cm^{-1} and 3300 cm^{-1}) stretching vibrations derived from poly(amic acid) were not observed in the FTIR spectra, implying the polyimides had been imidized completely.

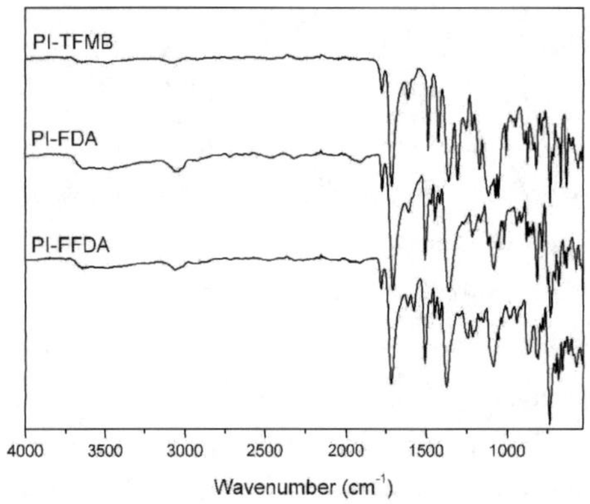

Figure 1. FTIR spectra of PI films derived from a-BPDA.

3.2 Solubility of the polyimides

The solubility of the polyimides was evaluated with several common organic solvents, as listed in Table 1. All of the polyimides could be dissolved in high boiling point polar solvents at room temperature, such as DMF, DMAc, NMP, m-cresol and DMSO. Besides, good solubility was observed in less strong solvents, such as dioxane, THF and chloroform. PI(DCBPDA/TFMB) was even soluble in ethyl acetate and acetone. It was noted that the solubility of polyimides was strongly dependent on the backbone structures, for polyimides based on a-BPDA, biphenyl-imide structures were highly twisted, this non-coplanar conformation disrupted the compact packing of the molecular chains and provided enhanced solubility.

Table 1. Solubility [a] of the polyimides.

polyimide	DMAc	DMF	Dioxane	NMP	m-cresol	THF	$CHCl_3$	EtOAc	Acetone	DMSO
PI-TFMB	++	++	++	++	++	++	++	++	++	++
PI-FDA	++	++	++	++	++	——	++	——	——	++
PI-FFDA	++	++	++	++	++	++	++	——	——	++

a, solubility was tested with 10mg of PI powder in 1mL solvent. (++) soluble at room temperature; (+—) soluble upon heating; (——) insoluble upon heating.

3.3 Thermal and mechanical properties of the polyimides

Table 2. Thermal properties of the polyimide films.

polyimide	Tg (°C)		T5% (°C)	CTE
	DSC	DMA		(ppm.K^{-1})
PI-TFMB	363	361	583	55.3
PI-FDA	429	430	572	54.4
PI-FFDA	423	421	583	58.9

Thermal stability of the polyimides was evaluated by TGA in N_2 atmosphere, T5%, which was the parameter of thermal stability, was listed in Table 4. All the PI samples demonstrated good thermal stability with T5% beyond 570°C, attributed to the fully aromatic chemical composition of the polyimides.

DSC and DMA were employed to measure the glass transition temperatures (Tg) of the polyimides. Due to the bulky fluorene unit of FDA and FFDA, the corresponding PI displayed superior high Tg.

Thermal dimensional stability of the PI films was examined by TMA, the in plane CTE data in the solid state of the films were listed in Table 2. The data indicated that the CTE was pretty large for the series of PI films, which could be interpreted from the viewpoint of twisted, non-coplanar conformation.

The tensile properties of PI films were tested and the data were shown in Table 3. The series of films had similar values, due to the twisted and bended feature of the molecular chain, the PI films displayed moderate mechanical properties.

Table 3. Mechanical properties of the PI films.

polyimide	Modulus (MPa)	Tensile strength (MPa)	Elongation at Break (%)
PI-TFMB	2950.2 ± 31.0	141.2 ± 9.3	12.1 ± 2.3
PI-FDA	2902.3 ± 35.5	139.7 ± 10.8	11.6 ± 1.9
PI-FFDA	2891.7 ± 41.1	133.5 ± 11.9	11.2 ± 2.7

3.4 Optical transparency of the PI films

Optical transparency of the PI films were studied by the use of UV-Vis spectroscopy and shown in Figure 2. Compared to pale yellow colored PI-FDA, PI-TFMB and PI-FFDA exhibited a more transparent and colorless feature, an obvious blue shift of the spectra were observed compared to PI-FDA, as displayed in Figure 2. The good optical transparency of PI-TFMB and PI-FFDA was attributed to the suppressed formation of intra- and inter-molecular charge transfer complex (CTC), due to the steric hindrance effect of 2,3,3',4'-biphenyl linkage combined with the fluoro content.

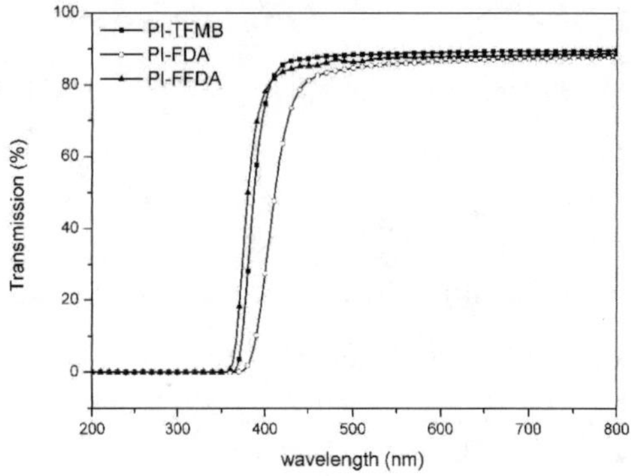

Figure 2. UV-vis spectra of PI films derived from a-BPDA.

4. Conclusion

The dianhydride monomer a-BPDA had distorted and twisted biphenyl linkage, the electronic conjugation along the polymer chain could be disrupted and the optical transparency of the PI films would be improved. Based on this designed feature, colorless polyimide films were developed and the physical properties were examined. The series of PI films displayed high thermal stability and high Tg, due to the fully aromatic structures. Besides, good optical transparency was obtained as expected. The dimensional stability was rather inferior due to the bended and twisted chain conformation, which should be improved in the future work.

5. References

[1] Hasegawa M 2017 *Polymers* **9** 520
[2] Hasegawa M, Watanabe Y, Tsukuda S and Ishii J 2016 *Polym. Int.* **65** 1063
[3] Hergenrother PM 2003 *High. Perform. Polym.* **15** 3
[4] Sroog CE 1991 *Prog. Polym. Sci.* **16** 561
[5] Hasegawa M and Horie K 2001 *Prog. Polym. Sci.* **26** 259
[6] Yi L, Huang W and Yan D 2017 *J. Polym. Sci. Part A: Polym. Chem.* **55** 533
[7] Ding M 2007 *Prog. Polym. Sci.* **32** 623

Preparation and luminescence of rare earth complex $Eu(DBM)_3(TPPO)$

Q Y Li, P S Tang, W B Shen, Y F Zhou, Z R Niu, Y L Wang and Y B Ding

Department of Materials Engineering, Huzhou University, Huzhou 313000, China
E-mail: tangps@163.com, 02566@zjhu.edu.cn

Abstract. Using Eu_2O_3 as source of europium, anhydrous ethanol as solvent system, methylene benzoyl(DBM) as first ligands, triphenyl phosphine (TPPO) as second ligands, the europium(III) ternary rare earth complex $Eu(DBM)_3TPPO$ has been successfully prepared. The samples were characterized by fluorescence spectrum(FS), Fourier transform infrared spectroscopy(FT-IR), thermogravimetry differential scanning calorimetry(TG-DS) and scanning electron microscope(SEM). The influence of TPPO on luminescence intensity, color temperature, color coordinates, morphology and thermal stability of the samples were analyzed. The results showed that the addition of second ligand could enhance the luminescence intensity of $Eu(DBM)_3TPPO$. $Eu(DBM)_3TPPO$ can emit strong red fluorescence when excited by 395nm ultraviolet light. The color coordinate of $Eu(DBM)_3(TPPO)$ is (0.63, 0.36). The $Eu(DBM)_3TPPO$ is a kind of red luminescent material with wide application prospect.

1. Introduction

The complex of europium has a half-peak width (about 10nm), a characteristic emission peak of about 612nm, and a pure red light (color coordinate 0.66, 0.33) [1], which is an important red LED material. Europium complexes have broad application prospects in fluorescent anti-counterfeiting field, fluorescent probe field [2], biomedical field [3] and optical field [4]. Therefore, it is of great value and significance to study europium complexes, especially europium ternary complexes. At present, the preparation methods of europium ternary complexes mainly include sol-gel method [5], hydrothermal method [6] and precipitation method [7]. The sedimentation method has the advantages of simple experimental process and convenient operation.

2. Experimental details

2.1 preparation of rare earth complexes

The purity of Eu_2O_3 purchased from Aladdin reagent company is 99.9%. Other chemical reagents are analytically pure. Typical experimental process, weigh a certain amount of Eu_2O_3 and dissolve it in concentrated hydrochloric acid, the white crystal, $EuCl_3 \cdot 6H_2O$, was obtained by evaporation under heating and stirring. $EuCl_3 \cdot 6H_2O$ was dissolved in ethanol with the first ligand DBM and the second ligand TPPO(molar ratio 1:3:2). And the pH was adjusted to 6~7 by ammonia reagent. After heating and stirring for 1 h, the precipitation was allowed to stand still, filtered, washed with ethanol for 3 times and dried to obtain the complexes of rare-earth europium.

2.2 reagents and instruments

The fluorescence quantum efficiency of Eu(DBM)$_3$(TPPO) was measured by Quantaurus-QY absolute quantum efficiency measuring instrument of Hamamatsu photonics trading (China) co. LTD. The fluorescence spectra of Eu(DBM)$_3$(TPPO) was determined by flouroSENS type steady state fluorescence spectrometer (FS) of Gilden. The infrared spectra of Eu(DBM)$_3$(TPPO) was determined by the Nicolet5700 Fourier transform infrared spectrometer of Thermo colliers. Eu(DBM)$_3$(TPPO) sample was measured by the American TA-SDTQ600 thermal analyzer. The morphology of Eu(DBM)$_3$(TPPO) was characterized by Hitachi S-3400N scanning electron microscope of Hitachi Company. The content of europium was determined by EDTA method.

3. Results and discussion

3.1 Infrared spectral analysis of Eu(DBM)$_3$(TPPO)

Figure 1 shows infrared spectra of ligands and ternary complexes. As can be seen from figure 1, the characteristic absorption peak of infrared spectrum of complexes and characteristic absorption peak of ligands are obviously different. After the formation of complexes, the P=O absorption peak of TPPO moves from 1190cm^{-1} to 1160cm^{-1}, indicating that TPPO uses its O atom to coordinate with Eu to form the Eu-O(TPPO) bond.The Eu-O absorption peak of complex was 541cm^{-1}. The significant reduction of wave number of carbonyl and unsaturated carbon-carbon bonds in complexes reflect the average of single and double bonds on the chelating ring. According to the literature, the metal and the complexes form the coordination structure of the six-membered cycle-ring, the carboxyl oxygen atoms of the free ligand DBM and the oxygen atoms of TPPO participate in the coordination.

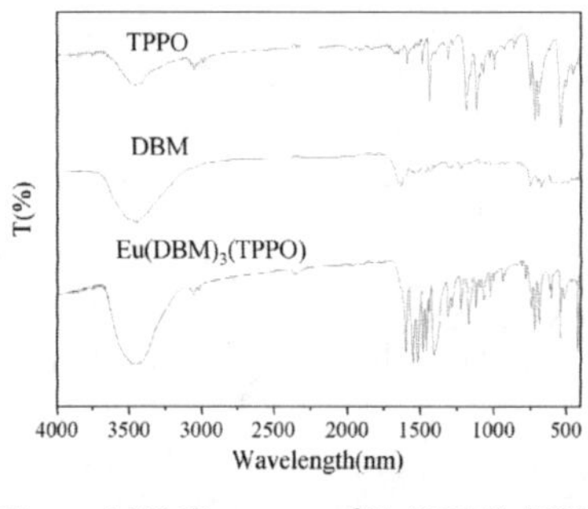

Figure.1 FT-IR spectra of Eu(DBM)$_3$(TPPO).

3.2 Excitation spectral analysis of Eu(DBM)$_3$(TPPO)

Figure 2 shows the excitation spectra of Eu(DBM)$_3$(TPPO). It can be seen from figure 2 that the excitation spectra of Eu(DBM)$_3$(TPPO) is strongly excited between 390-410nm and the maximum excitation wavelength is 396nm.

IOP Conf. Series: Materials Science and Engineering **733** (2020) 012036 doi:10.1088/1757-899X/733/1/012036

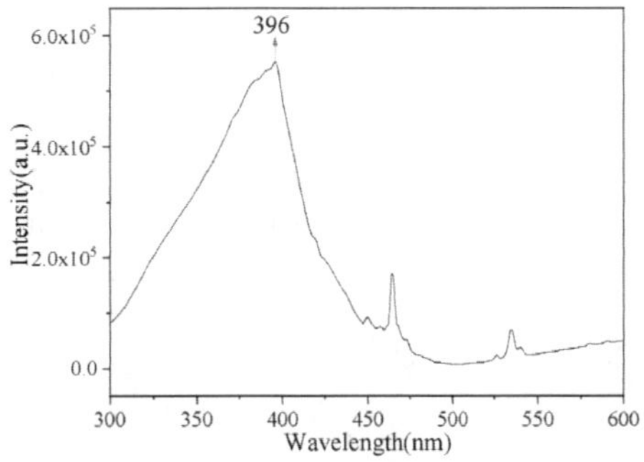

Figure.2 Excitation spectra of Eu(DBM)$_3$(TPPO)

3.3 Emission spectral analysis of Eu(DBM)$_3$(TPPO)

Figure 3 is emission spectrogram of Eu(DBM)$_3$(TPPO) measured with the excitation wavelength of 395nm. As can be seen from figure 3, the $^5D_0 \rightarrow {}^7F_2$ transition with the strongest emission wavelength of 612nm is the electric dipole transition of Eu^{3+}, with sharp peak shape, half peak width of 11nm and good monochromatic property. By comparing the emission spectrum of the complex Eu(DBM)$_3$(TPPO) with that of the ligand DBM, it can be seen that the ligand DBM has strong emission between 500-550nm, while the complex Eu(DBM)$_3$(TPPO) has no emission between 500-550nm, which indicates that after the formation of the Eu(DBM)$_3$(TPPO) complex, DBM transfers energy to Eu^{3+} central ion effectively.

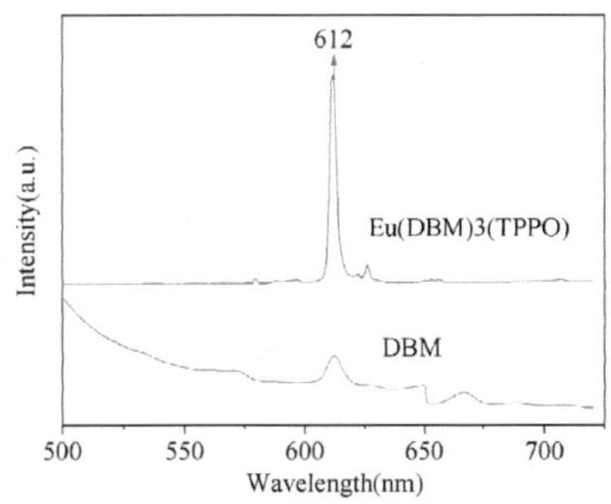

Figure.3 Emission spectra of Eu(DBM)$_3$(TPPO)

3.4 Color coordinates and color temperature of Eu(DBM)$_3$(TPPO)

Figure 4 is color coordinate diagram of Eu(DBM)$_3$(TPPO). It can be seen from figure 4 that the color coordinate of Eu(DBM)$_3$(TPPO) is (0.63, 0.36). This further indicates that Eu(III) ions coordinate with ligand DBM and ligand TPPO successfully.

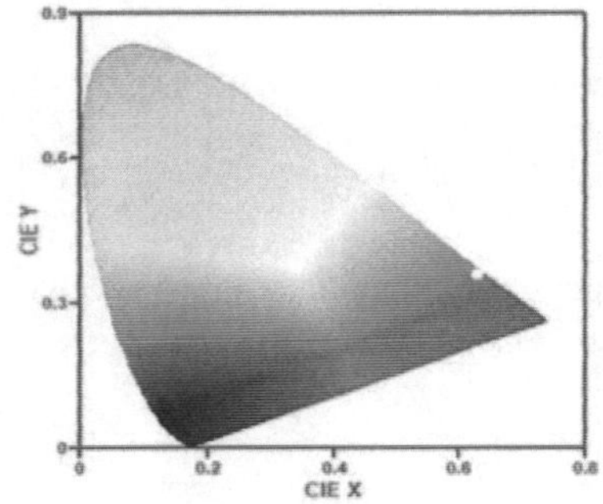

IOP Conf. Series: Materials Science and Engineering **733** (2020) 012036 doi:10.1088/1757-899X/733/1/012036

Figure.4 Emission spectra of $Eu(DBM)_3(TPPO)$

3.5 SEM image analysis of $Eu(DBM)_3(TPPO)$

Figure 5 is a scanning electron microscope image of complex $Eu(DBM)_3(TPPO)$. It can be seen from figure 5 that the particles of Eu (DBM) 3 (TPPO) are irregular flakes with uneven distribution. The average grain size is about 50-80 microns.

Figure.5 SEM image of $Eu(DBM)_3(TPPO)$

3.6 TG-DSC analysis of $Eu(DBM)_3(TPPO)$

The TG-DSC curve of $Eu(DBM)_3(TPPO)$ is shown in figure 6. According to the TG curve in figure 6,from room temperature 25°C to 300°C, the complexes have a loss of weight of about 4%, and there is a strong endothermic peak around 175°C on the DSC curve, which should be the removal of endothermic absorption of the complexes adsorbed solvents. The sample began to decompose when the temperature was greater than 300°C, and the surface sample has good thermal stability. From 300°C to 600°C, the total weight loss was about 68% and the exothermic peaks were obvious at 454°C, 514°Cand 577°C, indicating that the decomposition process of the complex was carried out step by step. At about 600°C, the complex has been decomposed, and the total weight loss is about 72%. The remaining 28% should be Eu_2O_3. This is consistent with the theoretical weight remaining from the thermal decomposition of $Eu(DBM)_3(TPPO)$, which also indicates that the coordination of ligand and europium is successful.

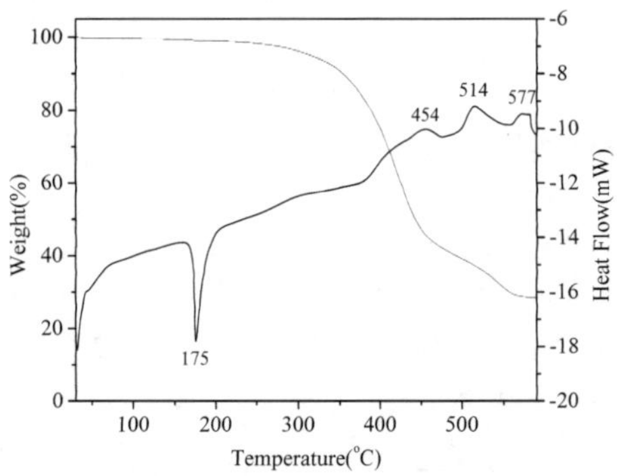

Figure.6 TG-DSC curves of $Eu(DBM)_3(TPPO)$

4. Conclusions

In ethanol solvent, Eu_2O_3 as europium source, DBM as the first ligand, TPPO as the second ligand, we use precipitation method to synthesize red europium (III) ternary rare earth complexes $Eu(DBM)_3TPPO$. $Eu(DBM)_3TPPO$ complex has good luminescence and thermal stability, and can emit pure red light.

$Eu(DBM)_3TPPO$ is a kind of red luminescent material with broad application prospects.

Acknowledgements
The authors acknowledge education leading talent project of Huzhou South Taihu special plan and financial supports from National Training Programs of Innovation and Entrepreneurship for Undergraduates.

References

[1] Su Y, Yang C L,Li Y B. Research progress on electroluminescent materials of europium complexes in red light . *Chinese journal of materials science and engineering, 2018,**36** (2):324-329.*

[2] Wei J Q, Sun C, Li Q H. The application of rare earth europium complexes in fluorescent anti-counterfeit printing ink . *Journal of Tianjin university of science and technology, 2012, **27** (4): 36-39.*

[3] Zhang Y. Application of europium-complexes based nano complexes in cancer diagnosis and treatment . *Lanzhou university, 2016.*

[4] You H, Fang J, Xuan Y, et al. Highly efficient red electroluminescence from stacked organic light-emitting devices based on a europium complex. *Materials Science & Engineering B, 2006, **131**(1-3): 252-255.*

[5] Zhang L, Yin G T, Zheng M L, et al. Synthesis and properties of gel-hybrid luminescent materials with new europium complexes. *Journal of biochemistry, 2018, **4**(3):102-104.*

[6] Hou X Y, Wang X, Fu F, Wang J J, Wang J W. Synthesis, characterization, fluorescence and thermal decomposition kinetics of new europium ternide complexes. *Henan science&technology, 2009, **27**(5): 530-535.*

[7] Zhang Y. Synthesis and properties of carboxylic acid europium complexes.*Dalian university of technology, 2016.*

Lightweight optimization design of arm frame for aerial work platform based on NSGA-II algorithm

Zhiyong Wang[1] and Xiping Liu[2]

1 Master, Yanshan university, Qinhuangdao,CHINA
2 Professor, Yanshan university, Qinhuangdao,CHINA
E-mail: mec29@mec.ysu.edu.cn

Abstract: In the previous lightweight design of the arm of aerial work platform, it is difficult to balance the relationship between lightweight and stiffness reduction. This paper proposes a lightweight design that minimizes the variation stiffness while reducing weight. Taking the arm mass and stiffness variation as the objective, which constrained by yield strength and stiffness of the material, the Pareto front of each arm section size is obtained by the second generation of elite strategy non-dominated sequencing genetic algorithm (NSGA-II). Finally, the finite element model is established by ANSYS to compare the static and dynamic performance of the arm before and after optimization. The results show that the weight of the optimized arm is reduced by the original 13.3%, lightweight effect is remarkable.

1. Introduction

As the main component of the aerial work platform, the arm structure occupies 58%-67% of the whole vehicle. Therefore, the lightweight optimization design for the arm is an effective way to reduce the overall mass.[1]. In recent years, topology optimization has been applied in the optimization of crane arm sections. Wang Xin et al[2] used the Optistruct module of the commercial software HyperWorks to optimize the topology of the crane arm section to obtain the optimal section topology. Zhang Xianhuan et al [3] applied the forward design method to reduce the weight reduction effect of the tower crane arm. Li Qianjin et al[4] use ANSYS to optimize the weight of the aerial work platform arm with the mass as objective, and succeeded in reducing the mass by 6.3%. In order to minimize the variation in stiffness of the arm while reducing weight, the NSGA-II is used to find the better size of the arm section. Taking the arm mass and stiffness variation as the objective, which constrained by strength and stiffness. The results show that the weight of the optimized arm structure is reduced by 871kg, accounting for 13.3% of the original mass. Finally, the statics and dynamics performance of the before and after optimized arm structure are compared under the typical working conditions. The results show that the lightweight effect is obvious, and the method can provide a reference for the lightweight design of the arm of aerial platform.

2. The finite element model and optimization model

In this paper, optimization model has five arms, which total weight is 6518kg, and section is rectangular. Using APDL language in ANSYS to establish parametric finite element model of each arm. The finite element model of the arm of aerial platform is shown in Figure 1.

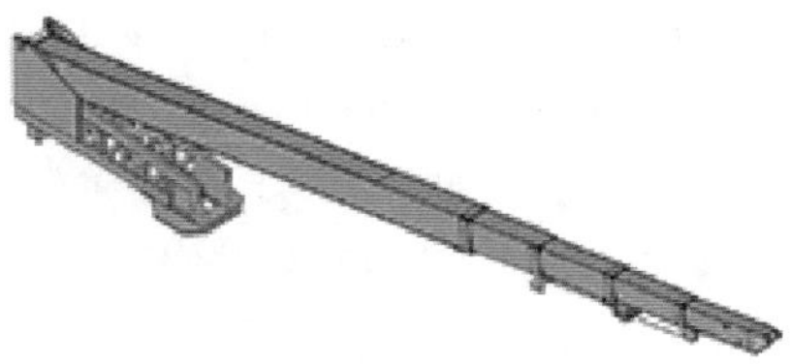

Fig. 1 Finite element model of arm frame for aerial work platform

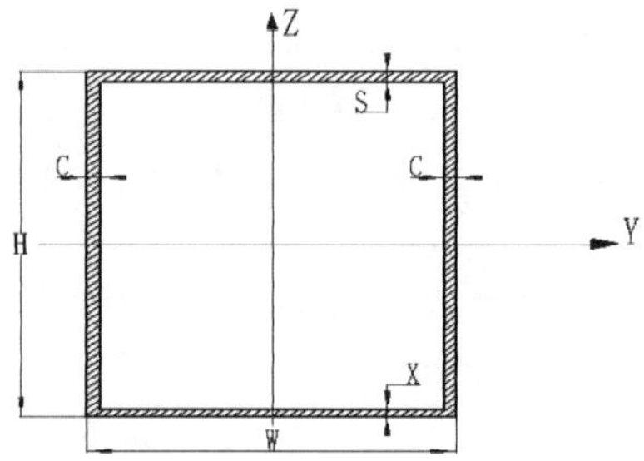

Fig 2 section of the arm

2.1 working condition

Six typical working conditions are selected for calculation, in which the dangerous working condition-4 is taken as the optimal calculation condition. Finally, the before and after optimized structure are compared under the last conditions. The typical working conditions are described in Table1

Tab. 1 Descript for working conditions

Working condition	1	2	3	4	5	6
Amplitude angle (°)	0	28	50	70	76	80
elongation (mm)	1739	2355	4082	8140	8750	8750
Arm head force (N)	265416	209385	199663	224958	244859	263870
Level hole force (N)	261444	211738	209863	237061	258001	277594

2.2 Optimization of arm section size

The arm section is rectangular, so the optimized sizes of the arm are flange plate and web, as shown in Figure 2. The size to be optimized in the figure is the section height H_i, section width W_i, section web thickness C_i, thickness of flange plate on the section S_i, and thickness of flange plate under the section X_i, i =1,2,3,4,5.

2.3 Optimization mathematical model

In the lightweight optimization design, the maximum stress of the optimized arm under all working conditions is required to be no more than the ratio of its yield strength and safety factor. Since the load of working condition has been enlarged to the safety factor in this paper, the stress constraint is no more than the yield strength of the material. In addition, the maximum displacement of the optimized arm head in the amplitude plane cannot exceed the allowable value. Finally, under the condition of satisfying the constraints of stress and displacement, the optimization objective is to minimize the arm mass and stiffness variation. The mathematical model is as follows:

$$\begin{cases} min \quad M,\ \Delta K \\ s.t. \quad \sigma_{max} \leq [\sigma] \\ \quad U_{max} \leq [U] \\ i \in [i-20, i+20], i=1,2,3,4,5 \\ j \in [j-10, j+10], j=1,2,3,4,5 \end{cases} \tag{1}$$

Where: $M, \Delta K$ are the arm mass and stiffness variation; σ_{max} is the maximum stress of the arm; $[\sigma]$ is the yield strength of the material; U_{max} is the maximum displacement of the arm in the amplitude plane; $[u]$ is the allowable displacement value; i is the arm section height and width dimension value; j is the arm section web and Flange plate thickness dimension value. Where ΔK can be transformed into the minimum displacement variation after optimization of the arm end, as shown in equation (2)

$$\Delta K = f - f_0 \tag{2}$$

Where, f_0 and f are the displacement before and after optimization of the arm head, which f can be divided into the static displacements fz and fy of the arm head in the amplitude plane and the rotation plane, as shown in formula (3). fz and fy are calculated by the amplification coefficient method, and the calculation formulas are (4), (5):

$$f = \left(fz^2 + fy^2\right)^{1/2} \tag{3}$$

$$f_z = \frac{1}{1-\frac{N}{0.9N_{cry}}}\left(f_{Wz} + \Delta_z \sum_{i=1}^{K-1} \frac{H_{i+1}}{l_{i+1}}\right) \tag{4}$$

$$f_y = \frac{1}{1-\frac{N}{0.9N_{crz}}}\left(f_{Wy} + \Delta_y \sum_{i=1}^{K-1} \frac{H_{i+1}}{l_{i+1}}\right) \tag{5}$$

Where: N is the axial pressure on the arm head; f_{Wz} and f_{Wy} are the arm head deflections caused only by the transverse load of the amplitude and the rotation plane when the axial pressure equal zero; N_{cry} and N_{crz} are the critical forces of the arm head on the amplitude plane and the rotation plane; Δ_z and Δ_y are the lateral gap between two adjacent arms, taking 1 ~ 3mm; H_{i+1} is the distance from the arm end to the head: l_{i+1} is the distance from the end of the $i+1$ arm to the head of the i arm; K is the number of telescopic arms. Then the optimization objective ΔK is transformed into f_z and f_y.

3. Multi-objective optimization

3.1 Optimization method

The second generation of non-dominated sorting genetic algorithm with elite strategy (NSGA-II)[5] is an improved algorithm of the traditional genetic algorithm. It continues the "survival of the fittest" criterion of the genetic algorithm, and then reduces the complexity of the non-dominated sorting genetic algorithm, with the advantages of fast running speed and good solution convergence. NSGA-II has the ability to find multiple Pareto optimal solutions in the optimization process, and can handle various types of objective numbers and constraints without imposing some mathematical relationships. So it can solve the problem of minimum mass and stiffness variation. The optimization flow of this paper is shown in Figure 3.

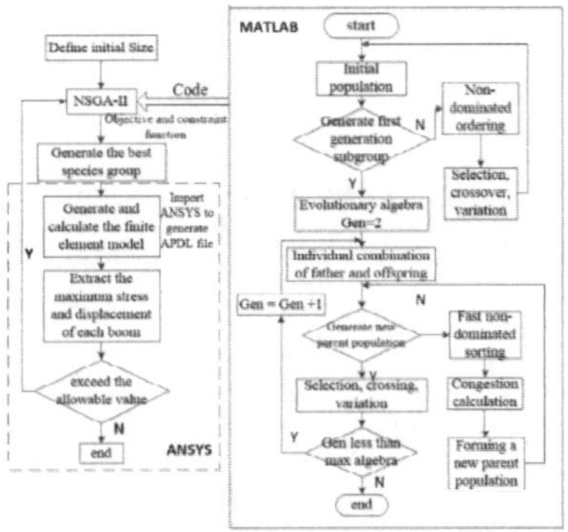

Fig 3 process of optimization solution

2.2 Design variable and Result process

For the design size of the initial input, firstly, the genetic algorithm transforms each variable into the form of binary coding, and then randomly generates the initial population. Then, the population is selected and evolved through crossover and mutation operations. The crossover algorithm uses the simulated binary crossover[6],and the following descendants are generated by the parents x_a and x_b, as equations (6) ~ (9) show. The variation uses the polynomial variation method. For details, see the literature[7]. In this paper, the probability of cross variation is 0.9, and considering the elite strategy, the offspring and the parent are combined to make the new species group more tend to be elite.

$$y_1(k) \leftarrow \frac{1}{2}[(1-\beta_k)x_a(k)+(1+\beta_k)x_b(k)] \tag{6}$$

$$y_2(k) \leftarrow \frac{1}{2}[(1+\beta_k)x_a(k)+(1-\beta_k)x_b(k)] \tag{7}$$

Where β_k is a random number generated by the probability density of formula (6) :

$$PDF(\beta) = \begin{cases} \frac{1}{2}(\mu+1)\beta^{\mu}, 0 \le \beta \le 1 \\ \frac{1}{2}(\mu+1)\beta^{-(\mu+2)}, \beta > 1 \end{cases} \tag{8}$$

Where, μ is any non-negative real number, $\mu \in [0,5]$, β can be obtained by (7)

$$r \leftarrow U[0,1], \beta \leftarrow \begin{cases} (2r)^{\frac{1}{\mu+1}}, r \le 0.5 \\ (2-2r)^{-\frac{1}{\mu+1}}, r > 0.5 \end{cases} \tag{9}$$

For the multi-objective optimization problem, there are n objective components $f_i(x)$, $i=1,2\cdots n$, and any two decision variables x_a and x_b are given. If following two conditions are true, x_a is said to dominate x_b :(1) For $\forall i \epsilon 1,2,\ldots,n$,always have $f_i(X_a) \le f_i(X_b)$;(2) $\exists i \epsilon 1,2,\ldots,n$, makes $f_i(X_a) < f_i(X_b)$.For a decision variable, if there are no other decision variables that can dominate him, then the decision variable is called a non-dominated solution. In this paper, the non-dominated solution set X_i, i=1, 2..., 25 of the section sizes of the arm is obtained by non-dominated sorting, furthermore, which is brought into ANSYS for modeling calculation, and the stress and displacement constraints will be adopted at the same time. Finally, the optimized structure is compared with the static and dynamic performance of the structure before optimization.

IOP Conf. Series: Materials Science and Engineering **733** (2020) 012037 doi:10.1088/1757-899X/733/1/012037

3 Analysis of Result

3.1 Lightweight results

Through many experiments, it is found that when the number of popular are 50 and the number of evolutionary iterations is 500, there is a better convergence effect. The optimization iterative process is shown in Figure 4. The 500th generation population is selected as the final optimized population, and the optimal solution is selected in this group. The change curve of arm displacement with mass is shown in Figure 5. Point A is the optimal solution point. The Pareto frontier obtained by optimization is shown in Figure 6.

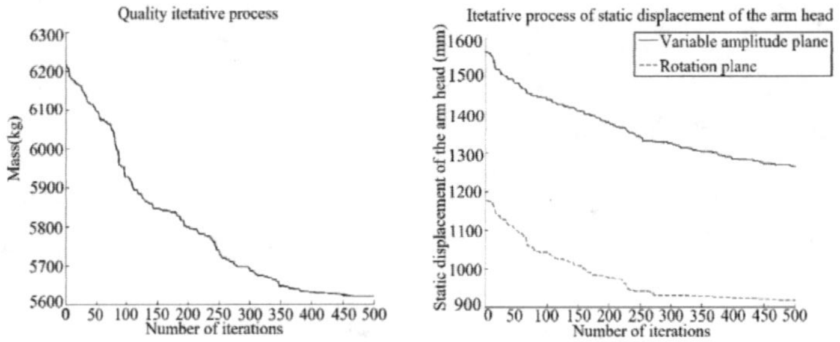

Fig 4 Optimized iterative process

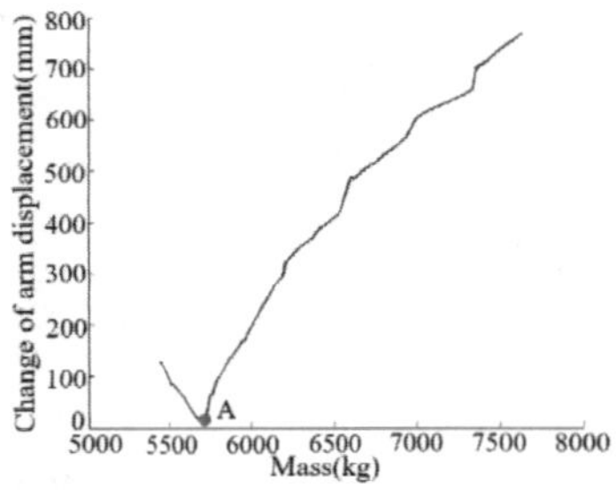

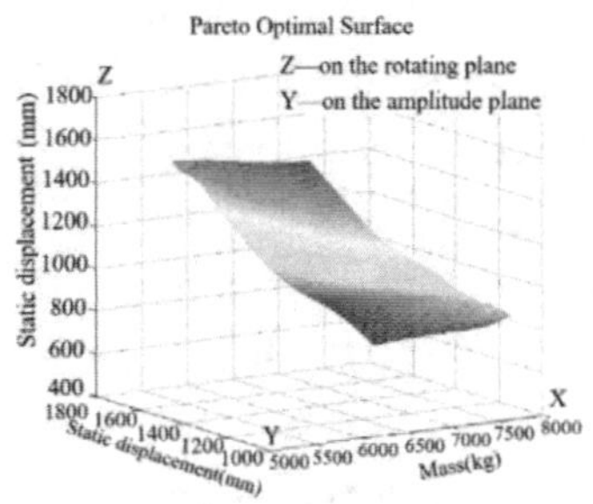

Fig 5 The curve of displacement with mass **Fig 6 Pareto frontier**

The optimized size has been rounded, and the maximum stress and displacement of each arm under the dangerous condition meet the requirements. The optimized result show that the weight after optimization is 5657kg , which is reduced 13.3% of the original mass, that is 6518kg.

3.2 Comparison of static performance

The strength and stiffness of the arm before and after optimization were analyzed and compared under six working conditions. The comparison results are shown in Table 2. The displacement and stress distribution under dangerous conditions are shown in Figure 7,8. According to the comparison data, it can be seen that, under all working conditions ,the maximum stress of the optimized arm is reduced, and the maximum displacement is increased, but it is still within the allowable range, so the static performance of the optimized structure is stable.

Tab2. strength and stiffness comparison

Working	Maximum stress /Mpa			Maximum displacement/mm		
condition	Initial	Optimized	Variation	Initial	Optimized	Variation
1	612.45	537.36	75.09	282.95	285.26	2.31
2	492.02	490.18	1.84	375.25	382.07	6.81

3	515.70	491.23	24.47	716.34	749.56	33.22
4	510.62	510.14	0.48	2116.33	2295.70	179.37
5	525.94	482.78	43.16	1996.00	2184.61	188.61
6	525.14	480.63	44.51	1634.59	1804.20	169.61

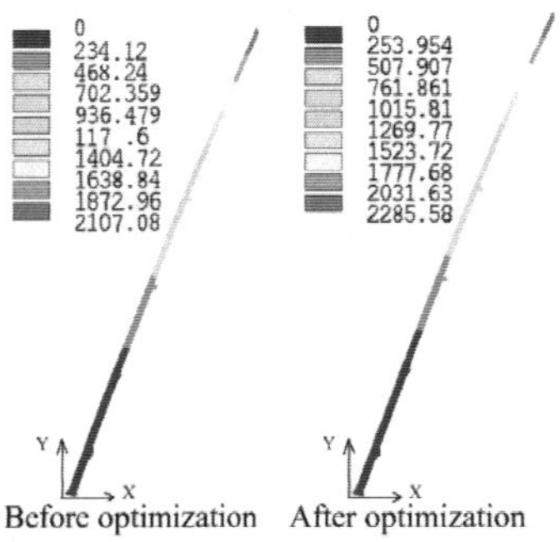

Fig 7 Comparison of displacement results

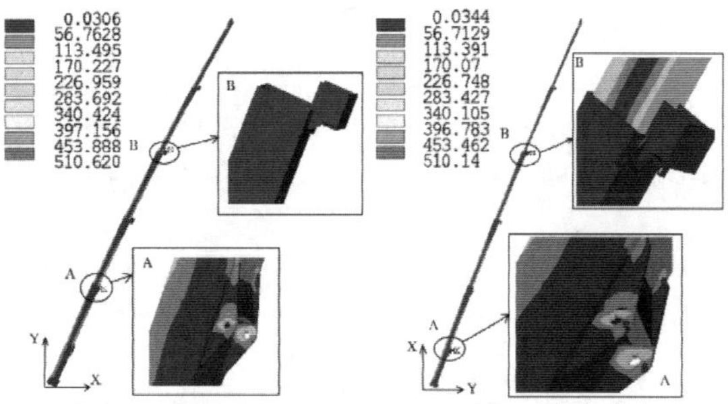

Fig 8 Comparison of stress results

3.2 Dynamic performance comparison

The modal analysis is carried out before and after optimization of the arm under dangerous conditions, and the first five modal frequencies and vibration modes of the arm are obtained. The modal comparison is shown in Table 3, and the main modal comparison is shown in Figure 9. Comparing the data, it can be seen that the vibration frequency difference before and after the first five-step optimization is small, and the vibration mode changes are not large. The dynamic performance of the optimized structure meets the requirements.

Tab 3 First 5 order Modals Frequencies of Optimal Arm (Hz)

Order	Initial	Optimized	Variation	Mode description
1	0.15156	0.16646	0.0149	arm Z direction first-order bending
2	0.43379	0.45008	0.0163	arm Y direction first-order bending
3	0.94420	0.91540	0.0288	arm Z direction second-order bending
4	1.6012	1.5787	0.0225	arm Y direction second-order bending
5	2.2967	2.2494	0.0473	arm Y direction third-order bending

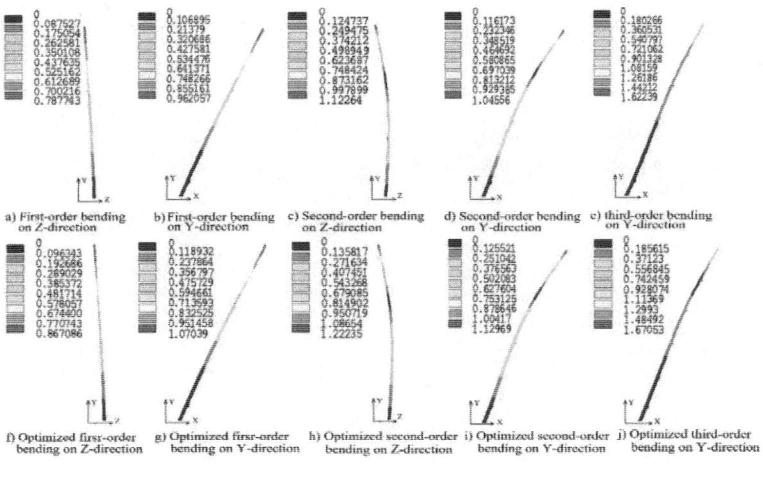

IOP Conf. Series: Materials Science and Engineering **733** (2020) 012037 doi:10.1088/1757-899X/733/1/012037

Fig 9 Comparisons of five Modal Vibration Modes Between Initial and Optimal Arm

4. Conclusion

In this paper, the second generation of non-dominated genetic algorithm (NSGA-II) is used to optimize the arm of aerial platform, which minimizes variation stiffness while reducing weight Meanwhile, the statics and dynamics performance of the arm after optimized are similar to that before optimization. The arm weight is reduced by 871kg, accounting for 13.3% of the original mass, which effect is obvious. The results can provide design suggestions for designers.

Reference

[1] Xia Xiufeng 2011 *on the development of aerial platform in China*(China: construction mechanization) chapter 9 pp 32-34

[2] Wang Xin, et al. 2009 *topology optimization of crane telescopic boom section*(China: Journal of Dalian University of technology) chapter 49(3) pp 374-379.

[3] Zhang xianhuan, Liu Zijian. Li Siming 2015 *research on forward design method of tower crane jib*(China: mechanical science and technology) chapter 10 pp1553-1559

[4] Li Qianjin, Zhang Shengli, Wang Zhiyong, Jiang Wenguang 2019 *lightweight design and improvement of boom of aerial platform*(China: construction mechanization) chapter 40(03) pp23-25

[5] P. A. Deb K , Agarwal S , et al. 2002 *A fast and elitist multiobjective genetic algorithm: NSGA-II* (American :IEEE Transactions on Evolutionary Computation) chapter 6(2) pp 0-197.

[6] R. B. Agrawal, K. Deb, and R. B. 2000 *Agrawal,Simulated Binary Crossover for Continuous Search Space*(Complex Systems vol 9(3)) pp 115-148.

[7] K. Deb and M. Goyal 1996 *A Combined Genetic Adaptive Search (GeneAS) for Engineering Design* (Computer Science and Informatics vol 26 (4)) pp 30-45.

Study on the influence of anticorrosive coating on chloride ion erosion resistance of concret

Tianyang Du[1]

1 Postgraduate, Shenyang JianZhu University, Shenyang, China
E-mail: 136964904@qq.com

Abstract. With the development of The Times and the progress of the society, more and more buildings have been built, more and more roads and bridges have been built, that is to say, the number of concrete structures is increasing.While concrete is convenient for our life, we also find that in the service of concrete structure, its durability sometimes fails to reach the designed service life. Especially, concrete buildings in coastal areas are corroded by chloride ions for a long time, which causes damage to the concrete structure.An effective way to solve this problem is to apply anti-corrosion coating on the concrete in service.This paper aims to introduce the influence of different anticorrosive coating systems on chloride ion erosion resistance of concrete.Through drilling samples of concrete specimens with dry and wet circulation in 10% NaCl solution.The content of chloride ion in concrete powder was tested at different depths.By comparing the experimental data of these three different anticorrosive coatings, we can find that polyurethane coating has the best chloride ion resistance, the next is polyurea coating.Both of them protect the concrete from chloride ion erosion.

1. Introduction
With the introduction of more and more concrete products in today's society, the durability of concrete has also come into people's eyes, especially the durability of concrete in coastal areas is a difficult problem. In many ways to solve the problem of concrete durability, concrete anticorrosive coating is the best way.We improved the traditional anticorrosive coating system and constructed two different anticorrosive coating systems:Epoxy primer,Epoxy intermediate paint and Acrylic polyurethane finish;Epoxy primer,Epoxy intermediate paint and Polyurea paint.These two coating systems can effectively solve the problem of chloride ion erosion resistance of concrete, so as to effectively improve the service life of concrete.

2. Expeiment

2.1. Experimental material
The cement used in this experiment is ordinary Portland cement P.O 42.5, which is produced in jidong cement co., LTD. All properties of this cement meet the requirements, and its physical properties are shown in Table 1;The water-reducing agent is basfaphthalene water-reducing agent sold in shenyang. It is white powder and soluble in water;The experimental water is urban water in Shenyang;The sand used in the experiment is medium sand;In the experiment, 5-10mm and 10-20mm gravel were used in the gravel field, and 5-20mm continuous grading gravel was synthesized according to the ratio of 3:7;The fly ash is grade I produced in shenyang, and the specific chemical composition is shown in

Table 2;The slag is S95 slag produced in Shenyang. The specific chemical composition is shown in Table 3;The strength of the concrete is C40 concrete.

Table 1. Physical properties of cement

Raw material	MgO (%)	Insolubles (%)	SO_3 (%)	Loss (%)	Initial setting time (min)	Final setting time (h)	Sulfate resistance	The hydration heat
The national standard	≤5.0	≤1.5	≤3.5	≤3.5	≥45	≤6.5	_	_
Measured value	3.0	0.57	3.5	0.2	60	6	Mid	Mid

Table2. Chemical composition of fly ash /%

component	CaO	SiO_2	Al_2O_3	Fe_2O_3	SO_3
content	2.86	59.67	28.31	3.92	0.62

Table 3. Chemical composition of slag /%

component	CaO	SiO_2	Al_2O_3	MgO	SO_3	Fe_2O_3	K_20	Impurities
content	51.28	27.56	9.99	5.79	1.91	0.64	0.59	2.24

2.2. Production of Specimens for Testing

(1) The cured concrete specimen is polished with sandpaper, and the defects such as empty drum, floating slurry and oil stains on the concrete surface are polished and removed to form relatively regular concrete test blocks for simple chamfering.

(2)Apply an epoxy sealant primer with a thin coating of about 50 μ m.After the primer coating is naturally dried, putty is used to fill the surface defects of the specimen to make it smooth and smooth.

(3)After 24h, paint the epoxy intermediate paint. After the first layer of epoxy intermediate paint is naturally dried, paint the second time. The thickness of each layer of dry film is about 110 microns, and the total thickness of the two layers of dry film is about 220 microns.

(4)After 7d, apply the polyurea (polyurethane) topcoat, and apply the second coat after the first coat is naturally dried. The thickness of each dry film is about 40 microns, and the total thickness of the two dry films is about 80 microns.

(5)After painting, put the specimen into the standard curing room for 7 days.

2.3. Test method

(1) Uncoated concrete, polyurea system coated concrete and polyurethane system coated concrete were immersed in 10% NaCl solution soaking tanks respectively.

(2) After completing the cycle every cycle, the coated specimens of different systems were taken out, that is, at day 8, 16 and 24, the concrete powder at 2mm, 5mm, 10mm, 15mm, 20mm and 25mm of the

specimens were respectively obtained with a high-precision coring machine and put into the sealed bag.

(3) The concrete powder samples taken in step (2) were weighed and taken 4g respectively for extraction test.Fill the reagent bottle with 40g of extract, and cover the bottle with the cap and oscillate vigorously until the concrete powder is fully integrated into the extract.

(4) Prepare the sswy-810 rapid determination instrument for chloride ion content: connect the equipment, clean the electrodes, open the test software, select 50ml 5.0×10^{-4}, 5.0×10^{-3}mol/L NaCl standard solution and place them in a pre-cleaned and dry beaker, add electrode stabilizer (1ml), and insert the electrodes into the standard solution in the order from thin to thick.After two calibration measurements, a straight line appears in the "calibration curve", which means the electrode calibration has been completed.

(5) Clean the calibrated electrode three times and wipe it dry with filter paper. Add electrode stabilizer (1ml) to each sample before test. Enter sample information into the software.

3. Result and discussion

3.1. Experimental results and discussion on chloride ion erosion of anticorrosive coated concrete

The corrosion form of chloride ion is one of the main causes of steel corrosion in concrete structure, which will also aggravate the freezing-thawing damage of concrete and adversely affect the durability of concrete.Chloride ion invasion of concrete is carried out in several ways, but under different circumstances, a certain invasion mode is the main one.The different ways of chloride ion erosion are actually permeated into concrete with water as the carrier. The permeability of concrete is closely related to its durability, so the impermeability is also an important index to evaluate the durability of concrete.

Table 4. The variation of chlorine ion content with the change of invasion depth (8d)

Concrete type	W（%）					
	d=2mm	d=5mm	d=10mm	d=15mm	d=20mm	d=25mm
Coating free system	0.0315	0.0115	0.0008	0.0001	/	/
Polyurea system	/	/	/	/	/	/
Polyurethane system	/	/	/	/	/	/

As can be seen from the data in Table 4, when the concrete is subjected to the dry-wet cycle chloride ion erosion test for 8 days after curing, the chloride ion content of the uncoated concrete at 2mm reaches 0.0315%, and the invasion depth has reached 15mm.Water-binder ratio is an important factor affecting the diffusion of chloride ions. Excessive water-binder ratio leads to poor compactness of concrete, increased capillary pores and decreased impermeability, leading to the advance of chloride ion invasion into concrete.At the same time, it can be seen that both polyurea system coated concrete and polyurethane system coated concrete were not corroded by chloride ions. It can be seen that the anticorrosive coating has an obstacle effect on the diffusion of chloride ions, indicating that the anticorrosive coating has a significant effect on the corrosion resistance of chloride ions.At the same time, it is not difficult to find that in the uncoated concrete, with the increase of depth, the content of chloride ion is gradually decreasing.Sections, subsections and subsubsections

Table 5. The variation of chlorine ion content with the change of invasion depth (16d)

Concrete type	W（%）					
	d=2mm	d=5mm	d=10mm	d=15mm	d=20mm	d=25mm
Coating free system	0.0998	0.0542	0.0338	0.0025	0.0010	/
Polyurea system	0.0085	0.0024	0.0010	0.0001	/	/
Polyurethane system	0.0045	0.0015	0.0001	/	/	/

As can be seen from the data in Table 5, when the concrete is subjected to the dry-wet cycle chloride ion erosion test, the data of 16 days of deterioration are as shown above: the invasion depth of uncoated concrete reaches 20mm, the invasion depth of polyurea system coated concrete reaches 15mm, and the invasion depth of polyurethane system coated concrete reaches 10mm.At the same invasion depth, the content of chloride ions in each sample ranged from large to small, including uncoated concrete, polyurea system coated concrete and polyurethane-coated concrete. In a short period of time, polyurethane system coating, compared with polyurea system coating, can better prevent the penetration of chloride ions and has better anti-chloride ion erosion effect.For example, at 2mm, the content of chloride in polyurea system coated concrete decreased by about 91.48% compared with that in uncoated concrete, and the content of chloride in polyurethane system coated concrete decreased by about 95.49% compared with that in uncoated concrete.It shows that both coating systems can effectively improve the chloride corrosion resistance of concrete, and both have a certain anti-corrosion effect, and polyurethane coating has the most obvious effect, followed by polyurea coating.

Table 6. The variation of chlorine ion content with the change of invasion depth (24d)

Concrete type	W（%）					
	d=2mm	d=5mm	d=10mm	d=15mm	d=20mm	d=25mm
Coating free system	0.1961	0.1554	0.0732	0.011	0.0042	0.0018
Polyurea system	0.0517	0.0112	0.0069	0.0041	0.0038	/
Polyurethane system	0.0215	0.0055	0.0001	/	/	/

According to Table 6, when the dry-wet cycle chloride erosion test of concrete deteriorated to 24 days, the invasion depth of uncoated concrete reached 25mm, the invasion depth of polyurea system coated concrete reached 20mm, and the invasion depth of polyurethane system coated concrete reached 10mm.At the same invasion depth, the content of chloride ions in each sample was uncoated concrete, polyurea system coated concrete and polyurethane system coated concrete from large to small.For example, at 2mm, the content of chloride in polyurea system coated concrete decreased by about 73.64% compared with that in uncoated concrete, and the content of chloride in polyurethane system coated concrete decreased by about 89.04% compared with that in uncoated concrete.It shows that both coating systems can effectively improve the chloride corrosion resistance of concrete in 24 days of dry and wet cycle, and both of them have a certain anti-corrosion effect, and polyurethane coating has the most obvious effect, followed by polyurea coating.

4. Conclusions

In this paper, experimental studies on the chloride ion permeability resistance of concrete with different anticorrosive coatings were carried out, and the following conclusions were obtained:

- With the increase of invasion depth (2mm-25mm), the chloride ion concentration in each sample decreased gradually.This shows that the distribution of chloride ion concentration from the concrete surface is a decreasing process;
- At the same invasion depth, the content of chloride ions in each sample was uncoated concrete, polyurea system coated concrete and polyurethane system coated concrete from large to small.The results show that the three coating systems can effectively improve the chloride corrosion resistance of concrete, and all of them have a certain anti-corrosion effect, and the polyurethane system has the most obvious effect, followed by the polyurea system coating.;
- At the same invasion depth, the chloride content of each sample increases with the increase of the time of dry-wet cycle corrosion.The increase rate of chloride ion content in uncoated concrete with time was significantly higher than that in urea-coated concrete and polyurethane coated concrete, among which the increase rate of chloride ion content in polyurethane coated concrete with time was the slowest.The protective effect of the two coatings did not decrease significantly with the increase of time, and the long-term protective effect was better.

Reference

[1]Pavla Halamickova,Rachel J. Detwiler,Dale P. Bentz,Edward J. Garboczi. Water permeability and chloride ion diffusion in portland cement mortars: Relationship to sand content and critical pore diameter[J]. Cement and Concrete Research,1995,25(4).

[2]A. Delagrave,J. Marchand,E. Samson. Prediction of diffusion coefficients in cement-based materials on the basis of migration experiments[J]. Cement and Concrete Research,1996,26(12).

[3]Sasan Parniani,Houssam Toutanji. Monotonic and fatigue performance of RC beams strengthened with a polyurea coating system[J]. Construction and Building Materials,2015,101.

[4]Zhe Shisun,Li Fazhi.Study on the corrosion resistance of coated concrete to hydrochloric acid[J].Shanxi building,2016,42(25):107-110.

[5]Shuo Zhang,Bo Wu. Effects of salt solution on the mechanical behavior of concrete beams externally strengthened with AFRP[J]. Construction and Building Materials,2019,229.

IOP Conf. Series: Materials Science and Engineering **733** (2020) 012039 doi:10.1088/1757-899X/733/1/012039

Preparation and Characterization of Tetraphenylporphyrin Dyes

Xinhui Li[1]
1 Postgraduate, Shenyang JianZhu University, Shenyang, China
E-mail: 2712126752@qq.com

Abstract: Porphyrin compounds exist widely in nature and are a key part of photosynthesis and photosynthesis. Because of its good photothermal stability and many unique physical and chemical properties, it has been widely used in the research and development of sensitized dye solar cells. As a new type of solar cell, dye-sensitized solar cell has attracted extensive attention and research due to its simple manufacturing process, relatively cheap price, high photoelectric conversion efficiency and environmental protection. In this paper, control experiments were carried out to synthesize tetrabenzoporphyrin with higher yield and purity by changing the solvent and washing solvent. The samples were characterized by uv-vis spectra. The results showed that the tetraphenylporphyrin crystals with better comprehensive effect could be obtained by using ethanol solution as sample solution and distilled water as washing solvent.

1.Introduction

Tetraphenylporphyrins were synthesized in 1935 by Rothemund under acidic conditions by the reaction of benzaldehyde and pyrrole in a sealed tube for 24 – 48 h under 150℃. But the yield obtained by this method is very low.

In 1967, Adlerimproved Rothemund's method. Tetraphenylporphyrin was prepared by the reaction of benzaldehyde and pyrrole in propionic acid at 141 °C for 30 min. After cooling and filtering , the filter cake with methanol and water respectively. The yield rate of tetraphenylporphyrin reached 20% .

The method of preparing tetraphenylporphyrins by condensation and cyclization of pyrrole with Aldehydes can also be used to synthesize simple meso-substituted porphyrins. This is a widely used method in the field of porphyrin synthesis.

2. Experiment

2.1First preparation

The The liquid paraffin was pour into the Constant temperature heating Magnetic stirrer, so that the liquid level reached two thirds of its volume. 180 mL xylene and 3.0 g salicylic acid were added into a 500 mL round bottlom three-port bottle, magnets, insert thermometer (measuring range 200 °C) , condensing tube and constant pressure drop funnel respectively, placeing the rounded bottom flask in the center of the Constant temperature heating Magnetic stirrer on a metal stand.

When the initial temperature was 16 °C and the reflux phenomenon began at 128 °C, the mixture of 20 mL xylene and 0.04 mol pyrrole was added to the reaction solution of round bottom flask through a constant pressure drop funnel. (within 5 min) The reaction liquid changed from light yellow to violet black very quickly. Continue reflowing for 3 h, stop heating, and cool to room temperature. Adding the same volume of distilled water, leaving for 8 h, washing with Water 4 times, the bright purple crystal was obtained. (net weight : 3.0 g) The yield rate was about 48.9%.

2.2*Second and third preparation*

1.0 g salicylic acid, 0.013 mol benzaldehyde and 60 mL xylene were added into the round bottom three mouth bottle of the same equipment, stiring and heating. The mixture of 0.013 mol pyrrole and 6.7 ml xylene were added through a constant pressure drop funnel after the reflux started (within about 2 minutes). The reaction liquid changed from light yellow to violet black very quickly. Continue reflow for 3 h, stop heating, cool to room temperature. After adding the same volume of ethanol, leaving for 8 hours, washing 4 times with water or washing 2 times with water and ethanol, bright violet crystal was obtained. (net weight 0.57 g or 0.6 g) The yield was about 27.7% and 29.3%.

2.3*Fourth preparation*

3.0 g salicylic acid, 0.04 mol benzaldehyde and 180 mL xylene were added into the round bottom three mouth bottle of the same equipment, stiring and heating. The mixture of 0.04 mol pyrrole and 20 mL xylene were added through a constant pressure drop funnel under agitation (within about 2 min) . The reactionliquid changed from light yellow to purple black quickly . Continue reflow for 3 h, then stopped heating and cooled to room temperature. After adding the same volume of ethanol and leaving it for 8 hours, washing twice with water and ethanol, the bright purple crystal was obtained (net weight : 1.2 g) . The yield rate was about 19.5% .

3. Result and discussion

3.1 The yield rate of tetraphenylporphyrin

Yield Rate = Actual Output / Shouldbe Output= Sample Weight / 6.14 g

Table 3-1 experimental results of porphyrin preparation by four different methods

	1	**2**	**3**	**4**
weight of salicylic acid	3.0g	1.0g	1.0g	3.0g
resting	water	ethanol	water	ethanol
washing	water（4 times）	water（4 times）	water（twice）+ ethanol（twice）	water（twice）+ ethanol（twice）
weight of sample	3.0g	1.7g	1.80g	1.2g
yield	48.9%	27.7%	29.3%	19.5%

The first preparation of porphyrin: The crystal was violet black, surface attached black flocs, yield was 48.9% ;

The second preparation of porphyrin: The crystal was bright purple, uniform thickness, yield was 27.7%;

The third preparation of porphyrin: The crystal was bright purple, but the size of the crystal was different, the yield was 29.3%;

The fourth preparation of porphyrin: The crystal was bright purple, uniform thickness, but the yield was 19.5%.

The conclusion drawn from the preparation of tetrabenzoporphyrin was as follows: The purified product can be obtained by washing with ethanol rather than distilled water alone; However, washing the sample with ethanol would lead to its dissolution in the solution, resulting in too little product,and yield was too low.

3.2 Detection of ultraviolet spectrum

Weigh 0.0003g of the product and dissolve it in dichloromethane, titrate to 100mL. The 4 kinds of solutions were blue and purple, with No. 4 being the darkest, No. 2 and No. 3 the second, and No. 1 the lightest. The four solutions were tested by uv-vis spectroscopy. The results were showed in Figure 1: UV spectra of 450-700nm.

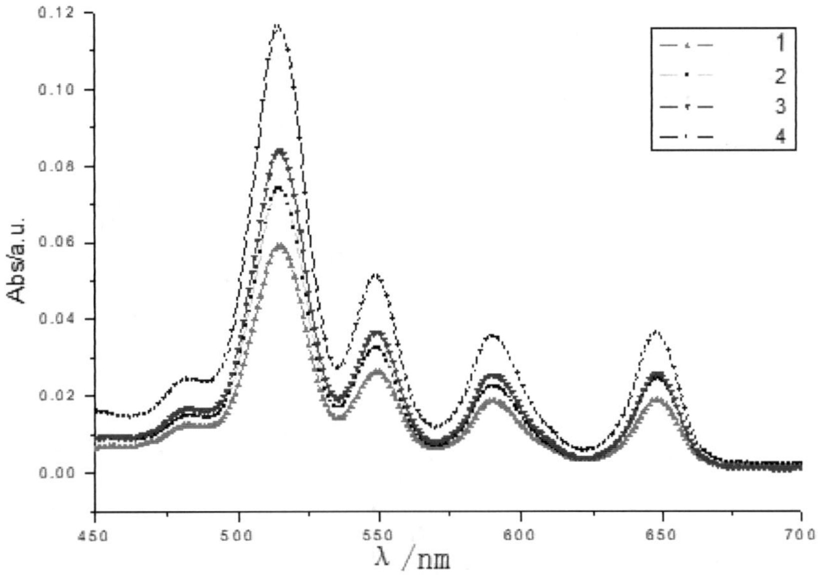

Fig 1 : UV spectra of 450-700nm

The typical UV-vis Absorption Spectra of porphyrins had a strong Soret band near 420 nm and four relatively weak q-band peaks between 450 nm and 700 nm. The maximum absorption peak of sample 1 – 4 was at 418 nm, and there were four relatively weak absorption peaks at 450 – 700 nm. The absorption intensity of sample 1 was the lowest, and that of sample 4 was the highest. It can be concluded that sample 4 has the highest purity.

4.Conclusion

In this thesis, tetrabenzoporphyrin with high yield and purity was synthesized by changing the sample static solvent and washing solvent. The samples were characterized by UV-vis Spectra. The effects of different solvents and concentrations on the photoelectric conversion rate of solar cells were studied. The results are as follows:

1. When tetraphenylporphyrin was prepared by four methods, the method of preparing tetraphenylporphyrin with absolute alcohol and washing with distilled water is the best, the yield can reach 19.5%;
2. The porphyrin samples were prepared by using anhydrous ethanol to rest and distilled water to wash the samples when the four samples were analyzed by UV.

Reference

[1] Lin P, Clegg W and R Harrington Synthesis and structures of 5-(pyridyl) tetrazole complexes of Mn(II) 2005 *J. Dalton Trans* **45** 2388–2394

[2] Zhao H and Xiong R G In situ hydrothermal synthesis of tetrazole coordination polymers with interesting physical properties2008 *J. Soc. Rev.* **37** 84–100.

[3] Wong C P, William D and Horrocks W New Metalloporphyrins: Thorium and Yttrium Complexes of T etraphenylporphin **1975** *J. Tetrahedron Lett.***16** 2637–2640.

[4] Timothy D L, Michael J H and John D S Conjugated Macrocycles Related to the Porphyrins. 21. Synthesis, Spectroscopy, Electrochemistry, and Structural Characterization of Carbaporphyrins **2002** *J. J. Org. Chem.* **6**74860–4874.
[5] Adler Alan D,Long Frederick R and William S Mechanistic investigations of porphyrin syntheses I: preliminary studies on ms-tetraphenylporphin 1964 *J. J. Am. Chem. Soc.* **86** 234–238.
[6] Gao J, Hide F and Wang H Efficient photodetectorsand photovoltaic cells from composites of fullerenes and conjugated polymers: Photoinduced electron transfer 1997 *J.* Syntheticmetals. **84** 979–980.

IOP Conf. Series: Materials Science and Engineering **733** (2020) 012040 doi:10.1088/1757-899X/733/1/012040

Preparation of TAT tri-layer insulating film on automotive glass using magnetron sputtering method

Yifan Chen[1]

1 Postgraduate, Material Science and Engineering, Shenyang Jianzhu University, Shenyang, China
E-mail: 356374012@qq.com

Abstract. In this paper, tri-layer structured films of titanium dioxide, silver and titanium dioxide were deposited on ordinary glass using radio frequency magnetron sputtering method to serve as thermal insulation films on automotive glass. The transmittance and thermal insulation of the prepared insulating films were changed by altering the deposition time of silver and titanium dioxide. The transmittance, absorbance and thermal insulation of the insulating films to visible light and infrared light were measured by ultraviolet-visible spectrophotometer and infrared spectrometer. The thermal conductivity of the films with different transmittances was tested by the thermal conductivity tester. The insulating films with lower thermal conductivity were screened out for analysis, and on this basis, the experiment was improved continuously. The total solar energy rejection rate was calculated subsequently. After summarizing the whole experiment process, the improvement measures were put forward to optimize the production of insulation films on automotive glass. The results show that the transmittance of TiO_2-Ag-TiO_2 tri-layer insulation film is 70.19%, and the solar energy rejection rate is 67.80%. The transmittance meets the national standard and the total rejection rate is also higher than many of the insulation films on the market.

1. Introduction

In this day and age, automobiles have gradually become the major means of transportation for people, thus the prospect of the auto film market is highly promising. In the next few years, the car parc in China will continue to increase by 20% annually. Therefore the development of automotive insulation film will be even more promising. Automotive insulation film refers to the film that is attached to the surface of automotive glass, which is also commonly known as automotive explosion-proof membranes. In addition to the functions of explosion-proofing, insulating and light-blocking, it also features unidirectional fluoroscopy and glare reduction. In this study, the performance of the automotive glass insulation film is optimised and improved. While ensuring high definition, a large amount of infrared rays are blocked, which can slow down the aging process of automobile interiors and reduce the workload of automobile air-conditioning. Therefore, the film performs high-definition, insulation and high UV-blocking functions while ensuring safety and preventing from explosion. At the same time, it provides high tear-proof and puncture-proof functions without sacrificing high transmittance. It plays an excellent role in cooling the interior of the car.

2. Experiment

In this experiment, reflective thermal insulating films were prepared using magnetron sputtering method. The experiment principle is that metal with high reflectivity such as silver (Ag) is deposited on a selected substrate using magnetron sputtering. After the film is formed, the deposited metal film will selectively reflect heat and various lights, mainly infrared, ultraviolet and visible light, from the sunlight on the surface to effectively block the heat. Metallic silver has very high reflectivity in all wavelengths, which is the main reason for choosing it as the target material for this experiment. Titanium dioxide (TiO_2) is applied on both sides of the silver layer to play a protective role. It is a wide bandgap semiconductor which can strongly absorb ultraviolet light and visible light. The magnetron sputtering film is uniform and transparent with no influence on the clarity of the substrate glass. Meanwhile, the favourable photocatalytic effect of titanium dioxide acts as a barrier to most of the light in the ultraviolet and visible wavelength ranges. The main work of this experiment is as follows:

(1) The Ag target material and the TiO_2 target material were deposited on a glass substrate of 20 mm×15 mm using radio frequency magnetron sputtering method. The effect of different deposition time of Ag target and TiO_2 target on the transmittance of film were studied. The parameter settings of the magnetron sputtering equipment applied in this experiment can be found in Table 1.

(2) The thermal conductivity of the prepared thermal insulation film was studied. After analysing comprehensively, the film samples with the lowest thermal conductivity and the highest visible light transmittance were selected, which provided the basis for the subsequent experimental studies and further optimisation of the experiment.

Table 1. Parameter settings of magnetron sputtering.

Vacuum on start-up	Reading of thermocouple	Amount of Ar	Sputtering frequency
2.4×10^{-4} pa	3 pa	10 sccm	120 w

3. Results and discussion

Figure 1 shows the transmittance of the film when TiO_2 target is deposited for 10 minutes, and Ag target is deposited for 5 seconds, 10 seconds and 15 seconds respectively. It can be seen that the transmittance in the visible light range is the highest when the Ag target is deposited for 5 seconds. The transmittance of 10 seconds is slightly better than that of 15 seconds, however, the gap between 5 seconds and 10 seconds is significantly larger than between 10 seconds and 15 seconds. The transmittance curve of the Ag film in the visible light wavelength range goes up first and then decreases. However, the highest transmittance within the visible light range for each time slot can be found around the wavelength of 400-500 nm. When the deposition time is between 5 seconds and 10 seconds, the transmittance rises first, then lowers and finally goes up. This may be caused by the fact that the deposition time for the Ag target material is too short and the silver film is too thin to be completely coated onto the surface of the glass substrate. As the deposition time of the Ag target increases, the average transmittance in the infrared light wavelength range decreases gradually, and the average transmittance in the visible light range increases first and decreases afterwards.

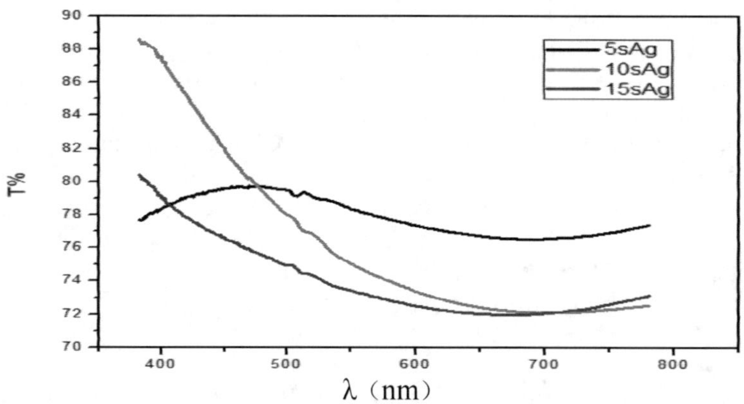

Figure 1. Curve of the effect of Ag target material deposition time on the transmittance of the insulation film.

As shown in figure 2, the average value of transmittance in this wavelength range is calculated by integrating the image in the visible light range (380 nm - 780 nm) using origin8 software and dividing by the interval 400 × 100 = 4000. Taking 15 seconds Ag target as an example, it is calculated as 28241.63/40000=73.54%, which shows the transmittance of this sample in the visible light range. Table 2 shows the transmittance in the visible light wavelength range of different silver deposition time.

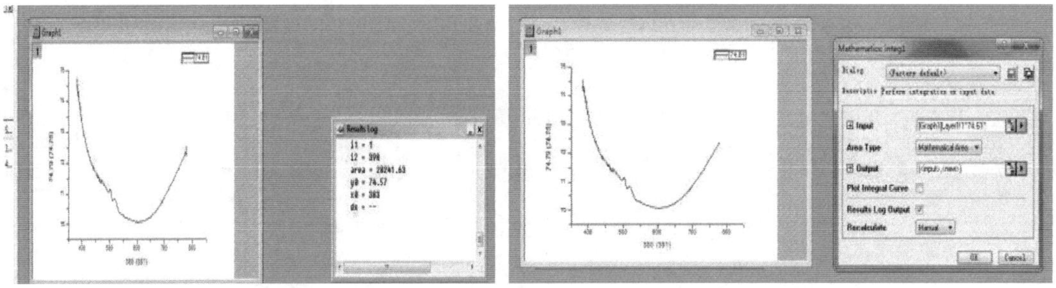

Figure 2. Average transmittance in the visible light wavelength range.

Table 2. Transmittance in the visible light wavelength range of different Ag deposition time.

Ag5s	Ag10s	Ag15s
79.88%	78.15%	73.54%

After studying the influence of silver deposition time on the transmittance and thermal insulation of the insulating film, 15 seconds Ag target material group was selected to conduct research on the impact of TiO_2 target deposition time on the transmittance of the insulation film. The Ag deposition time was kept as 15 seconds, and the time for TiO_2 was 20 minutes, 30 minutes and 40 minutes respectively as shown in figure 3.

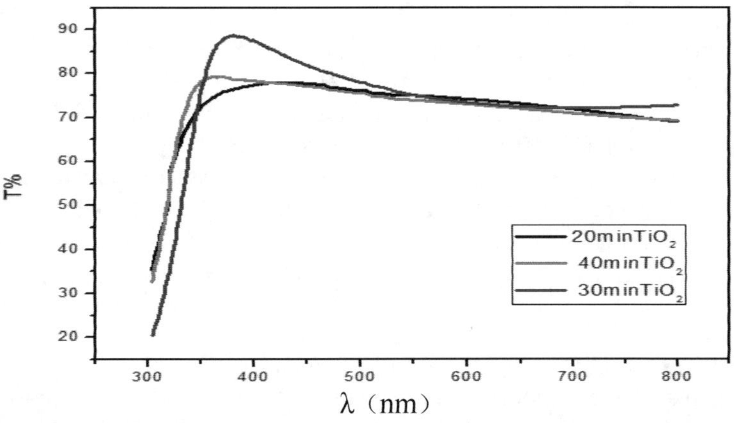

Figure 3. Curve of the effect of TiO_2 target material deposition time on the transmittance of the insulation film.

It can be seen from figure 3 that when the deposition time of TiO_2 target increases from 20 minutes to 30 minutes, the average transmittance of the prepared TiO_2-Ag- TiO_2 tri-layer film in the visible light range increases from 71.78% to 78.06%, and the average transmittance first increases and decreases afterwards. When deposition time is changed from 30 minutes to 40 minutes, the average transmittance of the prepared TiO_2-Ag- TiO_2 three-layer film in the visible light range decreases from

78.06% to 72.99%. After analysing the trend, two comparison groups, namely films with TiO_2 deposition time of 25 minutes and 35 minutes were chosen for further experiment, and the curve is formed accordingly. After calculation and analysis, it was found that when the TiO_2 deposition time is 25 minutes, the average transmittance of the prepared TiO_2-Ag- TiO_2 tri-layer film in the visible light range can reach its maximum value of 79.85%. This may be because the deposition time of the target material is increased, and hence the film deposited by the TiO_2 target is thickened, which indirectly improves the transmittance of the film. The visible spectrum features obvious selectivity, so the average transmittance of the prepared TiO_2-Ag- TiO_2 tri-layer film increases in the visible light wavelength range and the infrared wavelength range.

Table 3. Transmittance in the visible light wavelength range of different TiO_2 deposition time.

$TiO_2$20min	$TiO_2$25min	$TiO_2$30min	$TiO_2$35min	$TiO_2$40min
71.78%	79.85%	78.06%	74.37%	72.99%

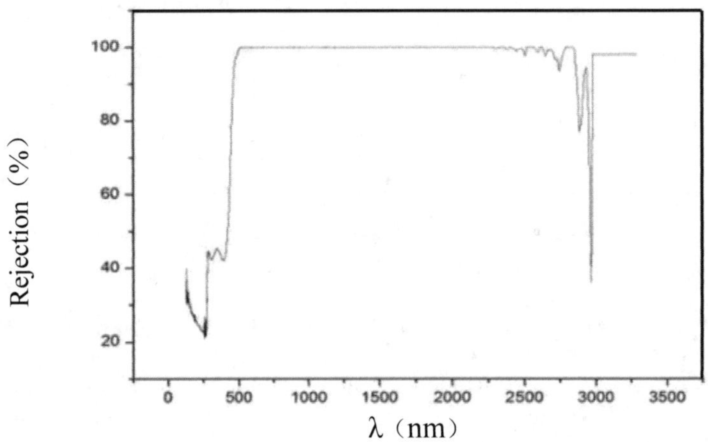

Figure 4. Curve of infrared rejection.

Figure 4 shows the curve of infrared rejection rate with TiO_2 25min + Ag 15s + TiO_2 25min. It can be seen through the experiment and analysis that the less the Ag deposition time, the higher the transmittance and the lower the insulation rate. The longer the TiO_2 film is deposited, the higher the insulation rate and the lower the transmittance. After analysis and calculation, it is found that the transmittance of all Ag samples with a deposition time of less than 15 seconds is higher than 70%. Finally, sample with a Ag deposition time of 15 seconds and TiO_2 time of 25 minutes was chosen. The calculation method of the infrared rejection rate is the same as the calculation method of the visible light transmittance. It is calculated that the infrared rejection rate was 96.68%.

4. Conclusions

After analysing and comparing the insulation and transmittance of each experimental control group of this subject, the sample with a Ag deposition time of 15 seconds and TiO_2 time of 25 minutes was selected. The final solar film thermal insulation rate was calculated based on the experimental data of 96.68% infrared rejection rate and 70.19% visible light transmittance. The total solar film rejection rate is calculated according to the following equation: infrared rejection rate × 53% (percentage of infrared in sunlight) + 44% (percentage of visible light in sunlight) × (1 - visible light transmittance) + 3% = 96.68% × 53% + 44% ×(1 - 70.54%) + 3% = 67.80%. Therefore, the prepared TiO_2-Ag-TiO_2 tri-layer automobile glass film has a transmittance of 70.54% and a solar energy rejection rate of 67.80%. The transmittance is up to standard and meets the national standard which regulates that the transmittance of the front windshield should be more than 70%. The total rejection rate is also higher than many of the insulation films on the market.

IOP Conf. Series: Materials Science and Engineering **733** (2020) 012040 doi:10.1088/1757-899X/733/1/012040

References

[1] H Zhang, R Liu, H Liu, et al 2010 Mn-doped Zn O transparent conducting films deposited by DC magnetron sputtering. *Materials Letters*. **64 (5)** 605-607

[2] Y Djaoued, K Ozga, A Wojciechowski, A H Reshak, J Robichaud and I V Kityk 2010 Photoinduced effect in TiO_2 nanocrystaline films with different morphology, *J. Alloys Compd.* **508** 599–605

[3] N Meshram, C Loka, K R Park and K S Lee 2015 Enhanced transmittance of ITO/Ag(Cr)/ITO (IAI) multi-layered thin films by high temperature annealing. *Materials Letters*. **145**

[4] J Hokim and H keyeollee 2015 Dependence of Optical and Electrical Properties on Ag Thicknessin TiO_2 /Ag/ TiO_2 Multilayer Films for Photovoltaic Devices. *Ceramics International.* **41** 8059-8063

[5] D F Rodríguez, P M Perillo 2015 TiO_2 nanopores with high sensitivity to ultraviolet light. *Optical Materials*, , **42** 52-55

IOP Conf. Series: Materials Science and Engineering **733** (2020) 012041 doi:10.1088/1757-899X/733/1/012041

Designing Core–Shell Materials Using Humic Acids

T S Skripkina, E M Podgorbunskikh and O I Lomovsky

Institute of Solid State Chemistry and Mechanochemistry SB RAS, Novosibirsk, 630128, Russia

Abstract. Due to the high content of complexing moieties, humic acids are promising compounds for designing heavy metal sorbents that can be used in natural environment objects. Experiments focused on designing sorbent particles containing a humic shell and a core (that is supposed to impart the desired performance characteristics to the particle) and applying humic acids onto various materials were conducted. It is demonstrated that the sorbent with a "lignocellulosic core–humic shell" structure can be produced mechanochemically. The efficiency of humic acid binding to lignocellulosic materials directly correlates with lignin content within lignocellulosic materials. A method for passing lignin from the surface of lignocellulosic biomass into the solution using humic acids is found. Unmodified highly lignified biomass is the optimal material for applying humic acids onto lignocellulosic materials.

1. Introduction

The interactions between humic acids and various materials are now being actively studied around the globe with different purposes. The sorption of humic acids onto such materials as nanosized zero-valent iron [1], nanosized inorganic oxides [2], graphene oxides [3], argillaceous minerals [4], chitosan granules [5], and polyacrylonitrile fibers [6] has been investigated. Humic acids are able to immobilize heavy metals partially due to precipitation of humic acid–metal complexes onto the fixed phase (argillaceous minerals, metal oxides, or silicates present in soils or bottom sediments [7]).

Among inorganic materials, carbon-based sorbents and silica gel varieties characterized by high surface area are also commonly used. Silica gel is subjected to pretreatment (typically with organosilanes). This process is conducted using organic solvents in anhydrous media at elevated temperatures. The data on sorption of humic molecules onto organic compounds or natural materials are very scarce because the analyzed system is very complex and there exist methodological challenges. However, it is widely recognized that the problem of searching for an abundant natural material that can adsorb humic acids is relevant today as these materials reduce the environmental load. Functional trimethylammonium-modified cellulose nanofibers were shown to efficiently bind humic acids. However, although this material is based on cellulose, it is a product of sophisticated multi-step synthesis [8]. Chitosan supported onto a clay substrate is another natural polymer for which the kinetic parameters and the mechanism of binding to humic acids have been studied [9].

It is reasonable to investigate the humic acid binding ability for a range of lignocellulosic materials of different chemical compositions (from isolated lignin and cellulose to lignocellulose varied different degree of chemical modification). Lignin is known to constitute up to 44% of peat and soil humic substances. The high affinity of humic acids for lignin is also attributed to the fact that these substances carry moieties of similar nature (mainly the orthophenolic ones). In addition to studying plant-derived materials having a natural structure, it is interesting to investigate the ability of lignocellulosic biomass with a modified structure to bind humic substances. Earlier, E.M.

Podgorbunskikh et al. [10] developed a method for increasing lignin content on the particle surface by mechanochemical pretreatment of the material in a laboratory-scale mechanochemical activator (attritor) at a temperature of lignin devitrification [10]. Under these pretreatment conditions, plant biomass lignin undergoes a phase transition to a rubber-like state, leaves the cell wall and is accumulated on the surface of the nascent particles. High lignin content on the surface of plant biomass, in combination with significant porosity, can impart an enhanced sorption capacity with respect to humic acids to the biomass. Verification of this hypothesis will substantially contribute to the development of complex sorbents based on humic acids.

The objective of this study was to design complex heavy metal sorbents with a core–shell structure.

2. Materials and Methods

A number of materials, both inorganic and organic, were selected for the experiments focused on supporting humic acids onto sorbents. The inorganic materials included activated carbon, mesoporous carbon adsorbent (MC), silica gel powder, and silica gel granules. Organic substrates were represented by a range of lignocellulosic materials of different chemical composition and morphology, including isolated lignin and cellulose, partially delignified wheat straw, lignocellulose after water extraction, and rice husk. The sorption of humic acids onto highly lignified materials (mechanically modified reed biomass and wheat straw) was also studied.

Humic acids manufactured by Aldrich were used in the experiments. The content of humic acids in the solution was determined spectrophotometrically on an Unico 2800 spectrophotometer (USA) at 465 mm in polystyrene cells with respect to water. Prior to each analysis, calibration curves were plotted using the points corresponding to humic acid concentrations of 250, 150, 100, 50, and 10 mg/l. The samples were preliminarily centrifuged and filtered using red ribbon filter paper.

The specific surface area (Ssp) of the test samples was measured using the nitrogen desorption method on a Sorbtometr-M instrument; the data were analyzed using the Bruner–Emmett–Teller (BET) equation. The morphology of the materials was studied by scanning electron microscopy on a Hitachi TM-1000 microscope and gold pre-deposition onto the surface (sputtering time, 1 min; ionic current, 30 mA). The composition of lignocellulosic materials was analyzed by successive extraction with hexane, ethanol, and water, followed by acid hydrolysis and delignification. The yields of the extracts were determined according to weight loss.

Sorption of Humic Acids onto Various Materials. One gram of each material was added to 50 ml of humic acid solution (250 mg/l, pH 7.0). The resulting suspension was thermostated at 25°C and shaken at a frequency of 130 min–1. Samples were taken at the initial instant and 48 h later to determine the content of humic acids remaining in the solution.

Desorption of Humic Acids. The materials with adsorbed humic acids were placed in a test tube containing 20 ml of Tris-HCl buffer (pH 9.0). The resulting suspension was thermostated at 25°C and shaken at a frequency of 120 min–1. After 5 h, solution samples were taken to determine the content of desorbed humic acids.

3. Results and Discussion

The chemical compositions of the materials used in this study as the "core" to support humic acids were preliminarily determined.

Water content in the materials was as follows: activated carbon, 3.5 ± 0.2 %; MC, 2.3 ± 0.1%; silica gel granules, 5.0 ± 0.2 %; and silica gel powder, 4.4 ± 0.2%. The series of plant-derived materials for supporting humic acids was originally selected so that the contents of the key components (lignin, cellulose, and hemicelluloses) were maximally diverse [11]. Lignin content was the highest for hydrolysis lignin (80.0 ± 0.9 %) and the lowest for microcrystalline cellulose (< 0.05%) and delignified wheat straw (3.0 ± 0.1 %). That is the key difference between delignified wheat straw and lignocellulose after water extraction, which has a similar composition. Cellulose content is minimal for hydrolysis lignin and maximal for microscrystalline cellulose and delignified wheat straw (89.0 ± 0.9 %), being slightly lower for lignocellulose after water extraction (74.0 ± 1.0 %).

The inorganic materials have similar morphology and differ only in terms of their particle size and ability to form agglomerates.

A morphological analysis of plant-derived materials demonstrated that they have a complex multi-level architecture, except for isolated lignin and microcrystalline cellulose. Lignin is a finely dispersed powder. The cellulose morphology is represented by fibrous fragments composed of fibrils up to 500 μm long and 10–20 μm wide. The morphology of delignified wheat straw is represented by fibers resembling those in the initial wheat straw and regions of the cross-linked structure with a large number of pores. The fibers are loosely packed and oriented parallel to each other so that particles have many large voids. The morphology of lignocellulose after water extraction is similar to that of delignified wheat straw but has higher pore content. Both samples have a well-defined pore structure (pore size, 3–6 μm). This happens because the rigid cellular framework starts to have vacancies as some soluble substances are dissolved, so humic substances can be bound via the mechanism of physical sorption.

Rice husk has an interesting morphology as its structure is rather complex. The outer side of the particles is covered by a cellulose silica membrane. The inner side consists of lignocellulose and has a smooth surface. Rice husk is a unique material due to its silicon dioxide content, which is extremely high for plant biomass, and exists in the amorphous form on the material surface. Silicon dioxide is accumulated near the outer surfaces of the plant tissue to form the cellulose silica membrane [12].

Table 1 shows the results of experiments on sorption/desorption of humic acids on inorganic materials.

Table 1. Parameters of sorption of humic acids onto inorganic materials and specific surface area values for the materials constituting the "core"

	Sorption (mg/g)	Desorption (mg/g)	Sorption capacity (mg/g)	Relative sorption capacity (mg/m^2)	S_{sp} (m^2/g)
Silica gel powder	6.3	1.7	4.6	0.014	309 ± 9
Silica gel granules	5.7	0.83	4.9	0.02	207 ± 7
Activated charcoal	6.9	0.00	6.9	0.02	280 ± 9
MC	13	0.25	13	0.04	350 ± 10

Carbon-based materials exhibited better sorption properties with respect to humic acids than silica gels did. The modified nano-carbon sorbent can remove the greatest amount of humic acids from the solution due to the lower size of its particles and higher surface area compared to the regular activated charcoal. Silica gel granules exhibited worse sorption properties than silica gel powder because silica gel particles had a higher specific surface area.

Simultaneously with being more efficient in removing humic acids from the solution, carbon-based materials showed better ability to retain humic acids. It should be mentioned that the degree of desorption from the surface of silica gel powder was the highest, which can be attributed to low porosity of this material. The desorption degree does not make any significant contribution to choosing the material for the "core." With allowance for the inaccuracy, the sorption capacities of silica gel granules and powder were equal; the maximum sorption capacity was observed for the modified nano-carbon sorbent.

Hence, among inorganic materials, carbon-based sorbents showed better results than silica gel did. One should take into account that activated charcoal is a much more accessible feedstock than modified nano-carbon sorbent, while the difference in their sorption capacity is not so significant. This fact, along with the convenience of working with activated charcoal, makes it the best material among inorganic ones.

3.1. Sorption of humic acids on unpretreated (native) organic materials.

The results of experiments on sorption/desorption of humic acids on lignocellulosic materials are shown in Table 2.

Table 2. Parameters of sorption of humic acids onto organic materials

Material	Sorption (mg/g)	Desorption (mg/g)	Sorption capacity (mg/g)	Relative sorption capacity (mg/m^2)	S_{sp} (m^2/g)
Cellulose	1.4	1.2	0.2	0.03	6.5 ± 0.3
Delignified wheat straw	2.4	1.5	0.9	0.06	16.0 ± 0.9
LC after water extraction	2.4	0.98	1.4	0.16	8.6 ± 0.5
Rice husk	2.0	0.66	1.3	0.10	13.1 ± 0.7
Lignin	4.6	1.3	3.3	0.33	9.9 ± 0.5

The sorption capacity of lignocellulosic materials increases in the cellulose–lignin series. Although the specific surface area of lignocellulose after water extraction is twice as low as that of the delignified biomass, their sorption degrees are identical due to an increased lignin content in lignocellulose after water extraction. This fact is indicative of the chemical nature of sorption. The delignified biomass is simultaneously characterized by maximum desorption.

The sorption capacity of lignin-containing materials is higher than that of cellulose. Sorption capacity with respect to humic acids depends linearly on lignin content in the biomass [11], which attests to the chemical mechanism of sorption according to which the functional groups of lignin form hydrogen bonds with respective functional groups of humic acids. Hence, the ability of plant biomass to bind humic acids depends on the lignin component. For the plant biomass under study, the maximum degree of sorption is 3.3 mg HA per g of dry sorbent.

The high affinity of humic acids for lignin is attributed to the fact that humic acids and lignins contain groups of similar nature (e.g., the orthophenolic moieties). This similarity is evolutionarily reasonable, since humic acids are the products of deep conversion of lignins and flavonoids. Binding between humic acids and lignin fragments takes place via polar interactions, cationic bridges, van der Waals forces, and hydrogen bonds [13]. Alcoholic hydroxyl groups of humic acids can interact with lignin by forming p-hydroxybenzyl esteric chemical bonds [13]. Hence, it has been demonstrated that when the sorbent particles with a "lignocellulosic core–humic shell" structure are designed, humic acids bind to lignocellulosic materials by forming chemical bonds with the lignin component of the biomass. The great variety of lignocellulosic materials that are formed as industrial wastes [14] can potentially be used to produce sorbents with a "lignocellulosic core–humic cell" structure.

3.2. Sorption of Humic Acids onto the Mechanochemically Modified Lignocellulosic Materials.
One of the stages in designing particles with a "lignocellulosic core–humic shell" structure is to search for a method to enhance the sorption capacity of lignocellulosic feedstock with respect to humic acids by performing mechanochemical pretreatment. This pretreatment allows one to conduct solid-phase chemical reactions without using solvents [15]. The most promising materials were those in which lignin content on the surface was increased by mechanochemical pretreatment in a laboratory-scale mechanochemical activator (attritor) at a temperature of lignin devitrification (180°C). Under these conditions, lignin undergoes a phase transition to a rubber-like state, migrates to the particle surface, and is accumulated there.

The high lignin content on the surface, in combination with high porosity, is expected to enhance the sorption capacity of plant biomass with respect to humic acids. Wheat straw and reed biomass (lignin content being 21.7 ± 0.1 and 38.5 ± 0.3 %, respectively) were selected as study objects. Table 3 shows the sorption capacity of the native and pretreated materials with respect to humic acids.

Table 3. Sorption capacity of wheat straw and reed before and after mechanochemical pretreatment

	Sorption (mg/g)	Desorption (mg/g)	Sorption capacity (mg/g)	Relative sorption capacity (mg/m^2)
Reed biomass without activation	7.3 ± 0.5	1.4 ± 0.1	5.9 ± 0.6	2.9 ± 0.3
Reed biomass (attritor, 100°C)	7.7 ± 0.5	1.5 ± 0.1	6.2 ± 0.6	2.7 ± 0.2
Reed biomass (attritor, 180°C)	0.5 ± 0.1	1.0 ± 0.1	Passing into solution	Passing into solution
Wheat straw (attritor, 25°C)	2.0 ± 0.2	0.9 ± 0.1	1.1 ± 0.3	0.6 ± 0.1
Wheat straw (attritor, 180°C)	2.4 ± 0.2	0.8 ± 0.1	1.6 ± 0.3	0.5 ± 0.1

The results demonstrate that the mechanochemical pretreatment of reed and wheat straw under different conditions did not significantly enhance the relative sorption capacity of these lignocellulosic materials with respect to humic acids. An increase in sorption capacity was observed for wheat straw, but this increase was found to be insignificant when the sorption capacity was recalculated per surface area. An opposite effect was observed for reed. In the experiments on sorption of humic acids from the solution, lignin was washed off the particle surface and passed into the solution, leading to an apparent increase in humic acid concentration in it. Hence, an interesting fact is that a method for bringing lignin from the surface of lignocellulosic feedstock to the solution using humic acids has been revealed. Unmodified highly lignified plant biomass is the optimal material to solve the problem of applying humic acids onto plant-derived materials.

The resulting core–shell materials exhibit high efficiency as heavy metal sorbents not only in the laboratory-scale experiments but also in full-scale ones [16]. It was demonstrated that core–shell materials can be repeatedly used up to five times; their sorption properties remain high after regeneration in the presence of 0.05 M HNO_3 [17].

4. Summary

Core–shell particles have been obtained, where humic acids act as a shell and a number of inorganic and lignocellulosic materials are used as the core. The efficiency of binding between humic acids and lignocellulosic materials directly correlates with lignin content within lignocellulosic materials. A method for passing lignin from the surface of lignocellulosic biomass to the solution using humic acids has been found. Unmodified highly lignified biomass is the optimal material for applying humic acids onto lignocellulosic materials. The resulting core–shell particles are promising heavy metal sorbents that can be used in natural environment objects.

5. References

[1] Giasuddin A B M, Kanel S R and Choi H 2007 *Environ. Sci. Technol.* **41, 6** 2022-27
[2] Eita M 2011 *Soft Matter* **7, 2** 709-715
[3] Chowdhury I, Duch M C, Mansukhani N D, Hersam M C and Bouchard D 2014 *Environ. Sci. Technol.* **48, 16** 9382-90
[4] Chen H, Koopal L K, Xiong J, Avena M and Tan W 2017 *J. Colloid Interface Sci.* **504** 457-467
[5] Dong C, Chen W, Liu C, Liu Y and Liu H 2014 *Colloids Surf., A.* **446** 179-189
[6] Deng S and Bai R 2004 J. *Colloid Interface Sci.* **280, 1** 36-43
[7] Metreveli G, Abbt-Braun G and Frimmel F H 2010 *Aquat. Geochem.* **16, 1** 85-100
[8] Sehaqui H, Schaufelberger L, Michen B and Zimmermann T 2017 *J. Colloid Interface Sci.* **504** 500- 506
[9] Chang MY and Juang R S 2004 *J. Colloid Interface Sci.* **278, 1** 18-25.
[10] Podgorbunskikh E M, Ryabchikova E I, Bychkov A L and Lomovskii O I 2017 *Dokl. Phys. Chem.* **473, 1** 49-51

[11] Urazova T S, Bychkov A L and Lomovskii O I 2015 *Russ. Chem. Bull.* **64, 5** 1183-88
[12] Park B D, Wi S G, Lee K H, Singh A P, Yoon T H and Kim Y S 2003 Biomass Bioenergy **25, 3** 319-327
[13] Popov A I 2004 *Humic substances: properties, structure, and formation.* (St. Petersburg: Publishing House of St. Petersburg State University)
[14] Sakovich G V, Skiba E A, Budaeva V V, Gladysheva EK and Aleshina L A 2017 *Dokl. Biochem. Biophys.* **477** 357-359
[15] Akopova T A, Demina T S, Cherkaev G V, Khavpachev M A, Bardakova K N, Grachev A V, Vladimirov L V, Zelenetskii A N and Timashev P S 2019 *RSC Advances.* **9, 36** 20968-75
[16] Smolyakov B S, Sagidullin A K and Chikunov A S 2017 *J. Environ. Chem. Eng.* **5, 1** 1015-20
[17] Smolyakov B S, Sagidullin A K, Bychkov A L, Lomovsky I O and Lomovsky O I 2015 *J. Environ. Chem. Eng.* **3** 1939-46

Acknowledgments
This study was supported by the Russian Science Foundation (project no. 16-13-10200).

IOP Conf. Series: Materials Science and Engineering **733** (2020) 012042 doi:10.1088/1757-899X/733/1/012042

Scaling up the Mechanochemical Technology for Producing Materials Enriched in Humic Acids

T S Skripkina[1], A L Bychkov[1,2], O A Rozhanskaya[3] and O I Lomovsky[1]

[1]Institute of Solid State Chemistry and Mechanochemistry SB RAS, Novosibirsk, 630128, Russia
[2]Novosibirsk State Technical University, Novosibirsk, 630073, Russia
[3]Siberian Federal Scientific Center of Agro-BioTechnologies, RAS, Krasnoobsk, Novosibirsk region, 630501, Russia

Abstract. Mechanochemical treatment is widely used in industry as it allows one to reduce the number of processing stages, either reduce or completely eliminate the use of solvents, automate the technology and ensure its mobility. Energy input is an important parameter of mechanochemical treatment. In this study, the technology of mechanochemical oxidation of brown coal was scaled up using the planetary mills of AGO-2 and APF types, as well as K-7 and CEM-20 flow-type vibrocentrifugal mills. The technology of mechanochemical oxidation of brown coal makes it possible to increase the availability of humic acids in brown coal and the content of oxygen-bearing groups capable of complexation, as well as to improve sorption characteristics of the resulting product. Materials enriched in humic acids can be used as heavy metal sorbents in water bodies and as agents for soil amelioration. The rotor speed of 1400 rpm and the feed rates of 15 kg/h and 1.8 kg/h for CEM-20 and for K-7 vibrocentrifugal mills, respectively, are the optimal modes from the point of view of chemical and economic efficiency. These modes increase the yields of alkali-soluble and water-soluble humic acids from 23 to 59 % and from <1 to 7%, respectively, which corresponds to the values reached using the laboratory-scale equipment.

1. Introduction

Humic acids are organic compounds that perform a number of functions in the biosphere, including the accumulative, transport, regulatory, protective, and other ones. Brown coal is one of the most important sources of humic acids. The world's brown coal reserves are estimated at billions of tons, with a significant portion being located in Russia. However, brown coal is poorly suited for use as a fuel because of its low calorific value and susceptibility to spontaneous combustion during transportation. Producing humic acids and humin-containing products from brown coal is a good alternative to its application as an energy source. The mechanochemical techniques make it possible to conduct solid-phase chemical reactions, thereby avoiding the use of solvents and increasing environmental friendliness and economical efficiency of the processes [1, 2]. Such advantages as the reduced number of processing stages and the mobility of mechanochemical activators make it possible to treat brown coal and brown coal waste directly on-site, which also solves the problems related to its transportation.

From the perspective of solid-state chemistry, brown coal is an interesting object for mechanochemical modification to change its chemical properties. Solid-phase treatment of brown coal with alkalies, which enhances the extractability of humic acids due to the formation of sodium humate via the acid–base mechanism, has been best studied. It has recently been found that mechanochemical treatment of brown coal in a laboratory-scale planetary-type activator in the presence of sodium percarbonate is accompanied by oxidation of the organic matter in brown coal and increases the content of phenolic and carboxyl groups in humic acids [3, 4]. Due to the high content of complexing groups, humic acids are rather promising substances for designing heavy metal sorbents, which can be used at natural environment sites [5]. A large number of reviews focused on determining the composition, structure simulation, and physicochemical properties of humic acids have been published. There also are publications devoted to binding of humic acids to organic and inorganic substances, as well as the methods for their modification [6].

The solid-phase methods for treatment and modification of humic acids within the humin-containing feedstock, including the mechanochemical ones, are currently used rather commonly in technology but remain insufficiently studied. Simultaneously, the research into mechanochemical reactions of humic acids typically focuses on laboratory-scale mechanochemical activators.

A laboratory ball mill and a planetary activator have been most widely described in the literature. These types of equipment are ideal for laboratory-scale experiments. The processes occurring inside the reactors and the exposure type have been carefully studied and simulated many times. The results of mechanical activation using these devices are well reproduced. However, only several dozen grams of a sample can be activated in a planetary mill per cycle. In order to develop the technology of mechanochemical modification of brown coal humic acids to produce heavy metal sorbents at a marketable volume, the processes under study need to be scaled up [7] using the flow-type mechanochemical equipment that has a similar type of mechanical impact.

The goal of this study was to determine the optimal conditions for mechanochemical treatment of brown coal in industrial-scale mills, which would allow one to attain the effects obtained in laboratory mills for larger-scale production of materials (heavy metal sorbents and ameliorants) enriched in humic acids.

2. Materials and Methods

Brown coal (alkali-soluble humic acids, 23.6 ± 0.4 %; ash, 10.7 ± 0.3 %) from the Itatskoe deposit was used in this study. The contents of alkali-extractable and water-soluble humic acids in all samples were determined in compliance with the International Standard ISO 5073:2013 (Brown Coals and Lignites – Determination of Humic Acids).

The first stage of technology scale-up was conducted using an APF water-cooled planetary-type mechanochemical activator equipped with two jars with steel grinding balls 8 mm in diameter (the total weight of the grinding bodies, 900 g; the weight of loaded substance, 100 g). Grinding body acceleration at an instant when it was detached from the grinding chamber walls was 200 m/s2; the duration of mechanical treatment was 2 min. The treatment conditions matched those for the laboratory-scale equipment: 5 wt.% sodium percarbonate additive; moisture content in the mixture, 3.5%. The maximum power consumption of the activator was 6 kW; the output was 6 kg/h.

The second scale-up stage was conducted using a K-7 flow-type vibrocentrifugal mill (Novits OJSC, Novosibirsk, Russia) with the output of 10 kg/h and the maximum power consumption of 2.2 kW. The rotor speed was varied from 420 to 1400 rpm. The mill was equipped with two cylindrical grinding bodies: the reagents were supplied through a feeder on one side, and the finished product was collected in a receiving hopper on the other side. The diameter and the operating length of a grinding body were 40 and 400 mm, respectively. The grinding bodies were steel balls 5 mm in diameter. The weight of the balls in a working body was 2600 g. The treatment conditions matched those for the laboratory-scale equipment: 5 wt.% sodium percarbonate additive; moisture content in the mixture, 3.5%.

The third scale-up stage was performed on a CEM-20 flow-type vibrocentrifugal mill, with the maximum output of 50 kg/h and the maximum power consumption of 2.2 kW.

In the experiments, cadmium(II), chromium(III), cobalt(II), and lead(II) ions were sorbed from solutions based on Tris-HCl buffer (pH 8.0); the concentration of each metal in the solution was 20 mg/kg for the first experimental series and 100 mg/kg for the second experimental series. An accurately weighed portion of the sorbent (1 mg sorbent per ml of the solution) was added. The mixture was thermostated at T = 25°C and shaking frequency of 130 min–1. Samples to determine the residual content of heavy metals were collected every hour, as well as at an instant when the sorbent was added. The residual content of heavy metals in the solution was determined by ICP-AES using a solution containing 0.5 m/l Sc(III) ions as an internal standard.

Evaluation of the effect of growth promoters on plants in field tests involved three stages: (1) setting up an experiment; (2) field observations and survey; and (3) laboratory analysis plant samples. A section of median traffic strip of a major highway in Novosibirsk (Russia) was selected as a test object. The 3 m wide grassplot is located between two lanes of the highway, on a slope being ~ 15° steep. The experiment was set up manually in four replicas; the plot length was 4 m; the area of the entire test ground was 204 m2. The test grounds of the same size without the product added to the soil were used as controls (five replicas).

3. Results and Discussion

The technology discussed in this paper consists in conducting solid-phase oxidation of the organic matter in brown coal in the presence of sodium percarbonate [3, 4, 8]. It was shown in the experiments performed using an AGO-2 laboratory-scale planetary-type activator that the optimal processing conditions were as follows: 5 wt.% sodium percarbonate additive; moisture content in the reaction mixture, < 12–13 %; exposure duration, 2 min; grinding body acceleration at an instant when they were detached from the jar walls, 20 g. Under these conditions it is possible to increase the content of alkali-soluble humic acids to 70 ± 3%, and the content of water-soluble humic acids from <1 % to 15 ± 1 %. The resulting product has an increased content of phenolic and carboxyl groups and better sorption characteristics compared to the original coal [9].

Proper equipment was selected at the first stage of technology scale-up. A review of the current market of semi-industrial mechanical activators was performed. The market segment of semi-industrial mechanochemical equipment (with the output of ~ 50 kg/h) included centrifugal elliptical ball mills (Novits OJSC, Novosibirsk, Russia; Xinxiang Unitfine Machinery Co., Ltd., China), jet mills (New technologies, St. Petersburg, Russia), finger disintegrator mills (Prodselmash OJSC, Novosibirsk, Russia; Desintegraator Tootmise, Estonia). It is worth mentioning that only the vibrocentrifugal mills can be used to perform a mechanochemical reaction (a prerequisite is that the grinding body has a simultaneous mechanical impact on the entire reagent mixture rather than on separate reagent particles).

At the first stage of scale-up, the mixture of brown coal and sodium percarbonate was treated in an APF planetary ball mill. The results obtained using the APF activator under conditions matching those used during treatment in an AGO-2 laboratory-scale activator (grinding body acceleration at an instant of detachment from the jar wall, 20 g; treatment duration, 2 min; moisture content, 3.5%) are well reproduced. The yields of alkali-soluble and water-soluble humic acids were increased to 65 ± 3% and 12 ± 1 %, respectively. The maximum output of the APF activator is 6 kg/h.

Although the maximum power consumption of this equipment is 6 kW, the actual power consumption with loaded jars does not exceed 3 kW. Unfortunately, this type of mechanochemical equipment is not flow-through. It requires careful maintenance by service personnel, as well as the time-consuming procedures of jar loading and unloading. So operation of this equipment is accompanied by high energy expenditure.

Table 1 summarizes the results of the second stage of mechanochemical technology scale-up, which was conducted using flow-type vibrocentrifugal equipment. The yield of water-soluble humic acids as an indicator of treatment efficiency was studied. The consumption of sodium percarbonate $Na_2CO_3 \times 1.5H_2O_2$ was estimated according to pH of the product suspension in distilled water.

Table 1. The results of mechanochemical treatment of brown coal in the presence of sodium percarbonate in a K-7 flow-type vibrocentrifugal mill

Rotor speed (rpm)	Yield of water-soluble humic acids (%)	pH
0	< 1%	10
420	< 1%	9.5
560	3.2 ± 0.1	8.5
840	3.6 ± 0.1	6.6
1120	5.8 ± 0.2	6.3
1400	6.8 ± 0.2	6.5

The pH values listed in Table 1 demonstrate that no mechanochemical reaction occurs between brown coal and sodium percarbonate at frequencies < 560 rpm, since sodium percarbonate is not consumed, remains in the final mixture, and increases pH. Since pH of the treatment product suspension for the samples obtained in the rotor speed range of 840–1400 rpm is not overestimated, it is fair to say that percarbonate was consumed and the samples obtained at these frequencies are the products of a mechanochemical reaction between the organic substance of coal and sodium percarbonate. When scaling up the technology of oxidative treatment of brown coal performed using the semi-industrial scale equipment, we managed to increase the extractability of water-soluble humic acids from 1 to 7 %. The technology implemented using the laboratory-scale equipment made it possible to increase this value to 14 %. This ratio between the efficiencies of brown coal processing using the laboratory-scale and semi-industrial equipment indicates that the technology scale-up was successful.

A pilot-scale batch (100 kg) was produced, with the CEM-20 industrial vibrocentrifugal mill (output, 50 kg/h; rotor speed, 1400 rpm) used at the stage of mechanical activation. The pH of the aqueous suspension of the product (1:10) is 6.5; the yield of water-soluble humic acids is 7.5 ± 0.3 %; the yield of alkali-soluble humic acids is 55 ± 2 %, being indicative that the organic matter of brown coal has been successfully oxidized with sodium percarbonate using the industrial-scale mechanochemical equipment. The resulting product was tested as a heavy metal sorbent under the laboratory and field conditions, as well as a soil ameliorant along a major highway in Novosibirsk.

Laboratory testing of the product as a sorbent. In order to assess the efficiency of brown coal subjected to mechanochemical modification in the presence of sodium percarbonate as a heavy metal sorbent, experiments focused on sorption from solutions containing a mixture of trace elements were conducted. Humic acid samples were added to a solution containing cadmium, chromium, cobalt, and lead ions (20 mg/l of each metal in the first experiment; ~ 100 mg/l in the second experiment). In all the cases, metal concentrations decreased dramatically after the first 30 minutes of product exposure to the mixture. One can see that the mechanochemically produced sorbent removed most of heavy metals from the solution due to sorption.

Table 2. The results of sorption of a metal mixture onto the products of mechanochemical activation of brown coal

Metal	$C_0(Me)$ (mg/l)	$C_{fin}(Me)$ (mg/l)	
		Initial coal	Coal subjected to MA in the presence of $Na_2CO_3 \cdot 1.5H_2O_2$
Pb^{2+}	130 ± 20	3.2 ± 0.4	0.69 ± 0.08
Co^{2+}	74 ± 8	1.1 ± 0.1	0.21 ± 0.02
Cr^{3+}	100 ± 10	1.4 ± 0.2	0.16 ± 0.02
Cd^{2+}	90 ± 10	10.0 ± 0.9	3.5 ± 0.4

Thus, brown coal mechanochemically oxidized using the scaled-up technology in the flow-type vibrocentrifugal mill efficiently removes heavy metals from the solution containing a mixture of trace elements (Cd^{2+}, Cr^{3+}, Co^{2+}, and Pb^{2+}). The residual concentrations of heavy metals in the solution are significantly lower than those in the control experiments, where the untreated brown coal was used as a sorbent. The product was also tested in field experiments on heavy metal sorption from the water area of the Novosibirsk reservoir [10]. It was applied onto a highly porous polymer matrix that ensured all the necessary performance characteristics: easy collection and regeneration and the possibility of its flow-through use. It was shown that the product could be used for removing Cd (II), Cu (II), Pb (II), and Zn (II) ions from a natural reservoir and was highly efficient.

Experiments were carried out to study the impact of adding the mechanochemically oxidized brown coal to the soil on the grassplot of the median strip of the Kamenskaya Highway in Novosibirsk. Table 3 lists the results of determining the state of the herbaceous cover by the end of the growing season. The percentage of soil surface covered with grass (the projective cover of grass stand) along the margins of the control plots was ≤ 30% and ranged from 60 to 80% across the experiment replicas. In variant 2, where 10 g/m^2 of the product was added, grass covered 70–80% of the soil. The results were similar in variant 3, where 30 g/m^2 of the product was added. In variant 4, where 10 g/m^2 of the product and a fertilizer were added and gramineous plants were undersown, the projective cover of grass stand was 80%.

Table 3. The average projective cover of grass stand on the test plot (% of the total soil area)

Variant	Total projective cover of grass stand	Including gramineous plants	Including weeds and motley grasses
1. Control	68	38	30
2. Product being added (10 g/m^2)	75*	40	35*
3. Product being added, 30 g/m^2	78*	48*	30
4. Product being added (10 g/m^2) + $N_{120}P_{60}K_{60}$ + seeding	82*	50*	32

*The difference with the control is reliable at the 95% level

The observed trend towards increasing projective cover of grass stand was accompanied by an increase in the gramineous component on experimental plots nos. 3 and 4 by 10% and 12%, respectively, compared to the control plot. The results of determining the height and productivity of grass stand on the experimental plot are presented in Table 4. Addition of the product significantly increased grass height and weight on the experimental plots compared to the control one. Undersowing of gramineous grasses did not lead to significant predominance of the gramineous component. The effect of the product increased with its dose. Fertilization and undersowing of gramineous grasses contributed to the increase in grass height and weight.

Table 4. Height and productivity of the grass stand on the experimental plot

Variant	Average height (cm)	Productivity (g/m^2) (abs. dry)
1. Control	12.8	94.8
2. Product being added (10 g/m^2)	14.4	114.0*
3. Product being added (30 g/m^2)	18.2*	117.6*
4. Product being added (10 g/m^2) + $N_{120}P_{60}K_{60}$ + seeding	22.8*	120.8*

*The difference with the control is reliable at the 95% level

Hence, the material produced using the semi-industrial flow-type vibrocentrifugal mill was efficient in the experiments on grass remediation on soil plots exposed to significant anthropogenic load. The treated plot is characterized by higher coverage, average grass height and weight, and a higher percentage of gramineous components compared to the control plot.

4. Summary

Proper equipment for scaling-up of the mechanochemical technology of brown coal oxidation has been selected. Three stages of scaling-up of mechanochemical technology of brown coal oxidation were used: from the larger-size equipment of the same type as that used in the laboratory-scale technology, to the semi-industrial-scale flow-type mechanochemical activator. The optimal technological modes of mechanical treatment of brown coal in a vibrocentrifugal mill giving rise to a product with elevated contents of water- and alkali-extractable humic acids were determined. The rotor speed of 1400 rpm and the feed rate of 15 kg/h and 1.8 kg/h for the CEM-20 for K-7 vibrocentrifugal mills, respectively, are the optimal modes from the point of view of chemical and economic efficiency. These modes increase the yields of alkali-soluble and water-soluble humic acids from 23 to 59 % and from <1 to 7%, respectively, which corresponds to the values reached using the laboratory-scale technology.

5. References

[1] Boldyrev V V and Tkáčová K 2000 J. Mater. *Synth. Process.* **8, 3-4** 121-132.

[2] Akopova T A, Demina T S, Cherkaev G V, Khavpachev M A, Bardakova K N, Grachev A V, Vladimirov L V, Zelenetskii A N and Timashev P S 2019 *RSC Advances.* **9, 36** 20968-75

[3] Skripkina T, Bychkov A, Tikhova V, Smolyakov B and Lomovsky O 2018 *Environmental technology & innovation.* **11** 74-82

[4] Skripkina T S, Bychkov A L, Tikhova V D and Lomovsky O I 2018 Solid Fuel Chem. **52, 6** 356-360

[5] Perminova I V and Hatfield K 2005 *Use of humic substances to remediate polluted environments: from theory to practice*, ed I V Perminova, K Hatfield and N Hertkorn (Dordrecht: Springer) pp 3-36

[6] Savel'eva A V, Yudina N V, Berezina E M and Petrova E V 2016 *Solid Fuel Chem.* **50, 2** 76-80

[7] Baibakova O V, Skiba E A, Budaeva V V and Sakovich G V 2017 *Catalysis in Industry* **9, 3** 257-263

[8] Urazova T S, Bychkov A L and Lomovskii O I 2014 *Russ. J. Appl. Chem.* **87, 5** 651−655

[9] Skripkina T S, Bychkov A L, Smolyakov B S and Lomovsky O I 2019 Water Resour. **46, 2** 242–248

[10] Smolyakov B S, Sagidullin A K, Romanov R E and Yermolaeva N I 2019 *Environ. Sci. Pollut. Res.* **26, 7** 6368-77.

Acknowledgments

Scaling-up of the mechanochemical technology with allowance for the electric intensity of the equipment was supported by the Russian Science Foundation (project no. 19-73-10074).

Modification of Manila Grass Activated Carbon for Reactive Dye Adsorption from Textile Printing Wastewater

P Charnkeitkong[1,*] and R Phoophuangpairoj[2]

[1]Department of Chemical Engineering, College of Engineering, Rangsit University, Pathumthani, 12000, Thailand
[2]Department of Computer Engineering, College of Engineering, Rangsit University, Pathumthani, 12000, Thailand

*Email: panida.s@rsu.ac.th

Abstract. The aim of this research is to focus on the activated carbon modification for the elimination of Novacron Brilliant Red EC-3GL which is a reactive red dye onto Manila Grass carbonaceous as an adsorbent with chemical activation. The Manila grass waste from the football field was obtained to carbonize at 400 °C. Three-types of chemicals, such as sodium chloride (NaCl), potassium hydroxide (KOH), and phosphoric acid (H_3PO_4) were used to soak for 1 h before pyrolysis at 700 °C for 2 h. The analysis of activated carbon properties and effects of dye adsorption, such as pH of the feed solution, time to adsorb, the amount of adsorbent, concentration and temperature were studied. Langmuir isotherm and Freundlich isotherm were applied to consider the adsorption mechanism. From the analysis, activated carbon with sodium chloride activation was a high surface area of 148.6 m^2/g and a low porosity of 25.72 °A. From the experiments were found that pH for adsorption of 2, absorption time 90 minutes, concentrations used in the adsorption of 50 ppm and the adsorption temperature 30 °C were optimum for dye adsorption. The NBR adsorption percentages from aqueous solution and textile printing wastewater were 77.87 and 85.52, respectively. The adsorption mechanisms of activated carbon were consistent with Langmuir isotherm for an aqueous solution and Freundlich for textile printing wastewater.

1. Introduction

Dyestuffs are separated from natural and synthetic dyestuffs. The most bleaching dyeing, printing, leather, finishing of textile fabric in industrials always apply with dyes synthetic A synthetic dyestuffs categorization are acid dye, azo dye, azoic dye, basic or cationic dye, direct dye, disperse dye, oxidation dye, mordant or chrome dye, sulfur dye, reactive dye and vat dye [2], especially, reactive dyes are widely spent since the high photolytic stability, high, resistance to microbial attraction and superior fastness to applied fabric are discovered. Dyes maintain in dyeing processes around 80 % and the remains are able to the mixture in the wastewater [3]. Therefore, an environment problem of wastewater is from the dyeing processes [4]. Various methods to eliminate dyes from wastewater, i.e., chemical precipitation, electrodialysis, ion exchange, chemical oxidation, chemical releasing have been applied [6,7]. Some techniques have operated more complications, high-cost for investment and low efficiencies [8]. Alternative the proficiency method for dye privation from wastewater is adsorption [9]. Therefore, the proficiency of dye adsorption and isotherm model onto Manila grass activated carbon were investigated in this work. The significant property of activated carbon produced is surface area and detected under a standard of American Society for Testing and Materials (ASTM). The objectives of this work were to focus on to remove Novacron Brilliant Red EC-3GL which is a reactive red onto Manila Grass carbonaceous modification with chemical activation and also study with Langmuir and Freundlich isotherms.

2. Materials and methods

Manila grass waste is a solid waste from the garden, forest, agricultural and biological waste. It was obtained from the decorating of the football field to remove other plants and dry with sunlight. The dried Manila grass was being carbonized in a furnace at 400 °C for 120 min. The Manila grass carbonaceous was cooled down to room temperature overnight and soaked in each solution 1 M H_3PO_4, 1 M KOH and 1 M NaCl for 60 min. The filtration was followed with a stainless screen as a filter and then activated in a furnace at 700 °C for 60 min. A cooled down to room temperature was used for the activated material. Manila grass activated carbon was washed many times with hot water until pH 7 and dried in an oven at 100 °C for 24 h. The activated carbon was obtained to the ground in a blender and sieve to a required particle size of 150-300 μm. It was kept in a desiccator for easier to use as an adsorbent for adsorption. MAC was called for an adsorbent in this work.

Novacron Brilliant Red EC-3GL (NBR) was dissolved with distilled water at 500 ppm for a stock solution preparation. The desired NBR concentrations were diluted from a stock solution. The pH of the working solutions was adjusted with the concentrated 6 M NaOH for requested pH higher 7.0 and 6 M HCl for requested pH lower. The textile printing wastewater was obtained from the Thai Textile Printing Public Co., Ltd.

The NBR adsorption was investigated in a batch adsorption technique. NBR solution of at the desired concentrations was put in Duran glasses Erlenmeyer flask containing 100 mL, the temperature was set with a thermostatic shaker. The effects of pH, adsorbed time, adsorbent dose, initial concentration, and temperature were investigated to determine for NBR adsorption in this work. The flasks were stirred at a vibrating rate of 110 rpm for 3 h to make sure to equilibrium adsorption. The solution in each flask was filtered with Whatman filter papers No.1 and 42. NBR remaining from the filtrate was measured with a UV-Visible spectrophotometer (HACH model DR 6000) at wavelength 610 nm. The data were reported in average values for all experiments were done in duplication. The NBR adsorption in percentage (%) was calculated from the equation (1).

$$\text{NBR adsorption (\%)} = \left(\frac{C_i - C_t}{C_i}\right)100 \quad (1)$$

where C_i is the initial NBR feed concentration (ppm), and C_t is the NBR concentration at time t (ppm). The optimization of the removal of an NBR adsorption system is significant to make the high proper correlation for the equilibrium system for each set of data. Langmuir and Freundlich isotherm models were employed to describe the relationship between the NBR adsorbed onto adsorbent and that attend in solution at equilibrium in linearization. Langmuir isotherm is based on adsorbing molecules in a monolayer and the forces of molecules interaction are insignificant. It can be shown in the non-linear form in equation (2) and modified in the following linear form in equation (3).

$$q_e = \frac{q_{max} K_L C_e}{1+K_L C_e} \quad (2)$$

$$q_e = \frac{1}{q_{max}} + \frac{1}{q_{max} K_L C_e} \quad (3)$$

where C_e is the concentration at equilibrium system(mg/L), q_e is the amount adsorption capacity at equilibrium (mg/g), q_{max} is the maximum capacity of adsorption and K_L is the Langmuir constant adsorption at equilibrium.

The significant parameter of equilibrium characteristics for Langmuir constant separation factor (SF) in dimensionless is shown in equation (4) as the following [10,11]:

$$SF = \frac{1}{1+K_L C_e} \quad (4)$$

Freundlich isotherm for the adsorption is an obvious model of molecules adsorbed onto the adsorbent surface in multilayer as shown in equation (5) as the following [10,11]:

$$q_e = K_F C_e^{1/n_F} \tag{5}$$

This linear expression can be shown in equation (6) as the following:

$$\log q_e = \log K_F + \frac{1}{n_F} \log C_e \tag{6}$$

where K_F is the Freundlich constant of the adsorption and means to the amount the solute molecules onto adsorbent for an equilibrium concentration.

3. Results and discussion

From the consideration properties of MAC with surface area analyzer (model: Quantachrome / Autosorb-1, Thermo Finnigan/ Sorptomatic 1990) are presented in table 1. Table 1 shows that the surface area from NaCl activation of MAC is higher than that of H_3PO_4 and KOH, respectively. The NaCl activation is normally used in activated adsorbent modification because of the high productive results and low cost in energies used has been found. The activations of encouraging bond disaffection from a salt catalyst are preferable acidic or alkaline catalysts. The higher NBR adsorption as a higher surface area results can be predicted with the adsorption sites increasing. Therefore, MAC activated with NaCl was selected as an adsorbent to study of NBR adsorption and mentioned to MAC-NaCl.

Table 1. Properties of MAC with chemical activation at 700 °C

Property	Manila grass activated carbon (MAC)		
	NaCl	KOH	H_3PO_4
Surface area (m^2/g)			
- Multiple points	148.6	83.91	120.9
- Single point	145.1	80.84	118.0
- Langmuir Surface Area	228.5	132.3	186.0
Total Pore volume (cc/g)	0.09925	0.06996	0.08401
Average Pore diameter (°A)	25.72	33.35	27.79

The determination of the organic functional groups at the surface structures of MAC by vibrational frequency changes can be measured with Fourier transform infrared spectroscopy (FTIR). Spectrum from the FTIR spectrometer was received by using the model: Perkin Elmer Spectum100. The consideration of the FTIR spectrum of before and after NBR adsorption onto MAC has been shown in figure 1. Many functional groups searched on the surface of MAC before and after adsorption was represented in The FTIR spectrum. Some peaks were shifted and a new peak has appeared in NBR adsorption. From figure 1, three the bonds –OH, -C=C- and –C-O- group showed the important frequencies at 3750, 1538, 1039 have been changed. The shifted of these frequencies in the spectrum can be shown the able concerning the specific functional groups which adsorbed on the MAC-NaCl surface from the NBR adsorption process.

IOP Conf. Series: Materials Science and Engineering **733** (2020) 012043 doi:10.1088/1757-899X/733/1/012043

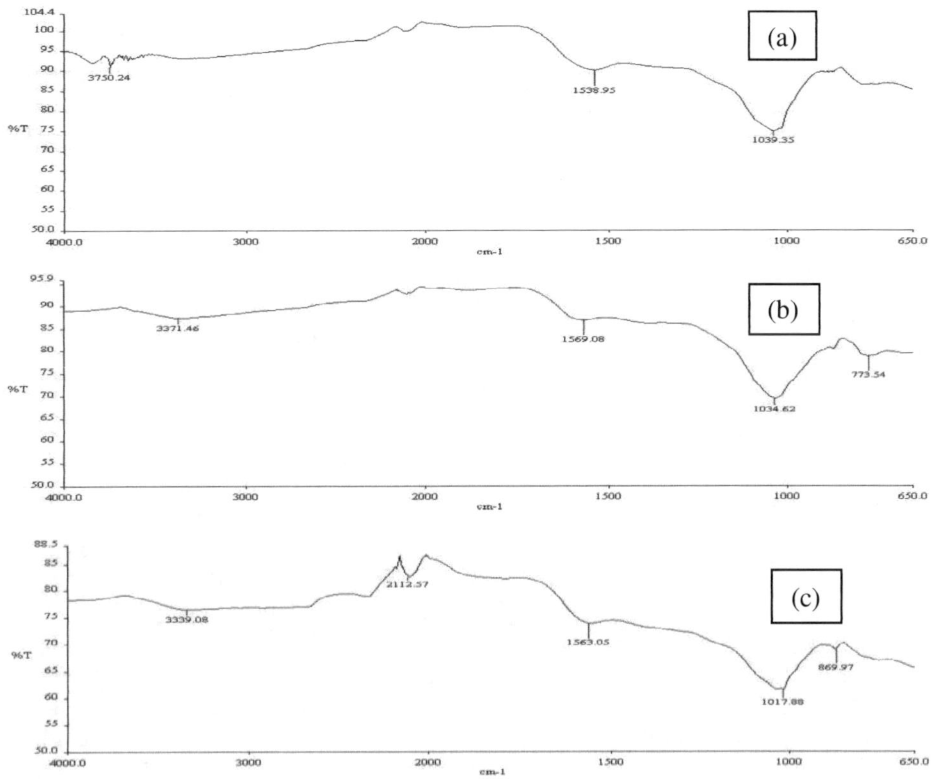

Figure 1. FTIR peaks (a) before and (b) after NBR adsorption from aqueous solution and (c) after NBR adsorption from textile printing wastewater on MAC-NaCl

3.1. Effect of pH for NBR adsorption on MAC-NaCl

The effects of pH solution on the surface adsorbent bonding with solute molecules were represented [11]. In this work, figure 2 has been showed the effect of pH in range 2-12 on MAC-NaCl. At the pH 2 was conducted for the maximum NBR adsorption. The decreasing of NBR adsorption with the increasing pH solution was observed. This action was similar to the textile dyes adsorption on agricultural waste [1]. The increasing of electron charges in the ionic form with pH solution decreased was affected to increase the electrostatic interaction with the adsorbent surface and NBR molecules. The percentage adsorption the aqueous solution was lower than that of textile printing wastewater.

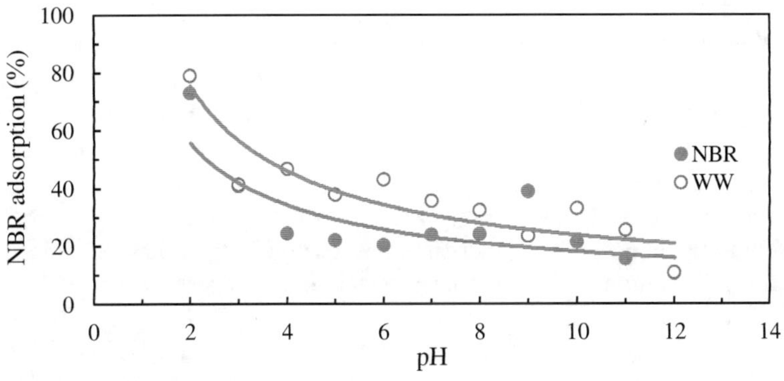

igure 2. Effect of pH on the [BR adsorption, initial oncentration 50 ppm, 1AC-NaCl 1.0 g in 100 mL, mperature 30 °C and dsorbed time 60 min

3.2. Effect of adsorbed time for NBR adsorption on MAC-NaCl

The effect of adsorbed time for NBR adsorption on MAC-NaCl is shown in figure 3. The operating time for 90 min was adsorbed to the maximum NBR adsorption. In the first stage, a rapid NBR adsorption has happened and after that, the adsorption decreased to equilibrium in finally. This result was similar to other works [10-12] that the higher adsorption rate was made from the greater and higher surface area of the adsorbent and some NBR adsorptions can be reversible from the adsorbent for slightly decrease adsorption at 90 min adsorption later. The percentage adsorption of the aqueous solution was also lower than that of textile printing wastewater.

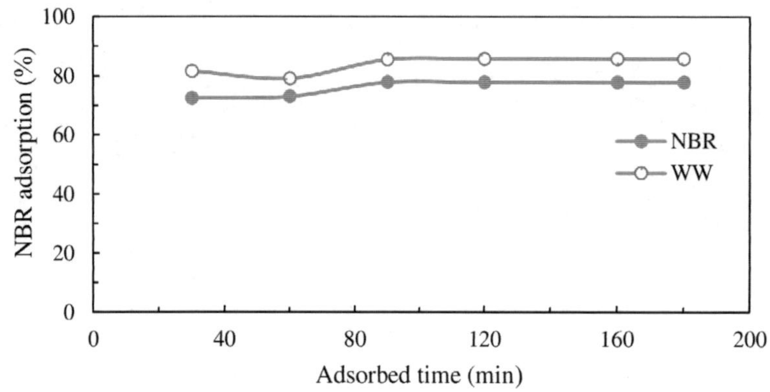

igure 3. Effect of lsorbed time on NBR lsorption, initial ncentration 50 ppm, AC-NaCl 1.0 g in100 mL, mperature 30 °C, and pH 0

3.3. Effect of adsorbent dose for NBR adsorption on MAC-NaCl

The determination of the adsorbent capacity at an initial feed concentration was influenced by the dose of adsorbent, which is a significant variable for adsorption. MAC-NaCl was affected on NBR adsorption has been shown in figure 4. From the results, an increasing of NBR adsorption with a high adsorbent dosage because of a larger surface area until saturation and after that, the decreasing adsorption has a possibility. The amount of adsorbate per unit mass decreasing should be decreased adsorption. Therefore, the adsorbent dose of 1.0 g in solution volume 100 mL was selected for these experiments.

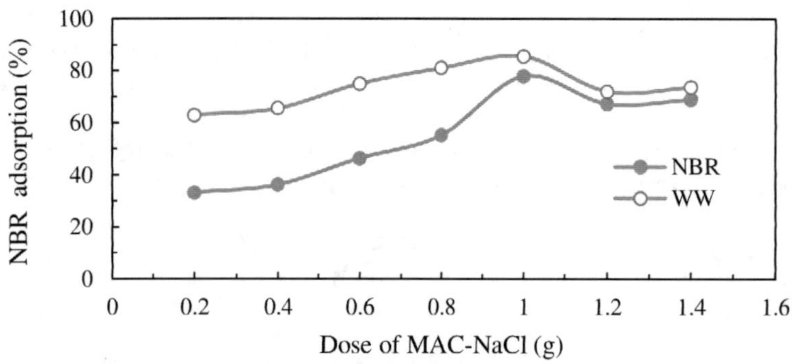

Figure 4. Effect of adsorbent dose on NBR adsorption, initial concentration 50 ppm, temperature 30 °C, pH 2.0, adsorbed time 90 min

3.4. Effect of initial concentration for NBR adsorption on MAC-NaCl

The effect of initial feed concentration for NBR adsorption on MAC-NaCl is shown in figure 5. The NBR adsorption was decreased with the concentration increases because the active sites of molecules and surface area of adsorbent would be saturated. It is similar to reactive dye adsorption on sugarcane bagasse pith activated carbon [12]. The percentage of NBR adsorption was 77.87% at initial feed aqueous concentration 50 ppm on 1.0 g of MAC-NaCl. There were many types of dyes to obtain from textile printing wastewater, the initial NBR containing in the wastewater of 50 ppm was diluted to lower concentration to study in this effect. The higher percentages of NBR adsorption from wastewater were found.

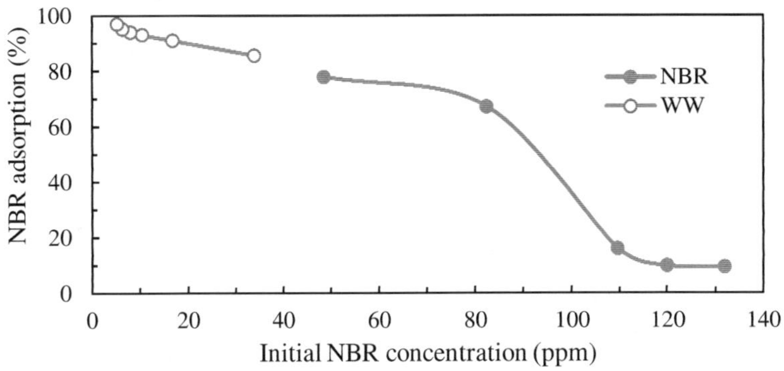

Figure 5. Effect of initial feed concentration on NBR adsorption, MAC-NaCl 1.0 g in 100 mL, temperature 30 °C, pH 2.0 and adsorbed time 90 min

3.5. Effect of temperature for NBR adsorption on MAC-NaCl

The effect of temperature was investigated at different temperatures 30, 35, 40, 45 and 50 °C as shown in figure 6. In most cases, removal very slightly decreased with the temperature increased. Therefore, the effect of temperature at equilibrium adsorption in this work seems no significant difference between each temperature.

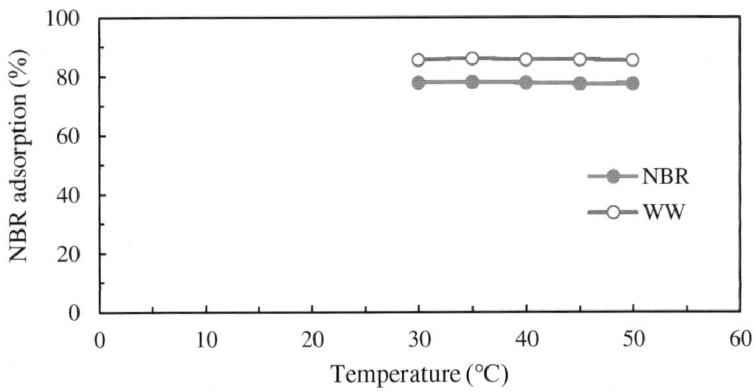

Figure 6. Effect of temperature on NBR adsorption, initial concentration 50 ppm, MAC-NaCl 1.0 g in volume 100 mL, pH 2.0 and adsorbed time 90 min

3.6. Adsorption isotherm

Adsorption is the interaction between adsorbate and adsorbent. The equilibrium adsorption isotherm models, i.e., Langmuir and Freundlich have been applied to explain the results obtained from these experiments. The correlation coefficient (R^2) is considered to look for the best-fit isotherm comparison. Langmuir adsorption isotherm can emerge on the monolayer of the homogenous surface adsorbent and non-interaction between adsorbed molecules. From the experimental data, the Langmuir isotherm model is a good fit for NBR adsorption from aqueous solution as shown in figure 7. The separation factor (SF) values concern with the adsorption process feature (SF = 0: Irreversible, $0 < SF < 1$: Favorable, SF = 1: Linear and SF > 1: Unfavorable). SF values of 0.3305 – 0.8459 were obtained to indicate in between 0 – 1 to verify the favorable NBR adsorption process. The experimental data gave a high R^2 of 0.9504 indicating the Langmuir model was satisfied to study for NBR adsorption.

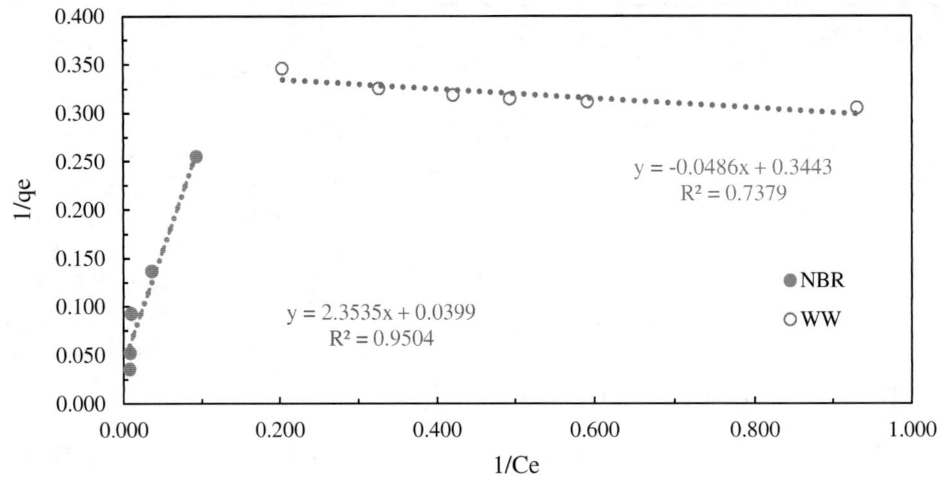

Figure 7. Langmuir isotherm of NBR adsorption from aqueous solution and textile printing wastewater on MAC-NaCl

Freundlich isotherm model for the interaction between the molecules adsorbed in multilayers was also used for the adsorption of NBR onto MAC-NaCl. The plotting of log q_e and log C_e is presented in figure 8.

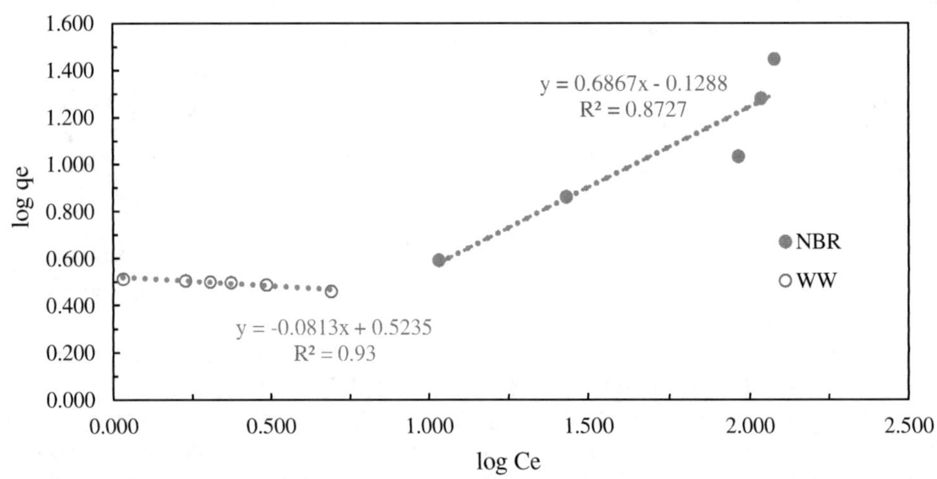

Figure 8. Freundlich isotherm of NBR adsorption from aqueous solution and textile printing wastewater on MAC-NaCl

K_F and $1/n_F$ values were obtained from the intercept and slope of the linear relation. The experimental data of NBR adsorption from aqueous solution, the Freundlich isotherm model showed a lower correlation coefficient (R^2 = 0.8924) than the Langmuir isotherm model (R^2 = 0.9504). It can be considered the non-acceptability of this isotherm model. Freundlich isotherm showed a good fit with NBR adsorption from textile printing wastewater (R^2 = 0.930).

4. Conclusions

The Manila grass waste from the football field was obtained to carbonize at 400 °C. The good activated carbon with sodium chloride activation by soaking for 1 h before pyrolysis at 700 °C for 2 h. was a high surface area of 148.6 m^2/g and low porosity of 25.72 °A for NBR adsorption. The pH of 2.0, absorption time 90 min, initial NBR in feed concentrations of 50 ppm and the adsorption temperature 30 °C were optimum for NBR adsorption. The NBR adsorption percentages from aqueous solution and textile printing wastewater were 77.87 and 85.52, respectively. The adsorption mechanisms of activated carbon were consistent with Langmuir isotherm for an aqueous solution and Freundlich for textile printing wastewater. NBR adsorption from aqueous solution was a monolayer. The adsorption constant (K_L) was 0.1965. The maximum monolayer of adsorption constant (q_{max}) was 25.06 mg/g and the correlation coefficient (R^2) was 0.9504. NBR adsorption from textile printing wastewater. NBR was a multilayer. The values of n_F and K_F were -12.30 and 3.381, respectively.

Acknowledgments
The author would like to thank the Thai Textile Printing Public Co., Ltd for the reactive red dye and wastewater supporting in this research.

References

[1] Baghapour M A, Pourfadakari S and Mahvi A H 2014 Investigation of reactive red dye 198 removal using multiwall carbon nanotubes in aqueous solution *J. Ind. Eng. Chem.* **20** 2912-2916
[2] Gupta V and Suhash J 2009 Application of low-cost adsorbents for dye removal – A review *J Environ. Manage.* **90** 2313-2342
[3] Mahvi A G, Ghanbarian M, Nasseri S and Khairi A 2009 Mineralization and discoloration of textile wastewater by TiO_2 nanoparticles *Desalination.* **239** 309-316
[4] Surana M, Mehta P, Pamecha K and Kabra B V 2011 Treatment of water contaminated with Reactive Red 198 (RR198) by Photo-Fenton Reagent *Der Chemica Sinica.* **2** 177-186
[5] Duan J, Liu R, Chen T, Zhang B and Liu J 2012 Halloysite nanotube-Fe_3O_4 composite for removal of methyl violet from aqueous solutions *Desalination.* **293** 46-52
[6] Mokhtari M, Ebrahimi A, Ghaneian M T, Jafari H, Talebi P and Farfani A S 2016 Efficiency of the combined chemical precipitation -Reduction process to remove dye and chromium from industry wastewater of home appliance *J. Health. Dev.* **5** 337-346
[7] Khataee A R and Kasiri M B 2017 Photocatalytic degradation of organic dyes in the presence of nanostructured titanium dioxide: Influence of the chemical structure of dyes *J. Mol. Catal. A Chem.* **328** 8-26
[8] Mishra A K, Arockiadoss T and Ramaprabhu S 2010 Study of removal of azo dye by functionalized multi walled carbon nanotubes *Chem. Eng. J.* **162** 1026-1034
[9] Sathishkumar P, Arulkumar M and Palvannan T 2012 Ultilxation of agro-industrial waste Jatopha curcas pods as an activated carbon for the adsorption of reactive dye Remazol Brilliant Blue R (RBBR) *J. Clean. Prod.* **22** 67-75
[10] Sampranpiboon P and Charnkeitkong P 2010 Equilibrium isotherm, thermodynamic and kinetic studies of lead adsorption onto pineapple and paper waste sludges *International J. Energy. Environ.* **4** 88-98
[11] Sampranpiboon P, Charnkeitkong P and Feng X 2014 Equilibrium isotherm models for adsorption of Zinc (II) Ion from aqueous solution on pulp waste *WSEAS Trans. Environ. Dev.* **10** 36-47
[12] Amin N K 2008 Removal of reactive dye from aqueous solutions by adsorption onto activated carbons prepared from sugarcane bagasse pith *Desalination.* **223** 152-161

The influence of terminal chain on the formation of chiral nematic phase of non-symmetric liquid crystal dimers containing fluorinated units

Zhu Chen[a], Rui Zhang[b], Mei Tian[c*]

Department of Chemistry, College of Science, Northeastern University,

Shenyang,China

[a]3111299695@qq.com, [b]2417115935@qq.com, [c]tianmei613@sina.com

Abstract. In this study, we have synthesized four fluorine liquid crystals (LCs), termed FA-PD-UBA, FA-PD-PBA, FA-PD-ABA and FA-PD-BA. Through the research on thermal properties and mesophase type of the dimers, we can draw a conclusion that the non-symmetric structure type with flexible terminal chain of appropriate length, may help the dimers to form cholesteric fluorinated liquid crystals.

1. Introduction

Cholesteric (chiral nematic) liquid crystals [1-5] have attracted considerable attention for their etc. Fluorinated liquid crystals [6-10] have gradually aroused the interest of the researchers for the special physicochemical properties. In previous study [8], the symmetric and non-symmetric fluorinated liquid crystal (LC) dimer containing chiral core display smectic phases. This may be due to the strong molecular interaction force. In this study, the four non-symmetric dimers were synthesized to find appropriate structure for preparing fluorinated cholesteric LCs. The effect of the terminal structure on the properties of fluorinated dimers is discussed.

2. Synthesis

The synthesis of liquid crystal (LC) the dimers is shown in Scheme 1. The synthesis of FA-PD-UBA was synthesized according to Ref. [8]. The synthesis of FA-PD-PBA, FA-PD-ABA and FA-PD-BA was similar. The synthesis of FA-PD-ABA is shown below.

A solution of ABA (17.819 g, 0.1 mol), DCC (20.06 g, 0.1 mol), DMAP (12.22 g, 0.1 mol), 20 mL pyridine and 140 mL THF were put into the flask. The THF solution of FA-PD (44.4 g, 0.1 mol) was put into the solution of ABA slowly. The solution was stirred at room temperature for 2 h and heated to 45 °C until the reaction is complete. Reaction mixture was filtrated to remove N,N′-dicyclohexylurea (DCU). The filtrate was concentrated, put into water and neutralized. The white product powder was obtained by filtration and recrystallisation. Yield: 50.0%. m.p.: 110.0 °C

FA-PD-ABA: Infrared spectroscopy (IR) (KBr, cm-1): 2975-2854 (-CH3, -CH2-), 1747, 1711 (C=O), 1642 (C=C), 1601, 1495 (Ar).

1H NMR (CDCl3, d, ppm): 1.457-1.470 (m, 3H, -CH3), 4.407-4.689 (m, 5H, -CH2-, -CH-), 5.347-5.364 (d, J=10.20 Hz, 1H, CH2=CHCH2O), 5.446-5.477(d, J=18.60 Hz, 1H, CH2=CHCH2O), 6.050-6.123 (m, 1H, CH2=CHCH2O), 6.895-6.911 (m, 2H, Ar-H), 7.254-7.261 (m, 2H, Ar-H), 7.337-7.356 (m, 2H, Ar-H), 7.677-7.721 (m, 4H, Ar-H), 7.809-7.826 (m, 2H, Ar-H), 8.111-8.155 (m, 2H, Ar-H), 8.351-8.361 (m, 2H, Ar-H).

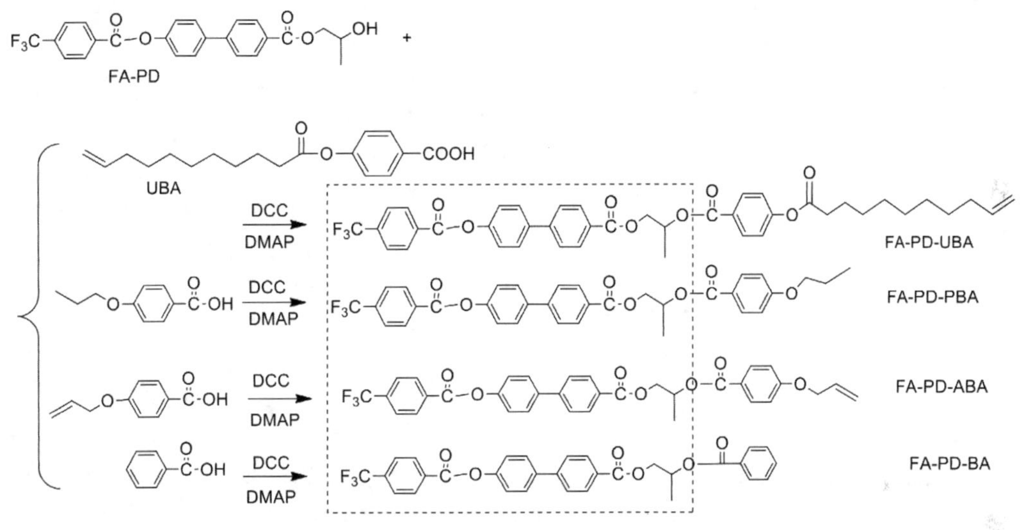

Scheme 1. Preparation route of the target products.

FA-PD-UBA: All data of FA-PD-UBA can be seen in [8].

FA-PD-PBA: All data of FA-PD-UBA can be seen in [11].

FA-PD-BA: Yield: 41.0%, m.p.: 154.9 °C.

Infrared spectroscopy (IR) (KBr, cm-1): 2984-2898 (-CH3, -CH2-), 1741, 1722 (C=O), 1601, 1410 (Ar), 1412-1326 (C-F).

1H NMR (CDCl3, d, ppm): 1.502-1.513 (d, J = 6.60 Hz, 3H, -CH3), 4.512-4.573 (m, 2H, -CH2-), 5.571-5.599 (m, 1H, -CH-), 7.332-7.347 (d, J = 9.00 Hz, 2H, Ar-H), 7.443-7.468 (t, J = 7.20 Hz, 2H, Ar-H), 7.561-7.568 (t, J = 7.20 Hz, 1H, Ar-H), 7.649-7.695 (dd, 4H, J1 = 19.20 Hz, J2 = 8.40 Hz, Ar-H), 7.805-7.819 (d, J = 8.40 Hz, 2H, Ar-H), 8.072-8.122 (m, 4H, Ar-H), 8.348-8.361 (d, J = 7.80 Hz, 2H, Ar-H).

3. Results and discussion

3.1.Thermal analysis. The LC performances of FA-PD-UBA, FA-PD-PBA, FA-PD-ABA and

IOP Conf. Series: Materials Science and Engineering **733** (2020) 012044 doi:10.1088/1757-899X/733/1/012044

FA-PD-BA were characterizated using DSC, POM and XRD. The thermal performance data are summarized in Table1. The results of DSC, POM and XRD are shown in Figure 1-3, respectively.

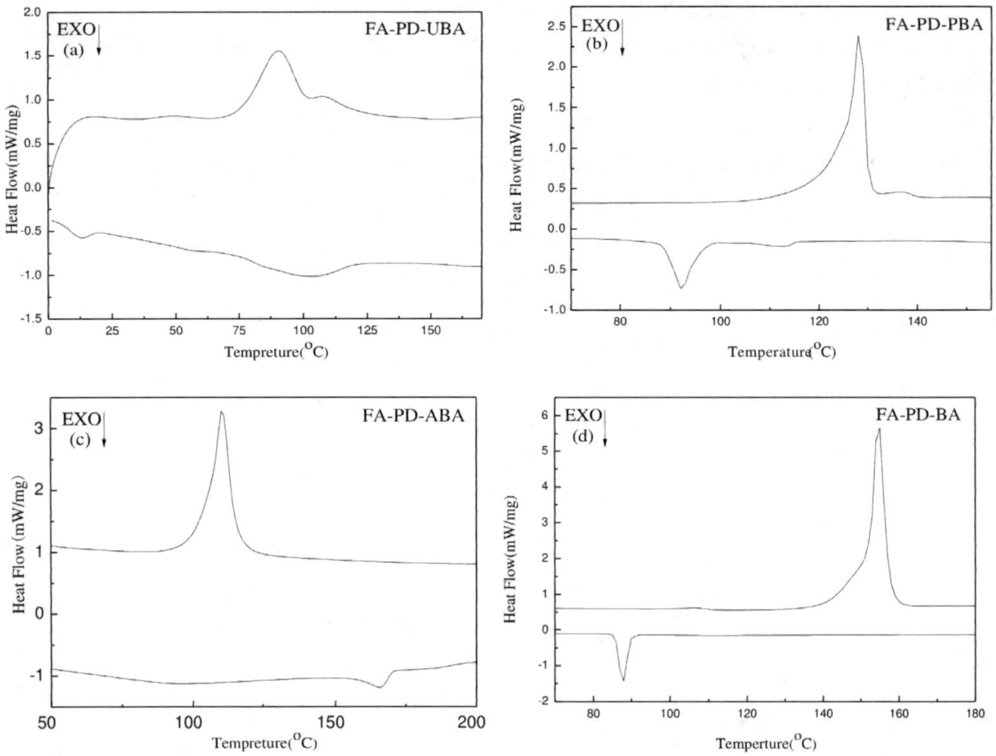

Figure 1. Thermal analysis results: (a) FA-PD-UBA; (b) FA-PD-PBA ; (c) FA-PD-ABA; (d) FA-PD-BA.

For FA-PD-UBA, FA-PD-PBA, FA-PD-ABA and FA-PD-BA, they have the same rigidity group, but the terminal flexible chains are different. In general, the flexibility of the molecule decreases with the length of terminal flexible chain decreasing, resulting the melting temperature increasing. For FA-PD-UBA, FA-PD-ABA and FA-PD-BA, the melting temperature increases from 90.6 °C to 110.0 °C to 154.9 °C with the length of the terminal flexible decreasing, while the mesomorphic region decreased (as shown in Table 1)

FA-PD-PBA and FA-PD-ABA displayed different thermal properties, although they have the similar terminal flexible chains which the only difference is the structure of the terminal bond. The terminal bond of non-fluorinated unit of FA-PD-PBA is C-C, while that of FA-PD-ABA is C=C. The melting point of FA-PD-ABA is lower than that of FA-PD-PBA because isolated double bond may decrease the rigidity of the molecule.

These results indicated that the structure of the terminal group played important roles on the performance of the target products.

Table 1. List of performance of target products.
[a]The transition temperature obtained from POM; SmA*, smectic A*; N, nematic; ch, cholesteric.

Product	Heating/cooling	T_m/T_k (°C)	$\Delta H_m/\Delta H_k$ (J.g^{-1})	T_i/T_c (°C)	$\Delta H_i/\Delta H_c$ (J.g^{-1})	ΔT_1 (°C)	Meso phase	SROT (°)
FA-PD-UBA	Heating	90.6	6.7	109.0	1.2	18.4	SmA*	-39.9
	Cooling	13.4	-2.2	103.2	-1.9	89.8	SmA*	
FA-PD-PBA	Heating	128.4	60.9	137.0	1.1	8.6	N	-
	Cooling	92.2	-36.3	112.6	-1.8	20.4	N	
FA-PD-ABA	Heating	110.0	55.9	-	-	-	-	-7.3
	Cooling	96.0[a]	-	166.0	-5.2	70.0	Ch	
FA-PD-BA	Heating	154.9	75.3	-	-	-	-	-6.6
	Cooling	87.9	-54.6	-	-	-	-	

3.2.LC properties analysis. FA-PD-BA showed no texture during heating and cooling process. FA-PD-BA is a non-LC compound.

FA-PD-UBA containing ten alkyl units in the terminal group displayed chiral smectic A phase [8]. The representative texture is presented in Figure 2 (a).

FA-PD-PBA displyed typical nematic schlieren texture with twofold and fourfold brushes (as presented in Figure 2 (b)).

FA-PD-ABA displayed fan-shaped texture (Figure 2 (c)). Figure 3 is the XRD result of FA-PD-ABA. These results state that FA-PD-ABA is a cholesteric LC. FA-PD-ABA only displayed cholesteric fan-shaped texture on cooling process. Thus FA-PD-ABA is a monotropic cholesteric LC.

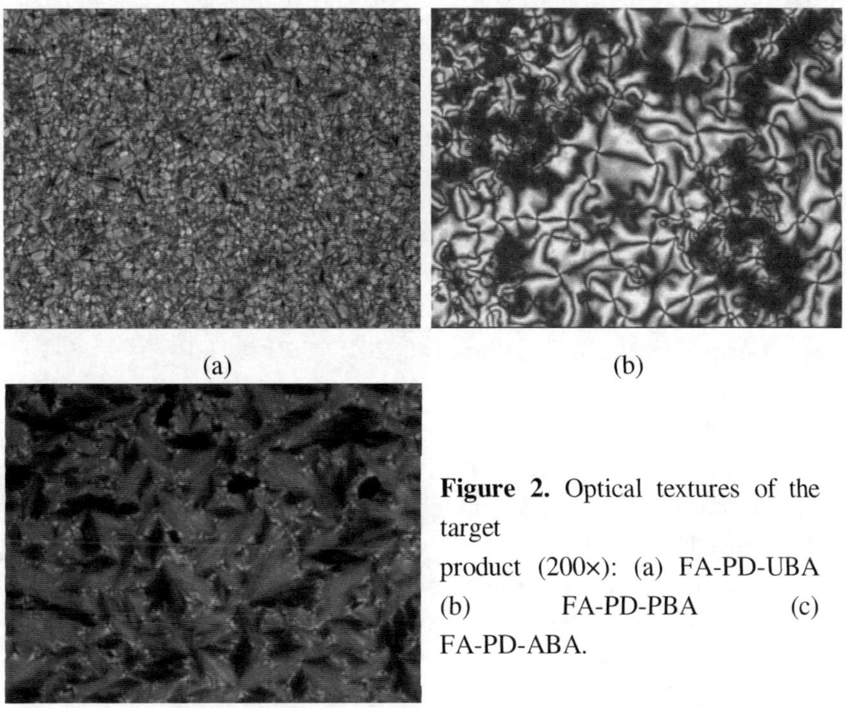

(a) (b) (c)

Figure 2. Optical textures of the target product (200×): (a) FA-PD-UBA (b) FA-PD-PBA (c) FA-PD-ABA.

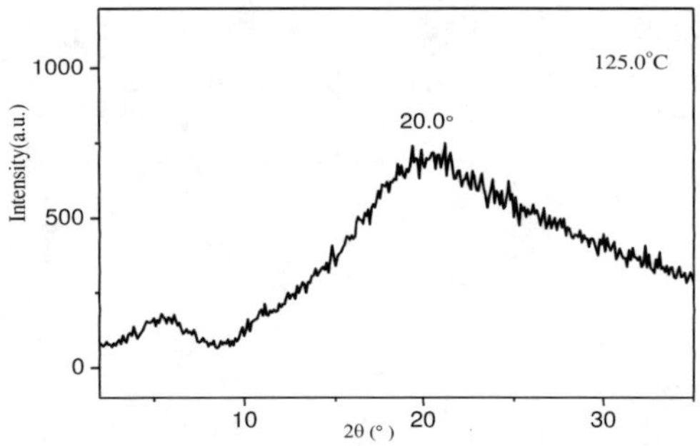

Figure 3. X-ray diffraction (XRD) pattern of FA-PD-ABA.

In this study, short alkyl terminal chain was adopted to obtain cholesteric dimer. The shorter flexible terminal chain successfully prevented the formation of lamellar phase. FA-PD-PBA and FA-PD-ABA containing three alkyl units in the terminal group did not display lamellar phase, while exhibited nematic phase and cholesteric phase, respectively. Although FA-PD-PBA and FA-PD-ABA all have three alkyl units in the terminal group, FA-PD-PBA and FA-PD-ABA displayed different thermal properties and mesophase type, suggesting that bonding type had an obvious influence in mesomorphic behaviour. FA-PD-BA is not a LC, which indicated that flexible terminal chain was necessary to form mesophase, which played a stabilizing effect on mesophase. The results indicated that the non-symmetric structure type, which contained fluorinated mesogenic unit and non-fluorinated mesogenic unit with flexible terminal chain of appropriate length, is advantageous to form cholesteric fluorinated liquid crystals.

4. Conclusion

Four non-symmetric dimers, FA-PD-UBA, FA-PD-PBA, FA-PD-ABA and FA-PD-BA, were synthesised. The structure of terminal chain of the non-fluorinated mesogenic unit had obvious effects on not only stabilizing mesophase but also the mesophase type. The non-symmetric structure type, which contained fluorinated mesogenic group and non-fluorinated mesogenic unit with flexible terminal chain of appropriate length, may help the dimers to form cholesteric fluorinated liquid crystals.

5. References

[1] Champagne PL, Ester D, Aldosari S, et al. Synthesis and comparison of mesomorphic behaviour of a cholesterol-based liquid crystal dimer and analogous monomers. Liq Cryst. 2018;45(8):1164-76.

[2] Coates D. Development and applications of cholesteric liquid crystals. Liq Cryst. 2015;42(5-6):653-65.

[3] Lesac A, Baumeister U, Dokli I, et al. Geometric aspects influencing N-N-TB transition - implication of intramolecular torsion. Liq Cryst. 2018;45(7):1101-10.

[4] Mitov M. Cholesteric liquid crystals in living matter. *J Soft Matter.* 2017,**13**,4176. doi: 10.1039/c7sm00384f.

[5] Lee SS, Seo HJ, Kim YH, et al. Structural Color Palettes of Core–Shell Photonic Ink Capsules Containing Cholesteric Liquid Crystals. *J Adv Mater.* 2017,**29**(23):1-8. doi: 10.1002/adma.201606894.

[6] Cha SW, Jin JI, Achard MF, et al. Anomalies of periodicity in TGB structures in new liquid crystal dimers. Liq Cryst. 2002,29(6):755-63. doi: 10.1080/02677822902 10129902.

[7] Wen JX, Tian RW, Dai XW. Synthesis and mesomorphic properties of four-ring fluorinated liquid crystals with trifluoromethyl group. Liq Cryst. 2017;44(10):1487-93.

[8] Xu ZX, Yao M, Feng WJ, et al. 1,2-Propanediol-linked chiral symmetric and non-symmetric liquid crystal dimers containing trifluoromethyl. Liq Cryst. 2016,43 (12):1846-1861. doi: 10.1080/02678292.2016.11197974.

[9] Tian M, Wu SJ, Tian XW, et al. Mesomorphic properties of chiral three-arm liquid crystals containing 1, 2, 4-butanetriol as core. *J Mol Struct.* 2016,**1107**:202-13. doi: 10.1016/j.molstruc.2015.11.040.

[10] Thompson M, Carkner C, Mosey NJ, et al. Tuning the mesomorphic properties of phenoxy-terminated smectic liquid crystals: the effect of fluoro substitution. *J Soft Matter.* 2015,**11**(19):3860-3868. doi: 10.1039/C5SM00473J.

[11] Dong L, Xu ZX, Tao HZ, et al. Non-symmetric chiral nematic liquid crystal dimers containing trifluoromethyl and 1, 2-propanediol. Liq Cryst. 2018; 45(12):1734-45.

Remark: Northeastern University Student Innovation Training Program Self-raised Project (191052) supported by “the Fundamental Research Funds for the Central Universities” and Project approval number (N182410001).

IOP Conf. Series: Materials Science and Engineering **733** (2020) 012045 doi:10.1088/1757-899X/733/1/012045

Evaluation of Uncertainty in Determining Average Grain Size by ASTM E112 Standard

Geping Bi[1,a*], Yong Guan[1], Xiaoyu Chen[1], Wenhua Tan[1], Wenniu Huang[1], Yuqing Gao[2] and Meimei Wang[3]

[1] China National Quality Supervision and Inspection Center for Knife, Scissors and Domestic Metal Tool, Yangjiang, 529539, China
[2]Heye Special Steel Co.,Ltd., Shijiazhuang, 050000, China
[3]Chinalco Materials Application Research Institute Co., Ltd., Beijing, 102209, China
[a*]bigeping@sina.com

Abstract. In this paper, the ASTM grain size number of aluminium alloy specimen was determined respectively by Abrams Three-Circle Procedure and Planimetric Procedure of ASTM E112-13 Standard, and the uncertainty of the result was evaluated. The results are as follows: 1) Using the Abrams Three-Circle Procedure determines G, G=4.13，U=0.26，k=2; Using the Planimetric Procedure determines G, G=3.94，U=0.15，k=2. 2) It is suggested that the ASTM E112 standard should provide that the errors of counting grain boundary intersection count or number of grains should be less than or equal to 2, and the ruler that determines the diameter of test circle or scale length should be with the scale that is less than or equal to 0.10 mm. On this condition, the uncertainties caused by them in determining G can be negligible. 3) The main sources of uncertainty for the determination of G by Abrams Three-Circle Procedure and Planimetric Procedure are repetitive measurement, which depends on the uniformity of grain size of specimen. 4) The %RA limit of ASTM E112 for determining G should be redefined according to the measurement uncertainty in order to be up to the precision requirement (±0.25).

1. Introduction

Metallographic examination is a very important testing field in metal material testing. In the field of metallographic examination, the semiquantitative comparison method is the main detection method, and the quantitative analysis items are very few. The ASTM E112-13 Standard test methods for determining average grain size[1] provides three methods for average grain size determination: comparison procedure, intercept procedures and planimetric procedure. The intercept procedures and planimetric procedure are very typical quantitative methods.

According to the ISO/IEC 17025-2017 Accreditation criteria for the competence of testing and calibration laboratories[2],Laboratories shall identify the contributions to measurement uncertainty. When evaluating measurement uncertainty, all contributions which are of significance, shall be taken into account using appropriate methods of analysis. A laboratory performing testing shall evaluate measurement uncertainty. Where the test method precludes rigorous evaluation of measurement uncertainty, an estimation shall be made based on an understanding of the theoretical principles or practical experience of the performance of the method.

In the field of chemical analysis and mechanical properties testing of metal materials, uncertainty evaluation of measurement results is widely used. In the two fields, there are many published papers about the uncertainty evaluation and standardization has been achieved [3-4]. In the field of metallographic examination, the uncertainty evaluation of measurement results is seldom used. The main reasons are as follows: 1) the semiquantitative comparison methods are difficult to be rigorous evaluated; 2) The quantitative metallographic examination methods are few; 3) there are hardly any quantitative metallographic results in product standards as a method of evaluating product eligibility. However, through the evaluating measurement uncertainty, the main contributions to measurement uncertainty could be identified. From the management point of view, the testing quality can be effectively controlled. From the technical point of view, it can not only further perfect the standard test method, but also improve the accuracy of the standard test method.
In this paper, the ASTM grain size number of aluminium alloy specimen was determined respectively by Abrams Three-Circle Procedure and Planimetric Procedure of ASTM E112-13 Standard. The test circletotal circumference used in Abrams Three-Circle Procedure is 500 mm, and the test circle area used in Planimetric Procedure is 5000 mm^2. And the uncertainty is evaluated and expressed according to the JJF 1059.2-2012 Evaluation and Expression of Uncertainty in Measurement[5].

2. Testing scheme

The test specimen is aluminium alloy (AlSi1). The polished surface is18 mm x 10 mm. The film on the surface of specimen is prepared by an odization. In the whole polished surface, the grain sizesare observed and the images are captured under polarized light randomly, and there are no overlapping fields. The magnification of objective lens is10×.The selection of magnification ensures that the grain boundary intersection counts are 50 to 100 and the numbers of grains are not less than 50.
The quantitative detection methods of G are as follows: 1) Abrams Three-Circle Procedure, the designed diameters of three test circles are 79.58 mm, 53.05 mm, and 26.53 mm. 2) Planimetric Procedure, the designed diameter of test circle is 79.80 mm. In the captured metallographic photographs, the scale and test circle are superimposed, and then the 100 ×photographs are printed. On the printed photographs, measuring the scale length and test circle diameter, counting the grain boundary intersection counts of three circles, and counting the numbers of grains completely within the test circle and the numbers of grains intercepted by the test circle.
Using the vernier calipers with 0.02 mm scale to determine the scale length or test circle diameter. The scale length is measured twice and the diameter of the circle is measured twice in the horizontal and vertical directions, and then the average scale length or average diameter of circle is calculated. When counting grain boundary intersection count or number of grains, appropriate labelling on the printed photographs is used to reduce the statistical counting error.
For the Abrams Three-Circle Procedure, the ASTM G is calculated from:

$$G = 6.643856\lg\left(\frac{M}{L}P\right) - 3.288 \qquad (1)$$

Where M is the magnification, L is the test circle total circumference in mm and P is the intersection counts.
For the Planimetric Procedure, the ASTM G is calculated from:

$$G = 3.321928\lg\left(\frac{M^2}{A}N\right) - 2.954 \qquad (2)$$

Where M is the magnification, A is the test circle area in mm^2 and N is the total number of grains.

3. Test result

Figure. 1a) is a metallographic photograph superimposed with the three circles and with a scale length of 200 μm for the Abrams Three-Circle Procedure. Figure. 1b) is a metallographic photograph superimposed with the circle and with a scale length of 200 μm for the Planimetric Procedure.

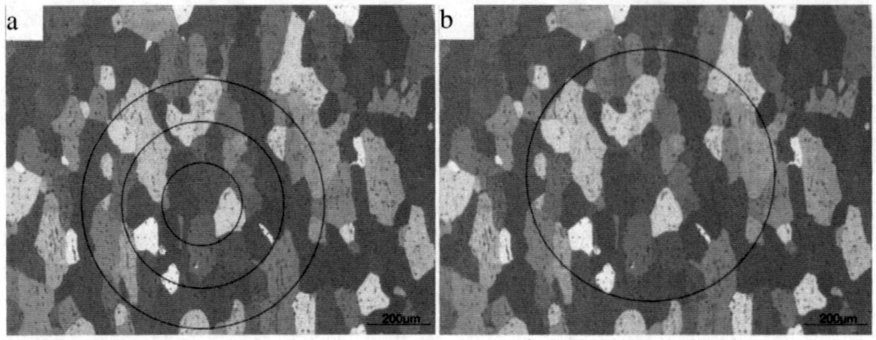

a) Metallographic photograph and three test circles; b) Metallographic photograph and test circle.

Figure 1. Metallographic photograph and test circle

The average scale length on Fig.1 is 23.30mm, and the range of the two measurements is 0.02 mm. The actual magnification of metallographic photograph is 101.50 x, that is, M=101.50 × in formulas (1) and formulas (2), and the magnification is within 2% of the permissible error of ASTM E112.
The average diameters of the three circles on Fig.1a are 79.48 mm, 52.96 mm and 26.32 mm, respectively. The range of the two measurements for each circle is 0.02 mm. The actual total circumference of the three circles is 498.76 mm, that is, L=498.76 mm in formula (1), and the relative error between actual total circumference and design total circumference is -0.25%.
The average circle diameter on Fig. 1b is 79.86 mm. The actual circle area is 5008.976 mm^2, that is, A = 5008.976 mm^2 in formula (2). The relative error between actual area and design area is 0.15%.
Table 1 shows the original data of average grain size determined by Abrams Three-Circle Procedure. In Table 1, P_L is the large circle grain boundary intersection count, P_M is the medium circle grain boundary intersection count, P_S is the small circle grain boundary intersection count, P_i is the total grain boundary intersection count on a metallographic photograph, and $P_i= P_L+ P_M+P_S$. $\sum P_i$ is cumulative grain boundary intersection count on several metallographic photographs, P is average grain boundary intersection count, and S_i is standard deviation of P_i. $t_{95\%}(v)$ is the quantile of t-distribution under the n fields (degree of freedom v = n-1) and confidence probability 95%. 95%CI is 95% confidence interval of several P_i. %RA is percent relative accuracy under 95% confidence probability, that is 95%CI/P. G is the ASTM grain size number obtained under %RA limit of several metallographic photographs. According to the requirements of ASTM E112, when %RA is less than or equal to10%, the G result is acceptable. After the tenth field has been detected, $\sum P_{10}$ = 653, P=65.3, 95% RA =9.42%≤10%, and G = 4.13. For the aluminium alloy (AlSi1) specimen, the ASTM grain size number is 4.13 determined by Abrams Three-Circle Procedure.

Table 1. The Abrams Three-Circle Procedure results

field	P_L	P_M	P_S	Pi	$\sum Pi$	P	S_i	$t_{95\%}(v)$	95%CI	%RA	G
1	37	28	14	79	79	79.0	-	-	-	-	4.68
2	33	23	13	69	148	74.0	7.07	12.706	63.53	85.85%	4.49
3	24	15	10	49	197	65.7	15.28	4.303	37.95	57.79%	4.14
4	29	21	11	61	258	64.5	12.69	3.182	20.19	31.30%	4.09
5	38	22	11	71	329	65.8	11.37	2.776	14.11	21.45%	4.15
6	35	18	9	62	*391*	65.2	10.28	2.571	10.79	16.56%	4.12
7	39	23	11	73	464	66.3	9.84	2.447	9.10	13.74%	4.17
8	28	19	11	58	522	65.3	9.57	2.365	8.00	12.27%	4.12
9	29	23	10	62	584	64.9	9.02	2.306	6.93	10.69%	4.11
10	33	23	13	69	*653*	*65.3*	8.60	2.262	6.15	*9.42%*	*4.13*

Table 2 shows the original data of average grain size determined by Planimetric Procedure. In Table 2, N_I is number of grains completely within the test circle, N_C is number of grains intercepted by the test circle, N_i is the total number of grains within the test circle on a metallographic photograph, and $N_i = N_I + 0.5\ N_C$. $\sum N_i$ is cumulative number of grains, N_{Pi} is average number of grains, and S_i is standard deviation of N_i within the test circle on several metallographic photographs. $t_{95\%}(v)$ is the quantile of t-distribution under the n fields (degree of freedom v = n-1) and confidence probability 95%.95%*CI* is 95% confidence interval of several N_{Pi}. %*RA* is percent relative accuracy under 95% confidence probability, which is 95%*CI* / N_{Pi}. *G* is the ASTM grain size number obtained under %*RA* limit of several metallographic photographs. According to the requirements of ASTM E112, when %*RA* is less than or equal to 10%, the *G* result is acceptable. After the thirteenth field has been detected, $\sum N_{13}$ =740, N_{Pi} =56.9, 95% *RA* =9.44%≤10%, and *G* = 3.94. For the aluminium alloy (AlSi1) specimen, the ASTM grain size number is 3.94 determined by Planimetric Procedure.

Table 2. The Planimetric Procedure results

Field	N_I	N_C	N_i	$\sum N_i$	N_{Pi}	S_i	$t_p(v)$	95%*CI*	%*RA*	*G*
1	59	39	78.5	78.5	78.5	-	-	-	-	4.40
2	38	33	54.5	133.0	66.5	16.97	12.706	152.47	229.28%	4.16
3	29	23	40.5	173.5	57.8	19.22	4.303	47.74	82.55%	3.96
4	38	29	52.5	226.0	56.5	15.92	3.182	25.32	44.82%	3.92
5	40	31	55.5	281.5	56.3	13.79	2.776	17.12	30.41%	3.92
6	44	29	58.5	340.0	56.7	12.37	2.571	12.98	22.91%	3.93
7	51	27	64.5	404.5	57.8	11.67	2.447	10.80	18.68%	3.96
8	30	36	48.0	452.5	56.6	11.35	2.365	9.49	16.77%	3.93
9	37	32	53.0	505.5	56.2	10.68	2.306	8.21	14.62%	3.92
10	44	32	60.0	565.5	56.6	10.14	2.262	7.25	12.83%	3.93
11	46	29	60.5	626.0	56.9	9.69	2.228	6.51	11.44%	3.94
12	40	30	55.0	681.0	56.8	9.26	2.201	5.88	10.37%	3.93
13	39	40	59.0	740.0	56.9	8.89	2.179	5.37	9.44%	3.94

4. Evaluation of measurement uncertainty

4.1. Evaluation of uncertainty in parameter measurement

4.1.1 Uncertainty evaluation of vernier caliper. Vernier calipers are used to measure the scale length and the diameter of test circle. According to GB/T 21390-2008 standard [6], the maximum permissible measurement error of vernier calipers with 0.02 mm scale is 0.03mm when measuring length is less than 150 mm. The uncertainty caused by the vernier caliperis as follows:

$$u_V = \sqrt{\left(\frac{0.03}{\sqrt{3}}\right)^2 + \left(\frac{0.02}{2\sqrt{3}}\right)^2} = 0.018$$

4.1.2 Uncertainty evaluation of magnification. The magnification of metallographic photograph is calculated from :

$$M = \frac{D}{l} \tag{3}$$

Where *M* is the magnification, *l* is scale length on the metallographic photograph (*l*=200 μm=0.2 mm), and *D* is the measured result of *l*.

The *D* is determined twice, and the range *R*=0.02mm and range coefficient *C*=1.13.The uncertainty caused by repeated measurement is as follows:

$$u_s = \frac{R}{C} = \frac{0.02}{1.13} = 0.018$$

The *D* is determined with the vernier calipers, the uncertainty caused by the vernier calliper has been described.

The combined uncertainty of D measurement results is as follows:

$$u_{D-M} = \sqrt{u_s^2 + u_V^2} = \sqrt{0.018^2 + 0.018^2} = 0.025$$

Ignoring the uncertainty of l itself, the uncertainty caused by magnification is as follows:

$$u_M = \frac{u_{D-M}}{l} = \frac{u_{D-M}}{0.2} = \frac{\sqrt{u_s^2 + u_v^2}}{0.2} = \frac{\sqrt{0.018^2 + 0.018^2}}{0.2} = 0.13$$

4.1.3 Uncertainty evaluation of test circle total circumference. The total circumference of three test circles is calculated from :

$$L = \pi d_1 + \pi d_2 + \pi d_3 \tag{4}$$

Whered_1, d_2, and d_3 are diameters of three test circle in mm.

The horizontal and vertical diameters of each circle are measured with vernier calipers, and the range R=0.02 mm and range coefficient C=1.13. In order to avoid the correlation, three circle diameters are measured by different vernier calipers.

The uncertainty evaluation of d_1, d_2, and d_3 is the same as the uncertainty evaluation of D. And the uncertainty caused byd_1, d_2, and d_3 is as follows:

$$u_{di} = \sqrt{u_s^2 + u_V^2} = \sqrt{0.018^2 + 0.018^2} = 0.025$$

The combined uncertainty of test circle total circumference is as follows:

$$u_L = \sqrt{(\pi u_{d1})^2 + (\pi u_{d2})^2 + (\pi u_{d3})^2} = \sqrt{3(\pi u_{di})^2} = 0.14$$

4.1.4Uncertainty evaluation of test circle area. The test circle area is calculated from:

$$A = \frac{1}{4}\pi d^2 \tag{5}$$

Where d is the diameter of test circle in mm.

The horizontal and vertical diameters of test circle are measured with vernier calipers, the average diameter is 79.86 mm, the range R=0.02 mm and range coefficient C=1.13. The uncertainty evaluation of d is the same as the uncertainty evaluation of D. And the uncertainty caused by$d$ is as follows:

$$u_d = \sqrt{u_s^2 + u_V^2} = \sqrt{0.018^2 + 0.018^2} = 0.025$$

The combined uncertainty of test circle area is as follows:

$$u_A = \frac{1}{2}\pi d \times u_d = \frac{\pi \times 79.86 \times 0.025}{2} = 3.2$$

4.1.5Uncertainty evaluation of repeated measurement. For Abrams Three-Circle Procedure, the G of the specimen was determined by using 10 fields, as shown in Table 1.For 10 fields, the average grain boundary intersection count is 65.3 and the standard deviation is 8.60, that is, P=65.3, S_{10}=8.60. The uncertainty caused by repeated P measurement is as follows:

$$u_{P1} = \frac{S}{\sqrt{n}} = \frac{8.60}{\sqrt{10}} = 2.7$$

According to the experience of statistical count, when an individualmetallographer counts the grain boundary intersection count, the counting error of every field is ±2. The uncertainty caused by counting error is as follows:

$$u_{p2} = \frac{2}{\sqrt{3}} = 1.2$$

The combined uncertainty of repeated measurement usingAbrams Three-Circle Procedure is as follows:

$$u_P = \sqrt{u_{P1}^2 + u_{P2}^2} = \sqrt{2.7^2 + 1.2^2} = 3.0$$

For Planimetric Procedure, the G of the specimen was determined by using 13 fields, as shown in Table 2. For 13 fields, the average number of grains is 56.9 and the standard deviation of $N_i$is 8.89, that is, N_{p13} =56.9, S_{13}=8.89. The uncertainty caused by repeated N measurement is as follows:

$$u_{N1} = \frac{S}{\sqrt{n}} = \frac{8.89}{\sqrt{13}} = 2.5$$

Also, according to the experience of statistical count, when an individual metallographer counts the number of grains, the counting error of every field is ±2. The uncertainty caused by counting error is as follows:

$$u_{N2} = \frac{2}{\sqrt{3}} = 1.2$$

The combined uncertainty of repeated measurement usingPlanimetric Procedure is as follows:

$$u_N = \sqrt{u_{N1}^2 + u_{N2}^2} = \sqrt{2.5^2 + 1.2^2} = 2.7$$

4.2. Evaluation of combined uncertainty

4.2.1 Abrams Three-Circle Procedure. For the aluminium alloy (AlSi1) specimen, the G is 4.13 determined by Abrams Three-Circle Procedure. In formula (1), M =101.50×, L = 498.76 mm, P=65.3.And according to formula (1), the combined uncertainty of G is calculated from:

$$u_{G-L-c} = \sqrt{\left(\frac{\partial G}{\partial M}\right)^2 u_M^2 + \left(\frac{\partial G}{\partial L}\right)^2 u_L^2 + \left(\frac{\partial G}{\partial P}\right)^2 u_p^2} \quad (6)$$

Where:

$$u_M = 0.13,\ \frac{\partial G}{\partial M} = \frac{2.88539}{M} = \frac{2.88539}{101.50} = 0.028$$

$$u_L = 0.14,\ \frac{\partial G}{\partial L} = -\frac{2.88539}{L} = -\frac{2.88539}{498.76} = -0.0058$$

$$u_P = 3.0,\ \frac{\partial G}{\partial P} = \frac{2.88539}{P} = \frac{2.88539}{65.3} = 0.044$$

The combined uncertainty is as follows:

$$u_{G-L-c} = 0.131$$

The expanded uncertainty are as follows:

U=0.26 k=2

The results of ASTM grain size number of aluminium alloy (AlSi1) specimen determined by Abrams Three-Circle Procedure are as follows:

4.13±0.26，k=2

4.2.2 Planimetric Procedure. For the aluminium alloy (AlSi1) specimen, the G is 3.94 determined by Planimetric Procedure. In formula (2), M =101.50×, A=5008.971 mm^2, N =56.9. And according to formula (2), the combined uncertainty of G is calculated from:

$$u_{G-A-c} = \sqrt{\left(\frac{\partial G}{\partial M}\right)^2 u_M^2 + \left(\frac{\partial G}{\partial A}\right)^2 u_A^2 + \left(\frac{\partial G}{\partial N}\right)^2 u_N^2} \quad (7)$$

Where

$$u_M = 0.13,\ \frac{\partial G}{\partial M} = \frac{2.88539}{M} = \frac{2.88539}{101.50} = 0.028$$

$$u_A = 3.2,\ \frac{\partial G}{\partial A} = -\frac{1.442695}{A} = -\frac{1.442695}{5008.971} = -0.00029$$

$$u_N = 2.7,\ \frac{\partial G}{\partial N} = \frac{1.442695}{N} = \frac{1.442695}{56.9} = 0.025$$

The combined uncertainty is as follows:

$$u_{G-A-c} = 0.069$$

The expanded uncertainty are as follows:

U=0.14 k=2

The results of ASTM grain size number of aluminium alloy (AlSi1) specimen determined by Planimetric Procedure are as follows:

3.94±0.14，k=2

5. Analysis and discussion

Using the Abrams Three-Circle Procedure determined G, G=4.13, U=0.26, k=2; Using the Planimetric Procedure determined G, G=3.94, U=0.15, k=2.The sources of uncertainty include: calipers for measuring circle diameter or scale length, magnification, three test circle total circumference, test circle area, errors of grain boundary intersection count or number of grains, repeated measurements.

In the table 3, U_P is the expanded uncertainty of G determined by the Abrams Three-Circle Procedure and U_N is the expanded uncertainty of G determined by the Planimetric Procedure.

Assuming that the permissible error of magnification changes from 0 to 2%and other conditions remain unchanged, the uncertainty caused by magnification is evaluated respectively according to uniform distribution and the uncertainty evaluation procedure in this paper. The U_P and U_N changing trends with the permissible error of magnification are shown in Table 3. In the table, Δ_R stands for magnification error.

Assuming that the scale of ruler changes from 0.01mm to 1mm,the maximum permissible measurement error of ruler meets JJG 30 standard [7]or JJG 1 standard [8], and other conditions remain unchanged, the uncertainty caused by length measurement is evaluated respectively according to uniform distribution and the uncertainty evaluation procedure in this paper. The U_P and U_N changing trends with the scale of ruler are shown in Table 3. In the table, CC stands for current calipers, SR stands for steel ruler, d stands for ruler scale, Δ_1 stands for permissible measurement error.

Assuming that the error of statistical counting changes from 1 to 5 and other conditions remain unchanged, the uncertainty caused by statistical counting is evaluated respectively according to uniform distribution and the uncertainty evaluation procedure in this paper. The U_P and U_N changing trends with error of statistical counting are shown in Table 3. In the table, Δ_2 stands for statistical counting.

Table 3. Sources of uncertainty and expanded uncertainty

Magnification			Length measurement					Statistical counting		
Δ_R	U_P	U_N	Ruler	d	Δ_1	U_P	U_N	Δ_2	U_P	U_N
0.0%	0.26	0.14	CC	0.01	0.03	0.26	0.14	1	0.25	0.13
0.5%	0.26	0.14		0.02	0.03	0.26	0.14	2	0.26	0.14
1.0%	0.26	0.14		0.05	0.05	0.26	0.14	3	0.29	0.15
1.5%	0.27	0.14		0.10	0.10	0.26	0.14	4	0.32	0.17
2.0%	0.27	0.15	SR	1	0.1	0.28	0.17	5	0.35	0.19

The extended uncertainty is up to two significant digits, usually one significant digit. If the extended uncertainty is rounded off according to the 1/3 rule, the U_P is 0.3 and the U_N is 0.2.From Table 3, it can be seen that the expanded uncertainty does not change with the magnification error and the ruler scale. It can be concluded that the contributions of the uncertainty caused by the magnification, the ruler, the scale length or test circle circumference, and test circle area to the uncertainty of G determined by the Abrams Three-Circle Procedure or the Planimetric Procedure can be negligible, when the magnification (≤2%) and the ruler (1mm scale) meet the requirements of ASTM E112 standard. It can be seen that the uncertainty of G is 0.02 different when the ruler scale is 1 mm and 0.10 mm. In order to reduce the determination risk, it is suggested that the ASTM E112 standard should provide that the ruler that determines the diameter of test circles or scale length should be with the scale that is less than or equal to 0.10 mm.

According to the experience of statistical count, when an individual metallographer counts the grain boundary intersection count or number of grains, the counting error of every field is ±2. ASTM E112 also requires control of 40 to 100 grain boundary intersection counts or 50 to 100 numbers of grains of every field. The lower limit number is required to ensure the boundary conditions of the mathematical model for calculating G, while the upper limit number is to ensure the accuracy of statistical counting and reduce the counting error. From Table 3, it can be seen that the expanded uncertainty changes greatly when the statistical counting error exceeds 2. So it is very important to control the counting error in the range of ≤2. In order to reduce the determination risk, it is suggested that the ASTM E112 standard should provide that the errors of counting should be less than or equal to 2. And for that, the

following methods can be adopted: 1) using counter; 2) labelling grain boundary intersection count or number of grains recommended by ISO 643:2012 standard [9]; 3) displaying method of specimen grain boundary; 4) ensuring gray difference between measurement grid or test circle and grain boundary.
Through the above analysis and discussion, it can be seen that the uncertainty of G determined by the Abrams Three-Circle Procedure or the Planimetric Procedure mainly comes from the repeated measurement of different field on the specimen, which is consistent with the repeatability requirement of ASTM E112 for controlling the detection precision. Repeatability measurements actually reflect the uniformity of the specimen grain size. Selecting measurement field randomly, fairly and representatively, may increase the number of measurement field. But this selecting method can reduce measurement uncertainty, improve accuracy and reduce the detection risk.
In Table 1, the field number is 10 when $\%RA \leq 10\%$. Formula (1) is used to calculate G of every field, and the standard deviation of G is 0.39. That is to say, under the condition of meeting the requirement ($\%RA \leq 10\%$) of ASTM E112 standard, the precision of G measurement is not up to the standard requirement (±0.25). In Table 2, the field number is 13 when $\%RA \leq 10\%$. Formula (2) is used to calculate G of every field, and the standard deviation of G is 0.22. That is to say, under the condition of meeting the requirement ($\%RA \leq 10\%$) of ASTM E112 standard, the precision of G measurement is up to the standard requirement (±0.25). It can be seen that the %RA limit of ASTM E112 for determining G should be redefined according to the measurement uncertainty in order to be up to the precision requirement (±0.25).

6. Conclusion

In this paper, the ASTM grain size number of aluminium alloy specimen was determined respectively by Abrams Three-Circle Procedure and Planimetric Procedure of ASTM E112-13 Standard. The test pattern total lengths used in Abrams Three-Circle Procedure are 500 mm, and the test pattern areas used in Planimetric Procedure are 5000 mm^2. And the uncertainty is evaluated. The results are as follows:

1) Using the Abrams Three-Circle Procedure determines G, G=4.13, U=0.26, k=2; Using the Planimetric Procedure determines G, G=3.94, U=0.15, k=2.

2) It is suggested that the ASTM E112 standard should provide that the errors of counting grain boundary intersection count or numbers of grains should be less than or equal to 2, and the ruler that determines the diameter of test circle or scale length should be with the scale that is less than or equal to 0.10 mm. On this condition, the uncertainties caused by them in determining G can be negligible.

3) The main sources of uncertainty for the determination of G by Abrams Three-Circle Procedure and Planimetric Procedure are repetitive measurement, which depends on the uniformity of grain size of specimen.

4)The %RA limit of ASTM E112 for determining G should be redefined according to the measurement uncertainty in order to be up to the precision requirement (±0.25).

References

[1] ASTM E112-13 Standard Test Methods for Determining Average Grain Size[S].
[2] ISO /IEC 17025-2017 Accreditation criteria for the competenceof testing and calibration laboratories[S].
[3] GB/T 28898-2012 Evaluation of uncertainty in chemical composition analysis for metallurgical materials [S].
[4] GB/T 4340.1—2009 Metallic materials - Vickers hardness test - Part 1: Test method [S].
[5] JJF1059.1-2012 Evaluation and Expression of Uncertainty in Measurement [S].
[6] GB/T 21390-2008 Vernier, dial and digital display height callipers [S].
[7] JJG30-2012 Current Calipers[S].
[8] JJG1-1999Steel Ruler[S].
[9] ISO 643:2012 Steels — Micrographic determination of the apparent grain size[S].

Preparation of silk-like zirconia and investigation of its heat-insulating properties

Chang Xi [1, a], Qinqin Xu [1, b], Xu Pang [1, c], and Tianchi Wang [1, d *]

1 School of Materials Science and Engineering, Nanjing University of Science and Technology, Nanjing 210094, China
Email: [a]veronica_xi@126.com, [b]1044643710@qq.com, [c]1175431703@qq.com, [d]tianchiwang@aliyun.com

Abstract. Many life forms possess the aesthetic and unique micromorphologies after long-term evolution and inspire researchers to fabricate different types of microstructures. In this study, silk was utilized as a template to produce silk-like zirconia fibers with excellent heat-insulating properties. We soaked silk in a solution containing Zr^{4+} and thereafter sintered the dried silk in air at temperatures below 1000 °C to obtain the zirconia fibers. The results showed that these zirconia fibers not only maintained the structures of silk but also exhibited hollowness, which rendered them with lower thermal conductivities and higher heat-insulating properties than conventional zirconia fibers.

1. Introduction

Compared with manufactured materials, natural materials often have unique morphology and excellent structural properties. People use various biological structures as templates to synthesize the required materials, so that these materials will have the morphology and structure of biological templates. We call this method of preparing materials "biomorphic method". The materials prepared by this method will possess excellent biological structure, thus improving the properties of materials. People have employed kinds of bio-structures as templates to synthesize the desired ceramics such as such as butterfly wings, radish, kapok and so on; these ceramics possess the morphologies and structures of the templates, which is the key factor that can give ceramics good and new properties [1-8].

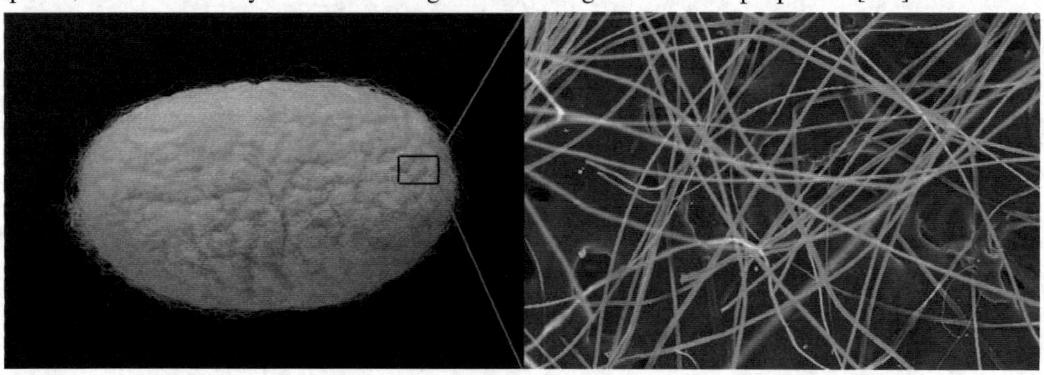

Fig. 1 Full-scale image of a pod and micromorphologies of the silk.

IOP Conf. Series: Materials Science and Engineering **733** (2020) 012046 doi:10.1088/1757-899X/733/1/012046

Zirconia (ZrO_2) fiber is an important product that is typically utilized for the heat insulator due to its good resistance to high temperatures and excellent heat-insulating capability [9-10]. Over recent decades, some precursor methods including the organic precursor, inorganic precursor, and sol-gel spinning technique have been applied to fabricate zirconia fibers [9-12]. The structures and shapes of these zirconia fibers have been artificially directed [11-12]. In contrast, silk is a kind of protein fiber produced by silkworm. Like wool, silk is one of the earliest animal fibers used by human beings. Silk is not only soft and breathable, but also green and healthy. Silk coating has good thermal insulation performance and can be used to make quilts and insect-proof clothing. The morphologies of silk and the pod are displayed in Fig. 1. In this study, silk was selected as the template to fabricate silk-like zirconia fibers, which possess the structures and shapes of the template. We expect that this method will be useful for the fabrication of zirconia fibers with excellent heat-insulating properties.

2. Experimental procedure

Dried and loose silk was used as the bio-template. It was first soaked in a solution containing 10% $Zr(NO_3)_4$ for 10 min. After drying at 70 °C for 12 h, it was moved to a muffle furnace and fired at 600 or 1000 °C for 2 h. Thus, we can obtain the zirconia fibers.

A scanning electronic microscopy (SEM, JSM-6380, JEOL, Japan) was used to detect the micromorphologies of zirconia fibers obtained in this study. The phase composition of ceramic fibers was examined by an X-ray diffraction (XRD, Cu-Kα, RigakuX 2038, Japan). The differential thermal analysis and thermal gravimetric analysis of the silk immersed in the solution of $Zr(NO_3)_4$ were performed by a thermal analyzer (TG-DSC, SDT Q600, TA Instruments, USA) in air. The range of the test temperature was from room-temperature to 1000 °C, and rate of heating was 20 °C/min. A steady-state plate diathermometer (PBD-02P, China) was employed to measure the thermal conductivities of the zirconia fibers. The apparent densities of all the specimens for thermal-conductivity testing were found to be 0.096 g/cm^3.

3. Results and discussion

The XRD patterns of the materials obtained by using the template of silk at different sintering temperatures are shown in Fig. 2. It is obvious that these silk-like fibers are composed of zirconia. These fibers were all made of monoclinic ZrO_2 or tetragonal phase ZrO_2.

When the sintering temperature is at 600 °C, the XRD peak position of ZrO_2 fiber are at $2\theta = 30.27°$ (011) and 50.38° (112), it represents the tetragonal phase, indicates that the crystal zirconia of a tetragonal phase can be obtained by sintering at 600 °C. It can be seen from the other curve in Fig. 2 that the peaks are at $2\theta = 28.08°$ (-111) and 31.32° (111), that is confirmed the presence of the monocline phase. The crystal zirconia can transform to the monoclinic phase after sintering at 1000 °C [10].

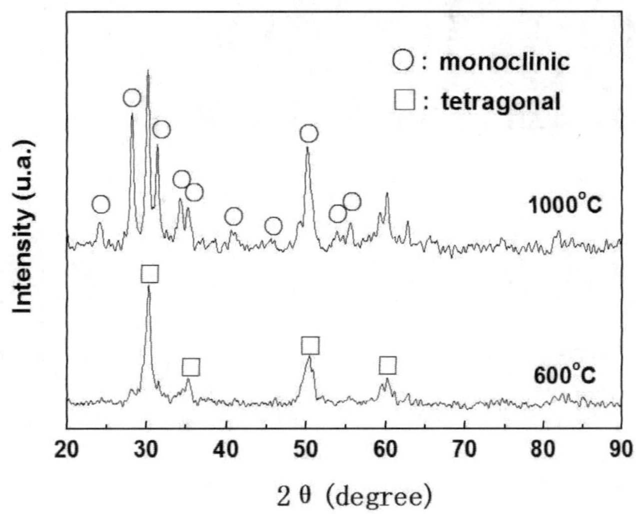

Fig. 2 XRD patterns of zirconia obtained by using the template of silk fibers sintered at 600 and 1000 °C.

Fig. 3a and b show the microstructures of the zirconia fibers sintered at 600 and 1000 °C. As can be seen from Fig. 1b, these zirconia ceramic fibers retain the fibrous microscopic morphology of the silk fibers well. The diameters of ceramic fibers are ranging from 5 to 10 μm. Different prepared using conventional methods from zirconia fibers, these fibers zirconia filamentous conformation is influenced by the material properties of the template. Microscopic observation found the presence of many hollow zirconia fibers. From Fig. 3c, it can easily be seen that the fiber is hollow. The inner diameters of these microtubules are ranging from to 3 to 6 μm. The reason for the formation of the hollow structure may be that when the natural silk fiber is immersed in the $Zr(NO_3)_4$ solution, the solute of $Zr(NO_3)_4$ adheres to the surface of the fiber, and when the fiber is dried, the solute is uniformly coated on the silk. Thereafter, during the heating and sintering, $Zr(NO_3)_4$ is decomposed into zirconia, which also covers the surface of the silk. When the temperature rises to a higher level, the silk template is completely burned, thus forming a hollow structure of zirconia fibers.

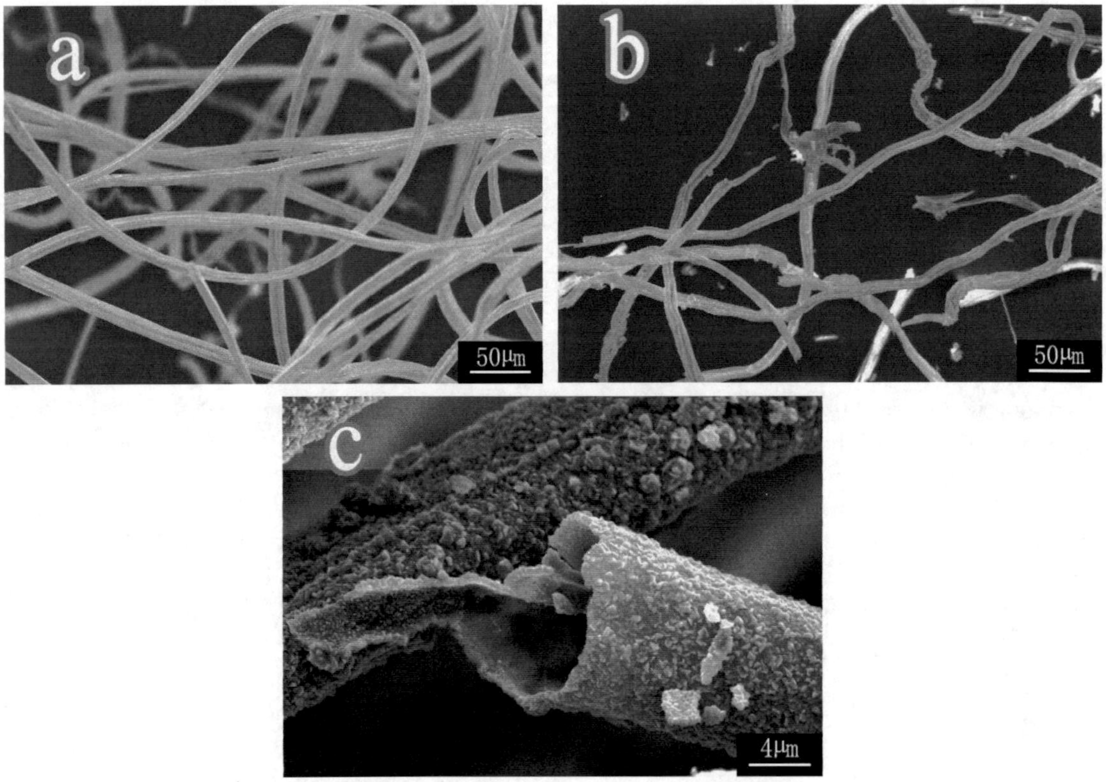

Fig. 3 SEM images of the silk-like zirconia: (a) sintered at 600 °C, (b) sintered at 1000 °C, and (c) tubular fiber

The differential thermal analysis of the silk immersed in the $Zr(NO_3)_4$ solution are shown in Fig. 4. An endotherm appeared at 80 °C, corresponding to 29% weight loss. This is ascribed to the dehydration of zirconium nitrate pentahydrate. Two exothermic peaks appeared at 196 ° C and 300 ° C, corresponding to 27% weight loss. The exothermic peak at 196 °C is caused by the solute decomposition reaction of $Zr(NO_3)_4$ solute to get zirconia, and the one at 300 °C can be ascribed to the preliminary crystallization of zirconia. An obvious weight loss is observed between 350 °C and 730 °C, along with the observation of several exothermic peaks at 404 °C, 515 °C, and 720 °C, which are ascribed to a series of actions that included the decomposition of amino acid in the silk and burning of the silk. As can be seen from the differential thermal analysis of Fig. 4, the formation of zirconia is earlier than the burning and decomposition of silk fibers. Therefore, this zirconia ceramic fiber retains the microscopic morphology of natural silk fibers.

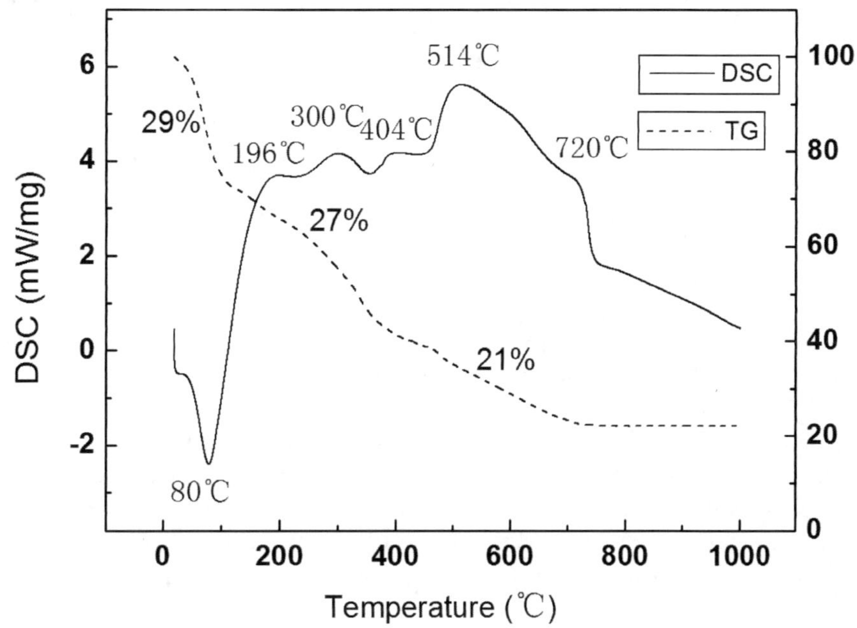

Fig. 4 The TG-DSC thermograms of the silk fibers immersed in the $Zr(NO_3)_4$ solution.

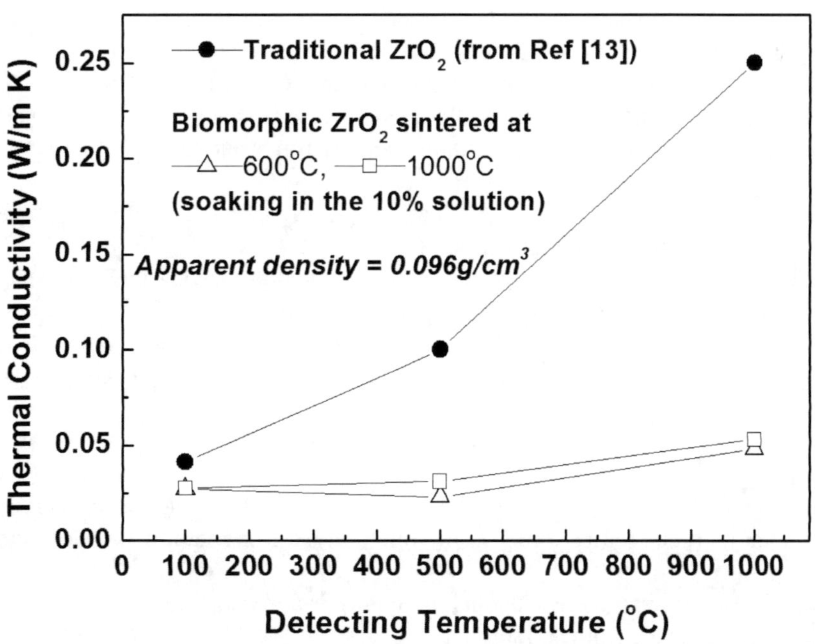

Fig. 5 Thermal conductivities of the silk-like and conventional zirconia fibers.

The thermal conductivities of the silk-like and conventional zirconia fibers are shown in Fig. 5. The data for the conventional zirconia comes from an article written by L. Hu et al [13]. The conventional zirconia fibers for testing have the same apparent density (0.096 g/cm^3) as the silk-like zirconia fiber. It can be clearly seen from the data in Fig. 5 that the thermal conductivity of the zirconia fiber is lower than that of the conventional zirconia fiber. When the detection temperature is 500 ° C and 1000 ° C, the thermal conductivity of the silk-like zirconia fiber is greatly reduced. This indicates that the silk-like zirconia fiber has better thermal insulation properties than the conventional zirconia fiber. The reason for this phenomenon is due to the hollow structure of the silk-like zirconia fiber. Hollow zirconia fibers (Fig. 3c) are more effective in preventing large areas of air flow in the sample and at the same time, the heat transfer mode of the fibers will be changed. The hollow structure breaks up the space into more parts, allowing more still air to stay in the fiber tube, and the flow of air will be trapped in each tiny hole, thereby hindering heat transfer heat transfer and heat convection heat transfer. Meanwhile, the presence of many tiny pores in the zirconia fibers provides a shorter mean free path for the photon, which can reduce the thermal radiation.

4. Summary

This study provides a new method for preparing heat-insulating ZrO_2 fibers using natural silk fibers as biological templates. We successfully immersed the natural silk fiber in the solution of Zr^{4+} and then sintered it at a high temperature to form a ZrO_2 ceramic fiber having a silk-like structure. The zirconia fibers we synthesized maintained the shapes of natural silk and exhibited hollow structures. TG-DSC tests show that zirconia formed earlier than the burning and decomposition of the bio-template such that the zirconia fibers could maintain the fibrous shape and micromorphology of silk. Compared with conventional zirconia fibers, the silk-like zirconia fibers have much better heat-insulating properties due to their hollow structure. Because our synthetic fibers have simpler process and exhibit better thermal insulation than conventional fibers, the way of utilizing a bio-template could give an innovative approach for the design of heat-insulating fibers.

5. References

[1] C. Zollfrank, R. Kladny, H. Sieber, P. Greil, Biomorphous SiOC/C-ceramic composites from chemically modified wood templates, J. Eur. Ceram. Soc. 24 (2004) 479-487.

[2] A. Zampieri, G.T.P. Mabande, T. Selvam, W. Schwieger, A. Rudolph, R. Hermann, H. Sieber , P. Greil, Biotemplating of luffa cylindrica sponges to self-supporting hierarchical zeolite macrostructures for bio-inspired structured catalytic reactors, Mater. Sci. Eng. C 26 (2006) 130-135.

[3] R. Sun, L. Sun, Y. Chun, Q. Xu, H. Wu, Synthesizing nanocrystal-assembled mesoporous magnesium oxide using cotton fibres as exotemplate, Microporous Mesoporous Mater. 111 (2008) 314-322.

[4] J. Kim, S. Myoung, H. Kim, J. Lee, Y. Jung, C. Jo, Synthesis of SiC microtubes with radial morphology using biomorphic carbon template, Mater. Sci. Eng. A 434 (2006) 171-177.

[5] G.J. Qiao, R. Ma, N. Cai, C.G. Zhang, Z.H. Jin, Mechanical properties and microstucture of Si/SiC materials derived from native wood, Mater. Sci. Eng. A 323 (2002) 301-305.

[6] P. Greil, E. Vogli, T. Fey, A. Bezold, N. Popovska, H. Gerhard, H. Sieber, Effect of microstructure on the fracture behavior of biomorphous silicon carbide ceramics, J. Eur. Ceram. Soc. 22 (2002) 2697-2707.

[7] T.C. Wang, T.X. Fan, D. Zhang and G.D. Zhang, "Fabrication and Wear Behaviors of Carbon/Aluminum Composites Based on Wood Templates", Carbon, 44 [5] 900-906 (2006).

[8] T.C. Wang, D.S. Xiong, T.L. Zhou, Preparation and wear behavior of carbon/epoxy resin composites with an interpenetrating network structure derived from natural sponge, Carbon 48 (2010) 2435-2441.

[9] A. Hartridge, M.D. Taylor, A.K. Bhattacharya, Synthesis and characterization of partially and fully stabilized zirconia fibers made from an inorganic precursor, J. Mater. Res. 16 (2001) 2384-2387.

[10] G. Yu, L. Zhu, X. Wang, H. Che, G. Zhang, Z. Sun, H. Fan, X. Liu, D.Xu, Fabrication of zirconia mesoporous fibers by using polyorganozirconium compound as precursor, Microporous and Mesoporous Mater. 119 (2009) 230-236.

[11] P.K. Chakrabarty, M. Chatterjee, M.K. Naskar, B. Siladitya, D. Ganguli, Zirconia fiber mats prepared by a sol-gel spinning technique, J. Eur. Ceram. Soc. 21 (2001) 355-361.
[12] R.C. Pullar, M.D. Taylor, A.K. Bhattacharya, The manufacture of partially-stabilised and fully-stabilised zirconia fibers blow spun from an alkoxide derived aqueous sol-gel precursor, J. Eur. Ceram. Soc. 21 (2001) 19-27.
[13] L. Hu, F. Gao, W. Chen, ZrO2 fiber and products, Journal of Synthetic Crystals 38 (2009) 265-270.

6. Acknowledgements

The authors wish to express thanks to National Natural Science Foundation of China (51672131), and the Scientific Research Training Program of Undergraduates.

AUTHOR INDEX